Dedicated to my parents

Vladimir Y. Rovenskii

Foliations on Riemannian Manifolds and Submanifolds

Birkhäuser
Boston • Basel • Berlin

Vladimir Y. Rovenskii
Mathematics Department/Geometry Chair
Pedagogical Institute
Krasnoyarsk 49
660049 Russia

Library of Congress Cataloging In-Publication Data

Rovenskii, Vladimir, 1953-
Foliations on Riemannian manifolds and submanifolds / Vladimir Rovenskii.
p. cm.
Includes bibliographical references and index.
ISBN-13: 978-1-4612-8717-9
(hardcover : acid-free)
1. Riemannian manifolds. 2. Foliations (Mathematics) I. Title
QA613.62.R68 1997 96-5737
516.3'73--dc20 CIP

AMS Classification: 52, 51, 53, 57

Printed on acid-free paper

Birkhäuser
Softcover reprint of the hardcover 1st edition 1998

ISBN-13: 978-1-4612-8717-9 e-ISBN-13: 978-1-4612-4270-3
DOI: 10.1007/978-1-4612-4270-3
Typeset by TEXniques, Inc., Boston, MA

9 8 7 6 5 4 3 2 1

CONTENTS

PREFACE

This monograph is based on the author's results on the Riemannian geometry of foliations with nonnegative mixed curvature and on the geometry of submanifolds with generators (rulings) in a Riemannian space of nonnegative curvature. The main idea is that such foliated (sub) manifolds can be decomposed when the dimension of the leaves (generators) is large. The methods of investigation are mostly synthetic.

The work is divided into two parts, consisting of seven chapters and three appendices. Appendix A was written jointly with V. Toponogov.

Part 1 is devoted to the Riemannian geometry of foliations. In the first few sections of **Chapter I** we give a survey of the basic results on foliated smooth manifolds (Sections 1.1–1.3), and finish in Section 1.4 with a discussion of the key problem of this work: the role of Riemannian curvature in the study of foliations on manifolds and submanifolds.

Chapter II contains local notions and results not only for a foliated Riemannian manifold, but also for the slightly more general situation of a pair of complementary orthogonal distributions. The latter is sometimes called an "almost-product structure." Beginning with the well-known "co-nullity operator" of D. Ferus, P. Dombrowski and others, we introduce in Chapter II a pair of structural tensors, which are similar to O'Neill's or Gray's pairs, but are more convenient for our purposes; these tensors also satisfy a Riccati type PDE with mixed curvature. Chapter II concludes with an overview of 1) integral formulas containing the mixed curvature along a compact total manifold or a compact leaf, 2) recent studies of Riemmanian foliations.

The next two chapters are devoted to the case which is local on transversal directions, i.e., a foliation is given on a neighborhood of one complete leaf.

Chapter III starts off with Ferus's result (1970) on the optimum (largest) dimension of a totally geodesic foliation with constant positive mixed sectional curvature on a given manifold. It contains a surprisingly deep relationship betweeen the Riemannian geometry of foliations and the topological notion of the number of vector fields on an n-sphere. In Section 3.4 we discuss Ferus's scheme: if the dimension of a (totally geodesic) foliation is "large," then there exists an L-parallel (Jacobi) vector field (i.e., induced by a foliation along some leaf geodesic) with a parallel initial value and first derivative. This is impossible in the case of constant curvature. From this point of view Chapter III is devoted to results on vector fields that are induced by a foliation or by a distribution given on a Riemannian manifold.

For non-constant positive mixed curvature, the dimension of a compact foliation is easily estimated by half of the totally geodesic dimension of the total manifold. The idea (by T. Frankel) is that two compact submanifolds (for instance, great spheres in a round sphere) in a Riemannian space of positive curvature must intersect if the sum of their dimensions is not less than the dimension of the total space. In Section 3.4 we extend this result to the case of positive partial mixed Ricci curvature.

The problem by V. Toponogov is to obtain a Ferus type estimate for the dimension of a foliation on a (compact) Riemannian manifold with positive mixed curvature. **Chapter IV** contains the author's results on the above problem. In the beginning of Section 4.1 a brief survey of foliations on space forms is given. In Section 4.2 we introduce a new variation procedure based on the *volume of an L-parallel vector field* and the *turbulence of a foliation along a leaf*, which allows us to obtain rigidity and splitting theorems for foliations with nonnegative mixed curvature, and in particular, to generalize Ferus's result. In Section 4.3 the Riccati equation procedure is extended to foliations. Combining some ideas, the integral inequality with mixed scalar curvature along a complete leaf is obtained. The case of when a structural tensor of a foliation is almost symmetric or skew-symmetric (i.e., the norm of the second fundamental form of the leaves or the norm of an integrability tensor for horizontal distribution is bounded above by some term with mixed Ricci curvature) is also studied.

In **Part 2** we consider a popular class of submanifolds equipped with the additional structure of a foliation whose leaves (generators) are totally geodesic, umbilic etc., in an ambient Riemannian space. The *ruled, canal* and *tubular submanifolds* (well-known in space forms) are intrinsically geodesic or umbilic foliations.

Systematic investigations of the local and global structure of submanifolds in (pseudo-) Riemannian spaces include a study of the relationships between intrinsic and extrinsic properties, the tests for totally geodesic, cylindrical submanifolds, estimates of codimension, etc.

Attention given to foliated submanifolds has increased due to studies of some special embeddings with degenerate second fundamental form: (strongly) parabolic, k-saddle, having nonpositive extrinsic curvature, small codimension, and others.

During the years 1960–1970 it was popular to examine submanifolds M in a Riemannian space with positive relative null-index $\mu(M)$. Different names for these submanifolds *with constant nullity, strongly parabolic, with constant rank, tangential degenerate*, and *k-developable* were introduced by geometers.

In the case of a *curvature invariant* submanifold (for example, when the ambient manifold is a space form) the *regularity domain* on M, which consists of the points with minimal value of relative null-index, has a *ruled developable* structure (with constant K_{mix} in the case of a space form), whose $\mu(M)$-dimensional rulings are tangent to the relative nullity distribution.

The *k-saddle* submanifolds and the submanifolds with *nonpositive extrinsic curvature* and small codimension are important examples of strongly parabolic submanifolds.

More general *parabolic* submanifolds, which were first introduced in space forms by A. Borisenko in 1972, have a large number of *rulings* under a certain condition on the curvature tensor of an ambient Riemannian space.

Part 2 begins with the bound (from above) on the dimension of a complete ruling on a ruled submanifold in a round sphere as half of the dimension of the ambient space; the best (Ferus's type) estimate holds in the case of a developable ruled submanifold. Then we continue the investigation of Toponogov's problem in the case of foliated submanifolds.

In **Chapter V** a detailed survey of all studies of ruled submanifolds in space forms and in arbitrary Riemannian spaces is given as well as the concepts and facts necessary for the subsequent chapters. We introduce the new (synthetically defined) class of *uniquely projectable submanifolds along generators* in a Riemannian space, which is deeply related to the class of ruled submanifolds with nonnegative (positive) mixed curvature. Also, in Section 5.3, we extend some recent results on submanifolds and foliations with nonpositive extrinsic sectional curvature for the case of extrinsic partial Ricci curvature.

The central result of **Chapter VI** is that a ruled submanifold in a sphere or in a complex projective space with a ruling of "large" dimension and a "small" norm of its second fundamental form is congruent to the Segre embedding; the latter plays the role of a "cylinder" in a space form of positive curvature. Combined with our method for the volume of an L-parallel vector field (from Chapter IV) we obtain the test for the Segre type decomposition of a ruled submanifold in a Riemannian space of positive curvature.

In **Chapter VII**, which is based on ideas from Chapter VI, we obtain for the first time the Decomposition Theorems for parabolic submanifolds (i.e., with degenerate second quadratic forms). The central result is that a complete parabolic submanifold in a sphere or in a complex projective space with "small" rank and a "small" norm of its second fundamental form is congruent to the Segre embedding. Combined with our method for a volume of an L-parallel vector field, as in Chapter VI, we obtain the test of Segre type decomposition of a parabolic submanifold in a Riemannian space of positive curvature.

Appendix A is devoted to Toponogov's conjecture: *if a complete simply connected Riemannian manifold with sectional curvature* ≤ 4 *and injectivity radius* $\geq \frac{\pi}{2}$ *has extremal diameter* $\frac{\pi}{2}$, *then it is isometric to CROSS* (i.e., compact rank one symmetric spaces). In Section A.1 the relationship between this problem and geodesic foliations of a round sphere (in particular, skew-Hopf foliations) is considered, but the proof of the conjecture along these lines is incomplete. In Section A.2 the proof is given, based on recent results, topological methods, and estimates of volumes for Blaschke manifolds.

Appendix B contains the author's results on submersions where some concepts of submanifolds with relative nullity are used.

In **Appendix C** we construct an even-dimensional Riemannian manifold with fibration on closed geodesics and $(\delta \approx 1)-$pinched mixed sectional curvature. This example is important in view of results in Chapters IV, VI, VII.

Acknowledgments. Victor Toponogov (Novosibirsk), for scientific and human help for many years, Professor Pawel Walczak and other colleagues from the Polish Academy of Sciences for discussions on the results from the Workshop "Foliations: Geometry and Dynamics" (1995, Warsaw), Robert Wolak (Krakow) for a useful discussion and improvements concerning the entire text, colleagues and friends from Novosibirsk, Kazan, Barnaul, Kharkov and Krasnoyarsk. Finally, I would like to warmly thank Ann Kostant for human support and help in the publishing process. My apologies for any remaining errors due to language and thanks to all the copyeditors who tried to make the book as readable as possible.

Vladimir Rovenskii
1997

CHAPTER I

FOLIATIONS ON MANIFOLDS

1.1. Definitions and examples of foliations

Intuitively, a foliation corresponds to a decomposition of a manifold into a union of connected submanifolds of the same dimension called leaves. The collection of leaves looks locally like pages of a book.

1.1 Definition. [Tam] A family $\mathcal{F} = \{L_\alpha\}_{\alpha \in A}$ of connected subsets of a manifold M^m is called a *ν-dimensional foliation* if

1) $\bigcup_\alpha L_\alpha = M^m$, 2) $\alpha \neq \beta \Rightarrow L_\alpha \bigcap L_\beta = \emptyset$,

3) for any point $p \in M$ there exists a C^r-chart (local coordinate system) (U_p, φ_p), such that $p \in U_p$ and if $U_p \bigcap L_\alpha \neq \emptyset$ the components of the set $\varphi_p(U_p \bigcap L_\alpha)$ are the following parts of parallel affine subspaces

$$A_c = \{(x_1, \ldots, x_m) \in \varphi_p(U_p) : x_{\nu+1} = c_{\nu+1}, \ldots, x_m = c_m\}$$

and the $c_i \in \mathbb{R}$ are constant (see Figure 1).

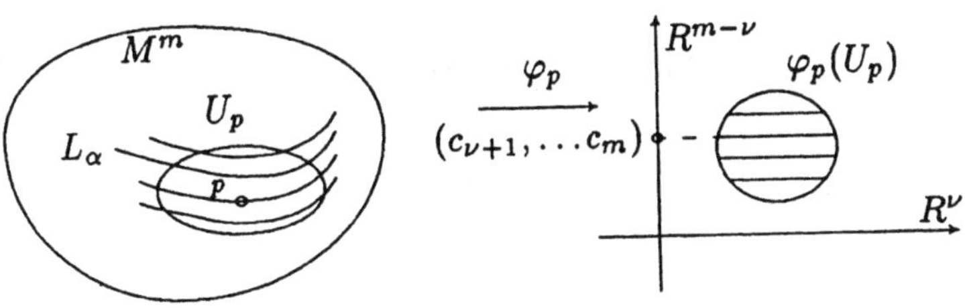

Figure 1. Foliation

Condition 3) does not follow from 1) and 2); this is obvious from the example *Sunrise* on $\mathbb{R}^2$ in Figure 2. The coordinate system (U_p, φ_p) in Definition 1.1 is called a *foliated chart*. The local picture of a foliated manifold is simply a space of the form $\mathbb{R}^\nu \times N$, where we regard sets of the form $\mathbb{R}^\nu \times \{n\}$ as leaves and N is a transversal as in Figure 3. To such a space there is canonically associated a ν-plane vector bundle $F_{|\mathbb{R}^\nu \times N}$ with fibers $F(x, n) \approx (\mathbb{R}^\nu)$. The integer $m - \nu = \dim N$ is called a *a codimension* of a foliation.

Foliations on manifolds have been studied in [CN], [God], [MS], [Tam], [HH], [VE], [Ree 1], [Rei 3], [Rem].

In [MS] a *foliated space* which generalizes the notion of a foliated manifold is considered. In this case M is a separable metrizable space with a collection of open sets $\{U_p : p \in M\}$ with $p \in U_p$ and homeomorphisms $\varphi_p : U_p \to \mathbb{R}^\nu \times N_p$. The continuous change of corrdinate maps has the form $t' = \varphi(t, n)$, $n' = \psi(n)$; moreover, a set in U_p of the form $\mathbb{R}^\nu \times \{n\}$ is sent to a set of the form $\mathbb{R}^\nu \times \{\psi(n)\}$ by a smooth map. The level surfaces coalesce to form maximal connected sets called *leaves*.

Each leaf is a smooth ν-dimensional manifold. An example of a foliated space that is not a foliated manifold is a *solenoid* with $\nu = 1$ and each N_p homeomorphic to a subspace of a Cantor set. Finally, M may be infinite dimensional.

In [Bou1, 2], [Žu 3], [Wol 4], [Mol], [Mol 2], and [Rei 3] *foliations with singularities* are studied. In this case the dimension of the leaves is not constant.

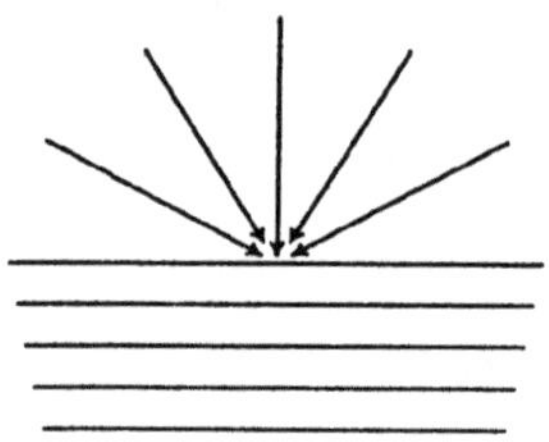

Figure 2. Sunrise

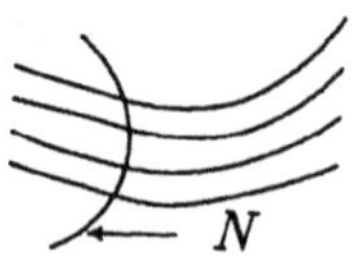

Figure 3. Transversal

Foliations arise in connections with such topics as *vector fields without singularities (when* $\nu = 1$*), integrable* ν*-dimensional distributions, submersions and fibrations, actions of Lie groups, and the direct constructions of foliations such as Hopf fibrations, Reeb foliations, etc.*

Each leaf of a foliation is a ν-dimensional connected manifold with a countable base. A leaf of a foliation is a component of the manifold M in the *leaf topology*, which is generated by all sets of the form $U \cap \varphi_p^{-1}(A_c)$, where U is an open subset of U_p. The injection mapping of a leaf into the manifold M with the original topology is continuous, but the image is not necessarily a closed set. A leaf L is called *closed* when it is a closed subset of M in the original topology. A leaf is called *proper* if both topologies coincide, as, for example, when a leaf is compact.

A manifold M with Euler characteristic $\chi(M) = 0$ admits a nonzero vector field, and the existence theory for ODE implies that M has a 1-dimensional foliation. The problem of the existence of a codimension-1 foliation is also solved: on every open smooth manifold such a foliation exists [Hae 2]; on a closed manifold it exists precisely when the Euler characteristic is zero (sufficiency was proved by W. Thurston). On the 2-dimensional torus there are vector fields that generate certain *line foliations*, each of whose leaves is dense in the whole manifold (see Figure 4).

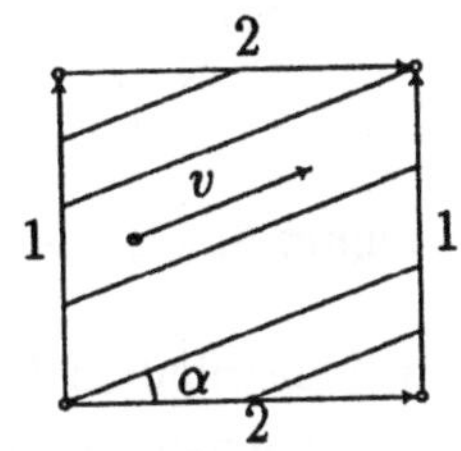

Figure 4. Line foliation on T^2

In this example the leaf topology on each leaf $L = \mathbb{R}$ is different from that induced by the original manifold topology.

1.2 Definition. A *saturated* (or an *invariant*) *set* on a foliated manifold is a subset which is the union of leaves. A *minimal set* is a closed saturated set which has no proper closed saturated subsets.

A particular case of a minimal set is a compact leaf. Note that any compact saturated set contains at least one minimal set and every foliation on a compact manifold has a minimal set.

1.3 Definition. A ν*-dimensional distribution* on a manifold M is a smooth field $\{T(p)\}$, $(p \in M)$ of ν-dimensional tangent subspaces, that is, a function whose value at each point $p \in M$ is a ν-dimensional subspace $T(p)$ of the tangent space T_pM. A distribution is called *integrable* if through each point $p \in M$ there exists a ν-dimensional *integral submanifold* which is tangent to the given distribution at each of its points. A 1-dimensional distribution is also called a *line field*.

Note that a collection of maximal integral submanifolds of an integrable distribution forms a foliation.

The *bracket (commutator)* of two vector fields x and y on M, denoted by $[x, y]$, is a vector field that acts on smooth functions by the rule $[x, y]f = x(yf) - y(xf)$. In a local coordinate system $[x, y]^i_m = (x^i \partial_j y^i - y^j \partial_j x^i)_m$. In particular, for basic vector fields we have $[\partial_i, \partial_j] = 0$. The bracket $[x, y]$ characterizes the non-commutativity of displacements along trajectories of the vector fields x and y (see Figure 5).

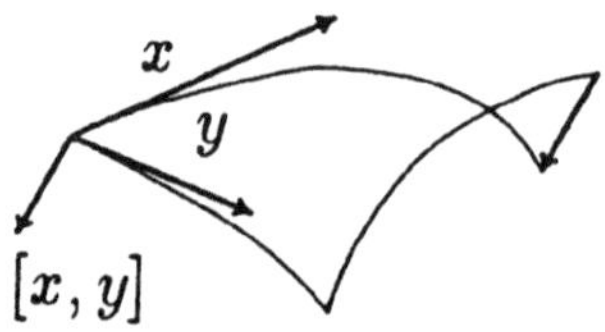

Figure 5. Bracket

By the Frobenius Theorem, a distribution T on M is integrable if and only if for any vector fields x, y on M which are tangent to the distribution, the vector field $[x, y]$ is also tangent to the distribution at each point. When $\nu = 1$, this condition is always satisfied, and so a line field is always integrable; it is equivalent locally, but not globally, to a vector field; see Example 1.11 below (a Reeb component on a torus with one compact leaf) and the examples in $\mathbb{R}^2 \setminus \{0\}$ in Figure 6.

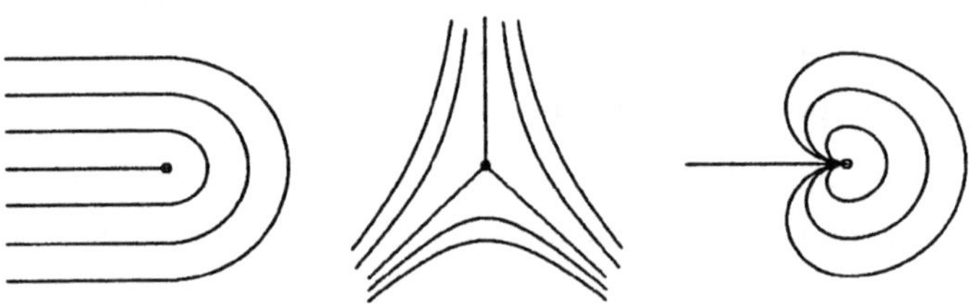

Figure 6. Foliations on $R^2 \setminus \{0\}$ — not vector fields

1.4 Example. [CN] Define a 2-plane field T on $\mathbb{R}^3$ by setting $T(x_1, x_2, x_3) = \{$ *the plane generated by* $X_1 = (1, 0, 0)$ *and* $X_2 = (0, \exp(-x_1), \exp(x_1))\}$. Then T has no integral surfaces: $[X_1, X_2] = (0, -\exp(-x_1), \exp(x_1)) \notin T$.

There are two orientations, $+1$ and -1, on $\mathbb{R}^n$ – two equivalence classes on the set of all bases.

Recall that a distribution T on a manifold M is *orientable* if it is possible to choose an orientation $\mathcal{O}(m)$ on $T(m)$ such that the map $m \to \mathcal{O}(m)$ is continuous. A foliation $\{L\}$ is *(transversally) orientable* if (some distribution complementary to) the tangent distribution TL is orientable [CN].

The foliation $L \times \{b\}$, where $b \in B$, in the direct product of manifolds $M = L \times B$, is called a *trivial fibration*. So locally, a foliation is a trivial fibration. For $\dim M = 2$ we have three trivial fibrations: $\mathbb{R}^1 \times \mathbb{R}^1$ on the plane, $\mathbb{R}^1 \times S^1$ on the cylinder and $S^1 \times S^1$ on the torus.

1.5 Definition. [CN] A smooth map $\pi : M \to B$ of differentiable manifolds M and B is called a *submersion* if for any point $m \in M$ the differential $\pi_*(m) : T_m M \to T_{\pi(m)} B$ is a surjective map of tangent spaces; M is called the *total space*, B the *base*, and $\{L_b = \pi^{-1}(b)\}$ the *fibers (leaves)*. A C^r-submersion $\pi : M^m \to B^n$ is called a *(locally trivial) C^r-fibration*, if the following is true:

1) For all $b \in B$ the set $\pi^{-1}(b)$ is a ν-dimensional submanifold which is C^r-homeomorphic to the fixed manifold L^ν,

2) For every $b \in B$ there exists a neighborhood U_b and a C^r-homeomorphism $\psi : \pi(U_b) \approx U \times L$ such that $\psi_b(\pi^{-1}(b')) = \{b'\} \times L$, $b' \in U_b$. (M is called the *total space* of the fibration, B the *base*, L the *fiber (leaf)*, and π the *projection*).

For a fibration, the following diagram is commutative ($\pi = P_1 \circ \varphi_b$)

$$\begin{array}{ccc} \pi^{-1}(U_b) & \xrightarrow{\ \varphi_b\ } & U_b \times L \\ & \overset{\pi}{\searrow} & \downarrow P_1 \\ & & U_b \end{array}$$

The simplest nontrivial fibrations in dimension two are flat *Möbius bands*, with fiber $\mathbb{R}^1$ and base S^1, and the flat *Klein bottle*, with S^1 as fiber and base. The 1-dimensional foliations that are complements to the above two fibrations are examples of a construction called the *suspension of homeomorphism*. On the product $B \times [0, 1]$ consider the equivalence relation $\sim$ defined by $(x, 0) \sim (f(x), 1)$, where $x \in B$ and $f : B \to B$ is a homeomorphism (diffeomorphism). The 1-dimensional foliation on the quotient space $B \times [0, 1]/_\sim$ corresponding to all $\{b\} \times [0, 1]$ is needed, see [CN].

The most famous fibrations are *vector bundles*, in particular, the *tangent bundle* $p : TM \to M$ of a manifold and the *normal bundle* $p : TM^\perp \to M$ of a submanifold $M \subset \bar{M}$.

A fibration with discrete fiber is called a *covering*; in this case, the dimensions of the total space and the base are equal. A popular example is the natural covering map $p : S^m \to RP^m$ from the sphere S^m onto a real projective space RP^m with two-point fiber (antipodal points of the sphere are identified).

Note that a submersion with compact leaves or with a compact total space and a connected base is actually a fibration [Tam]. The question of when a foliated manifold is actually a fibration, in particular, a product of manifolds, is studied in terms of an Ehresmann connection in Section 1.3.

A *Lie group* is a group G, which has a C^∞-differentiable manifold structure such that the map $(x, y) \in G \times G \to xy^{-1} \in G$ is C^∞-differentiable.

1.6 Definition. [CN] A *C^k-differentiable action of a Lie group G* (with identity e) on a manifold M is a C^k map $\varphi : G \times M \to M$ such that

$$\varphi(e, x) = x, \ \ \varphi(g_1 \cdot g_2, x) = \varphi(g_1, \varphi(g_2, x)), \ \ (g_1, g_2 \in G, \ x \in M).$$

The *isotropy group of a point $x \in M$* for the action φ is the subgroup of G defined by $G_x(\varphi) = \{g \in G : \varphi(g, x) = x\}$. The *orbit of a point $x \in M$* by the action φ is the subset $\mathcal{O}_x(\varphi) = \{\varphi(g, x) : g \in G\}$. We say that $\varphi : G \times M \to M$ is a *foliated action* if for every $x \in M$, the tangent space to the orbit of φ passing through x has a fixed dimension ν; if ν equals the dimension of G then φ is *locally free*.

Consider the equivalence relation on G: $g_1 \equiv g_2 \iff g_1^{-1} \circ g_2 \in G_x$. Let G/G_x be the quotient space of G by $\equiv$ and $p : G \to G/G_x$ the projection, which takes $g \in G$ to its equivalence class $p(g)$.

1.7 Theorem. [CN] *a) On G/G_x there is a unique differentiable structure that defines a fibration with fiber G_x.*
b) The orbits of a foliated action define the leaves of a foliation.

The action of the group $G = \mathbb{R}^1$ corresponds to a *dynamical system* on a manifold M. For example, the action of the group $O(2)$ by rotations on $\mathbb{R}^2 \setminus \{0\}$ defines a foliation with leaves that are concentric circles.

The action of the group S^1 on $S^3 \subset \mathbb{R}^4 \approx \mathbb{C}^2 = \mathbb{C}^1 \times \mathbb{C}^1$ by the isometries

$$\varphi_{p,q}(t)(z_1, z_2) = (\exp(2\pi i p t) z_1, \exp(2\pi i q t) z_2),$$

(where (z_1, z_2) is a pair of complex numbers with $z_1\bar{z}_1 + z_2\bar{z}_2 = 1$ and (p, q) is a pair of relatively prime positive integers), defines a foliation with equidistant leaves, which is not a fibration for most cases: the holonomy groups (see Section

1.3) of the leaves $z_1 = 0$ and $z_2 = 0$ are cyclic of orders q and p, respectively. The foliation is a Hopf fibration (see below) for $p = q = 1$.

The theory of foliations can be applied to obtain global information about the orbit structure of a locally free action of a group G on M.

1.8 Definition. [CN] The *rank of a manifold M* is the *maximum number of continuous pointwise linearly independent commuting vector fields* that the manifold admits. Equivalently, the rank of a manifold M is the greatest r for which there exists a locally free action of $\mathbb{R}^r$ on M.

This concept was introduced by J. Milnor. E. Lima [Lim] proved, that *every compact 3-dimensional manifold with finite fundamental group (for example, S^3) has rank one.* Moreover, *a compact 3-dimensional manifold with rank equal to two is a fibration over S^1 with torus fiber T^2* [CN].

Another similar topological invariant, *the number of continuous pointwise linearly independent vector fields on a manifold,* also has a deep relationship with foliations.

The exact upper estimate of the *number of vector fields on a sphere* is based on K-theory. We can write n as a product of an odd number and a power of two

$$n = (\text{odd})\, 2^{4b+c}, \quad (b \geq 0,\ 0 \leq c \leq 3).$$

Let $\rho(n) = 8b + 2^c$. Then $\rho(n) \leq 2\log_2 n + 2 \leq n$. Table 1 contains some values for the function $\rho(n) - 1$.

$n-1$	1	3	5	7	9	11	13	15	17	19	21	23	25	27	29
$\rho(n)-1$	1	3	1	7	1	3	1	8	1	3	1	7	1	3	1

Table 1.

1.9 Theorem. [Fu] *a) The maximal number of continuous pointwise linearly independent vector fields on the sphere S^{n-1} is equal to $\rho(n) - 1$.*

b) If $\nu \leq \frac{n}{2}$, then a continuous ν-dimensional distribution on S^{n-1} (and moreover a ν-dimensional foliation) exists $\Longleftrightarrow$ there exist ν continuous pointwise linearly independent vector fields on S^{n-1}.

The solution to the algebraic problem of the existence of orthogonal multiplications on $\mathbb{R}^n$ [Hus] allows us to construct $\rho(n) - 1$ orthogonal unit vector fields on the $(n-1)$-dimensional sphere: *for all* $\nu < \rho(n)$ *there exist* ν *orthogonal matrices* $\{B_i\}$ *of order* $n \times n$ *with the properties* $B_i^2 = -E$, $B_i B_j + B_j B_i = 0$, $i \neq j$.

For every $\nu \leq \rho(n) - 1$, the vector fields $\{w_i(x) = B_i x\}$, $(x \in S^{n-1}$, $1 \leq i \leq \nu)$ define a nonintegrable ν-dimensional *geodesic distribution* (see Chapter II) on the sphere S^{n-1}, that is, every great circle tangent to the distribution at one point is tangent to the distribution at every point.

Note that each vector field $\{w_i\}$ is *non-holonomic* (that is, its orthogonal hyper-distribution on S^{n-1} is non-integrable) and defines a Hopf foliation.

The sequence $\rho(n)$ plays an important role in the study of foliations with nonnegative mixed curvature (see Chapters IV–VII).

1.10 Example: Hopf fibration. Consider a round sphere $S^3 = \{x \in \mathbb{R}^4 : |x| = 1\}$ in a Euclidean space $\mathbb{R}^4$ with orthonormal basis $\{e_i\}$. A 2-dimensional subspace σ of $\mathbb{R}^4$ with orthonormal basis $\{a, b\}$ intersects S^3 in a *great circle* (geodesic) $\gamma(t) = a \cos t + b \sin t$, which plays the role of a straight line on a sphere. Two great circles in S^3 may not intersect, but are always *engaged*. For each great circle $\gamma \subset S^3$ and every point $x_0 \in S^3 \backslash \gamma$ with distance $dist\,(x_0, \gamma) < \frac{\pi}{2}$ there exist exactly two great circles $\gamma_1, \gamma_2 \subset S^3$ equidistant to γ (*Clifford parallelness*), i.e., for all $x \in \gamma_1 \cup \gamma_2$, $dist\,(x, \gamma) = dist\,(x_0, \gamma)$. The set $\gamma^\perp = \{x \in S^3 : dist\,(x, \gamma) = \frac{\pi}{2}\}$ of maximal distance is also a great circle; it is *orthogonal* to γ (see Figure 7).

In particular, the surface equidistant to γ $T(\gamma; s) = \{x \in S^3 : dist\,(x, \gamma) = s\}$ with $0 < s < \frac{\pi}{2}$ has a locally Euclidean metric and two (orthogonal when $s = \frac{\pi}{4}$) foliations by great circles; $T(\gamma; \frac{\pi}{4})$ is called a *Clifford torus*.

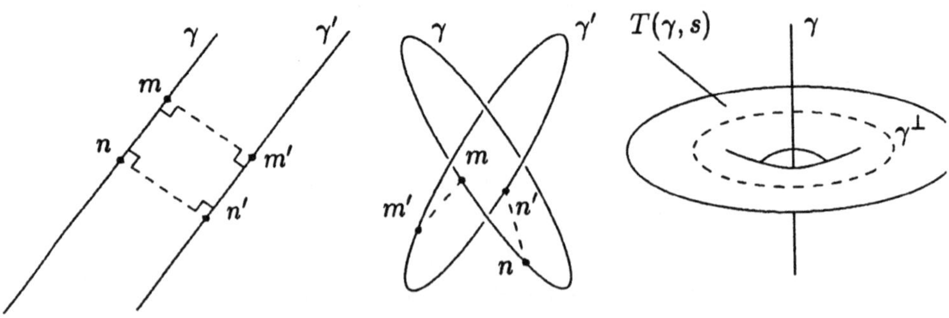

Figure 7. Clifford parallelness

Each of the two continuous partitions of S^3 by great circles, which are

Clifford parallel to the given γ, are called *Hopf fibrations*; their base is S^2 [Be 2].

The Hopf fibration, introduced in [Hop 1,2] about sixty years ago, had a powerful effect in **topology** because it provided the first example of a homotopically nontrivial map from one sphere to another of lower dimension. It also greatly influenced **geometry** because the fibers of a Hopf fibration are equidistant geodesics and form a *Riemannian foliation* on a round sphere. (Also see Appendix A.)

The Hopf fibration can be defined in the following (equivalent) ways:

1. The Hopf fibration $p : S^3 \to S^2 \approx CP^1$ for complex variables $z_1 = x_1 + ix_2, z_2 = x_3 + ix_4$ in $\{\mathbb{C}^2 = \mathbb{C}^1 \times \mathbb{C}^1\} \cap S^3$ has the form

$$p(z_1, z_2) = (|z_1|^2 - |z_2|^2,\ 2\Re(z_1\bar{z}_2),\ 2\Im(z_1\bar{z}_2));$$

for real variables the Hopf fibration $p : (x_1, x_2, x_3, x_4) \to (y_1, y_2, y_3)$ has the form

$$y_1 = x_1^2 + x_2^2 - x_3^2 - x_4^2,\ y_2 = 2(x_1x_3 + x_2x_4),\ y_3 = 2(x_2x_3 - x_1x_4)\ .$$

Note that $y_1^2 + y_2^2 + y_3^2 = (x_1^2 + x_2^2 + x_3^2 + x_4^2)^2$.

2. Introduce a complex variable z on S^2 via stereographic projection: $z = \frac{y_1+iy_2}{1-y_3}$. Then the above p can be written in the compact form $z = \frac{x_1+ix_2}{x_3+ix_4}$, as the usual map from the unit sphere in $\mathbb{C}^2 = \mathbb{R}^4$ to the complex projective line $\mathbb{C}P^1 = S^2$ [GW].

3. Consider the action φ of the group $\mathbb{R}^1$ on S^3

$$\varphi(t)(z_1, z_2) = (\exp(2\pi it)z_1, \exp(2\pi it)z_2).$$

The orbits of this foliated action form a Hopf fibration.

4. A Hopf fibration can be given as the collection of intersections of S^3 with all *holomorphic* 2*-planes* $\{\sigma = x \wedge Jx\}$, where J is any *complex structure* in $\mathbb{R}^4$ – a linear operator, given for some orthonormal basis $\{e_i\}$ by the rule

$$Je_1 = e_2,\ Je_2 = -e_1,\ Je_3 = e_4,\ Je_4 = -e_3. \tag{1.1}$$

Note that the multiplication by J gives an orientation of fibers.

The *projective space* KP^n, $(K = \mathbb{R}, \mathbb{C}, \mathbb{H})$ is the set of orbits for the right action of the group $K^* = K \setminus \{0\}$ on $K^{n+1} \setminus \{0\}$, that is, $x \sim y \iff$ there exists a $\lambda \in K^*$ such that $x = y\lambda$. Such a factor-space is not defined for a (non-associative) Cayley algebra $\mathbb{C}a$.

For $K = \mathbb{C}$ or $\mathbb{H}$ the orbits are equidistant $(\dim K - 1)$-dimensional great spheres. We define the *distance* between points on KP^n to be equal to the distance between corresponding orbits on a round sphere $S^{n \dim K - 1}$ and obtain a *canonical metric* on the projective spaces $\mathbb{C}P^n$ and $\mathbb{H}P^n$. In particular, a canonical metric on $\mathbb{C}P^1$ makes it into a round 2-sphere of radius $\frac{1}{2}$. The generalized *Hopf fibrations* (see [GWZ 1])

$$S^1 \subset S^{2n-1} \to \mathbb{C}P^{n-1}, \; S^3 \subset S^{4n-1} \to \mathbb{H}P^{n-1}, \; S^7 \subset S^{15} \to S^8 \qquad (1.2)$$

and a corresponding complex fibration [Es 3]

$$\mathbb{C}P^1 \subset \mathbb{C}P^{2n-1} \to \mathbb{H}P^{n-1} \qquad (1.2')$$

are Riemannian submersions. Moreover the fibrations in (1.2), except possibly the last one, are unique Riemannian submersions (that is, the leaves are equidistant) of Euclidean spheres [GG 2].

The following example plays an important role in the development of foliation theory.

1.11 Example: Reeb foliation. The *Reeb component* in the strip $S = \{(x, y) \in \mathbb{R}^2 : |x| \leq 1\}$ is the following foliation (see Figure 8): $\{L_a\}$, $(a \in R \cup \pm\infty)$

$$L_a = \{(t, a + f(t)) : |t| < 1\}, \; L_{\pm\infty} = \{(\pm 1, t) : |t| < \infty\},$$

where for example, $f(t) = \exp(\frac{t^2}{1-t^2}) - 1$ leads to a C^∞-foliation [Tam].

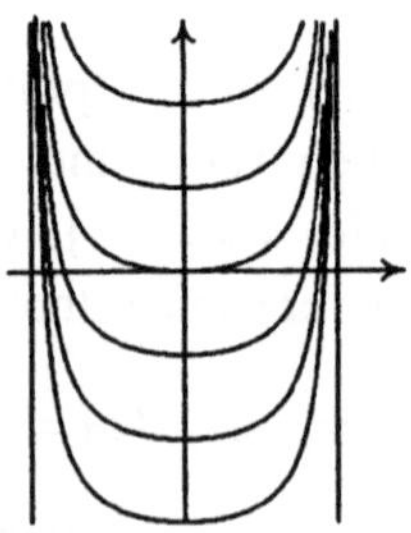

Figure 8. Reeb component

Reeb components can also appear on surfaces. By factoring a strip using a translation along axes OY, we obtain the following Reeb components: a Möbius band and its double covering on a cylinder (*Reeb annulus*), see Figure 9. The Reeb annulus with boundary lines each to the other forms a foliation on a torus or on a Klein bottle. Both these foliations have exactly one compact leaf, but on the torus such a foliation cannot be obtained from a nonsingular vector field.

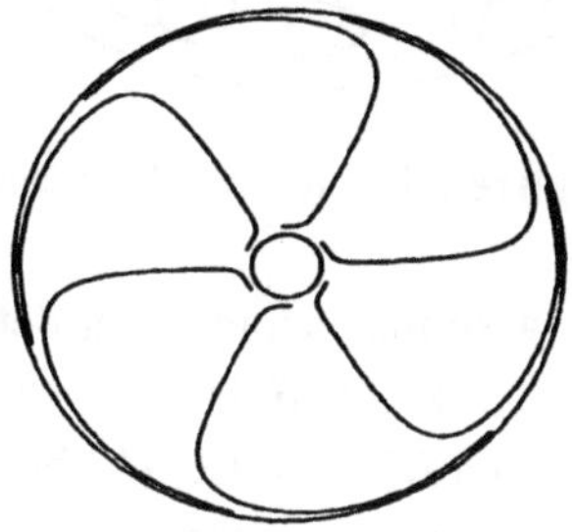

Figure 9. Reeb annulus

Any field of curves transverse to the Reeb annulus $\{L\}$ must have a limit cycle by Poincaré-Bendixson. But if $\{L\}$ were *geodesible*, i.e., there exists a complete Riemannian metric for which all leaves are geodesics, then its orthogonal trajectories would be a constant distance apart, measured along the curves of $\{L\}$, and hence could contain no limit cycle. Hence, the *Reeb component in a cylinder (or in the Möbius band) is not geodesible.*

H. Gluck studied the question of *what could prevent a foliation on M from being geodesible*? and proved the following theorem.

1.12 Theorem. [Gl 2] *On a closed surface any smooth curve field without Reeb components is geodesible.*

For simple examples of geodesic fields, consider parallel lines in Euclidean space, and parallel circles or winding lines on a flat torus as in Figure 4.

By rotation of a strip around the OY-axis, we obtain the Reeb component on a solid cylinder $D^2 \times \mathbb{R}$ (and hence on a solid torus, see Figure 10), which is also defined by a submersion $f : D^2 \times \mathbb{R} \to \mathbb{R}$:

$$f(x_1, x_2, x_3) = a(r^2)\exp(x_3), \tag{1.3}$$

where $r = \sqrt{x_1^2 + x_2^2}$ and $a(r) = \exp(-\exp(\frac{1}{1-r^2}))$.

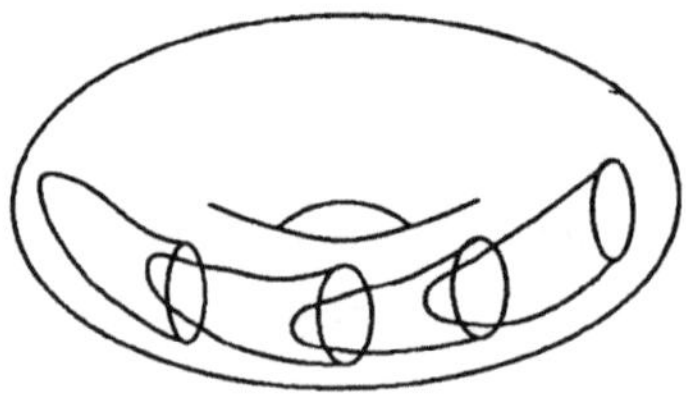

Figure 10. Reeb solid torus

Let $f_1 : \mathbb{R}^3 \to \mathbb{R}$ be a similar submersion defined on $\mathbb{R}^3$ by the same formula (1.3), where now $a : \mathbb{R} \to \mathbb{R}$ is a C^∞-function such that $a(1) = 0$, $a(0) = 1$ and if $t > 0$, then $a'(t) < 0$. Let $\{L\}$ be the foliation of $\mathbb{R}^3$ whose leaves are the connected components of the submanifolds $f^{-1}(c)$, $(c \in \mathbb{R})$. The leaves in the interior of the solid cylinder $C = f^{-1}([0,1]) = \{(x_1, x_2, x_3) : x_1^2 + x_2^2 \le 1\}$ are all homeomorphic to the plane $\mathbb{R}^2$. The boundary of C, $\partial C = f^{-1}(0) = \{(x_1, x_2, x_3) : x_1^2 + x_2^2 = 1\}$, is also a leaf – a cylinder. Outside C the leaves are all homeomorphic to cylinders (see Figure 11). *This submersion is not a fibration.*

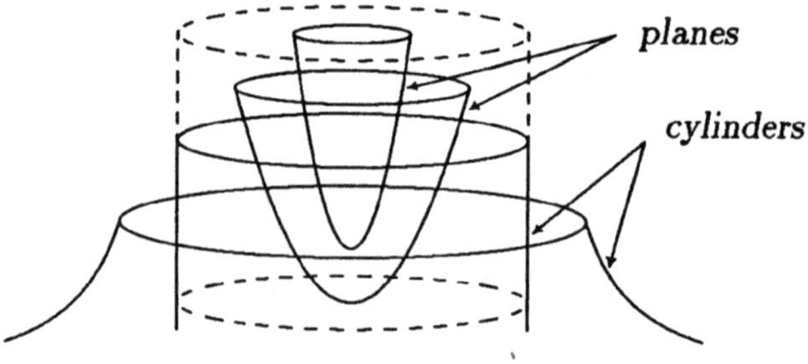

Figure 11. Submersion — not a fibration

As in the 2-dimensional case, we can factor the Reeb component in a solid cylinder, thereby obtaining a foliation in a solid torus $D^2 \times S^1$ by copies of $\mathbb{R}^2$; these accumulate only on the compact leaf, the boundary torus $\partial(D^2 \times S^1) = S^1 \times S^1$. *This foliation is not defined by a submersion.*

Finally, from two Reeb components in $D^2 \times S^1$ we obtain a C^∞-foliation of a 3-sphere, because $S^3 = \{x = (x_1, x_2, x_3, x_4) \in \mathbb{R}^4 : \sum_{i=1}^4 x_i^2 = 1\}$ can be considered as the union of two solid tori $S^3_+ = \{x \in S^3 : x_1^2 + x_2^2 \ge \frac{1}{2}\}$ and $S^3_- = \{x \in S^3 : x_1^2 + x_2^2 \le \frac{1}{2}\}$ identified along the boundary $T^2 = \{x \in S^3 : x_1^2 + x_2^2 = \frac{1}{2}\}$. This *Reeb foliation on* S^3 has exactly one compact leaf. Thus it is optimal according to the Novikov Theorem (see [CN], [Tam]), which states:

Every 2-dimensional C^2-foliation of a compact 3-dimensional manifold M with finite fundamental group has a compact leaf that is homeomorphic to a torus in the case of $M = S^3$.

1.2. Holonomy

The behavior of leaves near a fixed leaf L can be studied by the first returns of leaves to a small transverse section passing through a point $x \in L$. The notion of holonomy is a generalization of Poincaré's first return map from flows to foliations and is essential for diffeomorphism and structural questions about foliations.

1.13 Definition. [CN] Let X, Y be topological spaces and $x \in X$. On the set of maps $f : V \to Y$, where V is a neighborhood of x, consider the equivalence relation $f \sim g \iff$ there is a neighborhood W of x such that $f_{|W} = g_{|W}$. The equivalence class of f is called the *germ* of f at x.

When $X = Y$ the set $\Gamma(X, x)$ of *germs of local homeomorphisms which leave x fixed* is a group with the natural multiplication. The example that is important for foliations is Γ_n^k, the *group of germs of* C^k*-diffeomorphisms of* $\mathbb{R}^n$ *leaving* 0 *fixed.*

The mappings of cross-sections associated to curves lying in leaves allow us to obtain a representation of $\pi_1(L, x)$ (the fundamental group of the leaf L) by the germs of diffeomorphisms of transversal manifolds at x.

1.14 Lemma. [Rei 3] *Let L be a leaf of a foliation of codimension n, $x \in L$, and h an embedding of $\mathbb{R}^n$ transverse to the foliation such that $h(0) = x$. If the foliation is of class C^k, then h is also assumed to be of class C^k. Then there is a homomorphism $f(L, x, h)$ taking $\pi_1(L, x)$ into Γ_n^k, which is determined up to conjugacy by the foliation and the leaf.*

Idea of Proof. There exists at least one Riemannian metric on any smooth manifold M [GHL]. If one of these metrics is fixed, then the orthogonal distribution $TL^{\perp}$ of the foliation $\{L\}$ may be identified with the *normal bundle,* which is defined algebraically as the factor-bundle TM/TL. Then the local embedding h may be given by $\exp(TL^{\perp})$, the exponential mapping along geodesics orthogonal to L.

Let the loop $c : [0,1] \to L(x)$, $(c(0) = c(1) = x)$ represent an element of $\pi_1(L, x)$. Then the vector bundle $T_c L^{\perp}$ (restriction along c) admits a 1-dimensional foliation transverse to the fibers (having the zero section c as a leaf), and h induces an embedding $\tilde{h}$ of a neighborhood of 0 in $\mathbb{R}^n$ into the total space $T_{c(0)} L^{\perp}$. If $\pi : T_c L^{\perp} \to M$ is a projection, then $\pi^{-1} * c$ makes each leaf into a path defined on some open subinterval of $[0,1]$. Since a zero section is a leaf, paths having $(\pi^{-1} * c)(0)$ close to 0 will be defined on all of $[0,1]$. The map taking the initial point of each such path to its final point is a diffeomorphism of some neighborhood of 0 in the fiber $T_x L^{\perp}$ to some other neighborhood of 0; it still induces a uniquely defined germ of a diffeomorphism at the origin in $\mathbb{R}^n$. This germ is taken to be $f(L, x, h)(c)$. □

1.15 Definition. [Rei 3] The *holonomy homomorphism* of a leaf is the homomorphism $Hol\,(L) : \pi_1(L) \to \Gamma_n^k$ occurring in the statement of Lemma 1.14. The *holonomy group* of a leaf is the image of the holonomy homomorphism, and the *holonomy covering* is the covering whose fundamental group is the *kernel of the holonomy homomorphism.*

A simply connected leaf of a foliation has trivial holonomy. It is clear that if the holonomy group of a leaf L is small, then the recurrence of leaves near L will be small. If $k > 0$, the foliated structure in the neighborhood of a compact leaf is determined up to diffeomorphism by the holonomy homomorphism [Hae1].

The holonomy of the core circle γ on a flat Möbius band foliated by circles equidistant to γ is the group $\mathbb{Z}_2$ generated by the diffeomorphism $f(x) = -x$ (see Figure 12a). In the Reeb foliation of S^3 (see Section 1.1) the fundamental group of the unique compact leaf (torus T^2) is isomorphic to $\mathbb{Z} \oplus \mathbb{Z}$. Its holonomy group (see Figure 12b) is the same and can be represented by two C^∞-diffeomorphisms f, $g : \mathbb{R} \to \mathbb{R}$ with $f(0) = g(0) = 0$ such that

$$f(x) \begin{cases} < x, & \text{for } x > 0 \\ = x, & \text{for } x \le 0 \end{cases}, \qquad g(x) \begin{cases} = x, & \text{for } x \ge 0 \\ < x, & \text{for } x < 0. \end{cases}$$

For more details on holonomy for codim $L = 1$ see [HH].

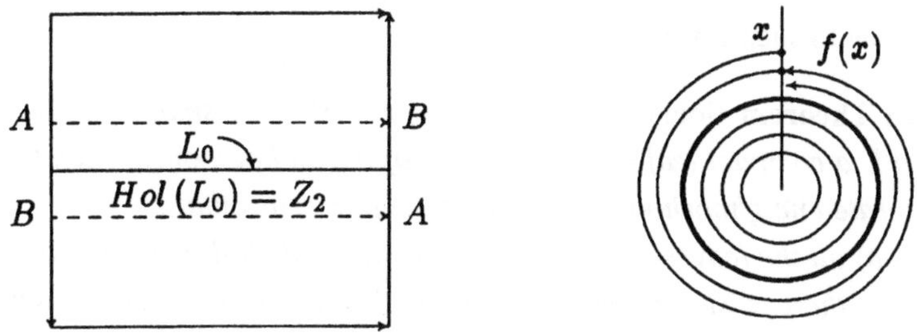

Figure 12a,b. (a) Holonomy of Möbius band;
(b) Holonomy of Reeb foliation

1.16 Lemma. *[EMT],* [Rei 3] *The union of all leaves with trivial holonomy is a dense set which is the intersection of countably many open sets.*

Idea of Proof. Let M be covered by a finite or a countable number of foliated charts. For x from some chart U_α (with submersion $\pi_\alpha : U_\alpha \to \mathbb{R}^n$ constant along components of leaves) let a loop $c : [0,1] \to L$, $(c(0) = c(1) = x)$ represent an element of $\pi_1(L, x)$. The loop c is covered by a finite number of foliated charts, which define the local diffeomorphisms $\psi_i : \mathbb{R}^n \to \mathbb{R}^n$; these difffeomorphisms have $\pi_\alpha(x)$ as a fixed point. There are countably many (finite sequences of given foliated charts) of such diffeomorphisms ψ_i. Decompose $\mathbb{R}^n = G_i \cup \mathring{A}_i \cup \partial A_i$, where $G_i = \{y \in \mathbb{R}^n : \psi_i(y) \neq y\}$ is an open set of moved points, $\mathring{A}_i$ is an open set (the interior of the set of fixed points, A_i) and the boundary ∂A_i is a closed nowhere dense set. A holonomy of a leaf through x is non-trivial if and only if $\pi_\alpha(x) \in \partial A_i$ for some loop c. But the union of a countable number of nowhere dense sets $\{\partial A_i\}$ is also a nowhere dense set. The transition from U_α to all of M is obvious. □

The set $\{m \in M : Hol(m) \neq 0\}$ may have positive measure. (For related problems see [Lan 2].)

1.17 Example. [MS] Let K be a Cantor subset of the unit circle of positive measure and $f : S^1 \to S^1$ the homeomorphism which has K as its fixed point set. The associated foliation (using the *suspension of homeomorphism* construction) on the torus has closed leaves corresponding to each point of K and each of these leaves has nontrivial holonomy.

All leaves of a fibration are diffeomorphic; for foliations the sufficient conditions for this are given in the following stability theorems. Holonomy is the key to such results.

1.18 Theorem (local stability). [CN] *Let M be a smooth foliated manifold and L a compact leaf having a) finite, b) trivial holonomy. Then there exists a saturated neighborhood U of L with the properties that :*

a) all the leaves in U are compact with finite holonomy groups,

b) there exists a diffeomorphism $\psi : L \times \mathbb{R}^n \to U$, which preserves leaves.

The Global Stability Theorem for $\operatorname{codim} L = 1$ was obtained by Reeb. For the Local Stability Theorem for a Riemannian foliation with singularities see [Žu 3].

From Lemma 1.16 and Theorem 1.18 it follows that for a foliation with compact leaves, the union of all leaves with trivial holonomy is an open dense set of M.

1.19 Theorem. *[Epd],* [Rei 3] *For a foliation with compact leaves, the following statements are equivalent (other similar conditions are in Section* 1.4*): a) the holonomy group of the leaf L is finite, b) the space of leaves is Hausdorff in a neighborhood of L, c) the volume of leaves in a neighborhood of L is uniformly bounded in some (and hence every) Riemannian metric on M, d) the leaf L admits a fundamental system of saturated neighborhoods.*

For a foliation with compact leaves: 1) if either $\operatorname{codim} L = 1$ or $\operatorname{codim} L = 2$ and M is compact, then according to [Ree 1] and [EMS] all leaves have finite holonomy groups, 2) if either $\operatorname{codim} L \geq 3$ or $\operatorname{codim} L = 2$ and M is not compact, then the assertion in 1) need not be true, see [Bes 1].

1.20 Definition. [Rei 3] The *graph (holonomy groupoid)* $\Gamma(M, L)$ of a foliated manifold M is the collection of all triples $(x, y, [\alpha])$, where x, y belong to the same leaf L, α is a piecewise smooth path from x to y in L and $[\alpha]$ is the holonomy equivalence class of α. Two triples $(x, y, [\alpha])$ and $(x', y', [\beta])$ are equivalent iff $x = x'$, $y = y'$ and the holonomy of the curve $\alpha\beta^{-1}$ is trivial.

The notion of holonomy groupoid was introduced by [Ehr 1]; it was later studied in some detail by [Win 1], who was interested primarily in Riemannian foliations (the graph in this case is Hausdorff) and geometric properties of the graph. The study of its analytic properties was started by [Con]. The importance of the graph also comes from the fact that it is the starting point of the construction of the C^*-algebras associated to this foliation.

The following are associated canonically to $\Gamma(M, L)$: a *diagonal embedding* $\Delta : x \in M \to (x, x, [0]) \in \Gamma(M, L)$, an *involution* $i : (x, y, [\alpha]) \in \Gamma(M, L) \to (y, x, [\alpha]) \in \Gamma(M, L)$, and the mappings $p_1 : (x, y, [\alpha]) \to x$ (the *source projection*) and $p_2 : (x, y, [\alpha]) \to y$ (the *target projection*) which define two submersions of the graph onto M (see [Wol 4] for local coordinates). The fibers of the submersions p_1, p_2 are the holonomy coverings of leaves of the given foliation.

This locally compact groupoid is primarily a measure of the behavior of the foliation near a leaf.

The set M with the leaf topology is not connected, but each component L (leaf) separately gives a manifold of dimension $2 \dim L$ in the graph. $\Gamma(M, L)$ has a locally Euclidean topology of dimension $\dim M + \dim L$, but in general this topology is non-Hausdorff. Moreover, if the elements of the holonomy pseudogroup are determined by their jets, then the graph is a Hausdorff topological space.

1.21 Remark. In other words, the holonomy groupoid $\Gamma(M, L)$ is a quotient of the fundamental groupoid of M, when M has the leaf topology. As a set, the *fundamental groupoid* $\Gamma(X)$ of the topological space X is the quotient of the set of paths in X by the relation of homotopy with fixed endpoint. It is a groupoid under the composition and inversion induced by the composition and inversion of paths. The topology of $\Gamma(X)$ is a compact open topology on the space of paths, and with this topology $\Gamma(X)$ is a topological groupoid.

In fact there are two groupoids associated with a foliation, the homotopy groupoid and the holonomy groupoid, sometimes called the graph.

The *homotopy groupoid* of a foliation is defined in a similar way (as the holonomy groupoid). We identify leaf curves which are homotopic relative to their ends in the corresponding leaf. Likewise, the homotopy groupoid is a manifold, but not necessarily Hausdorff. There is a natural submersion from the homotopy groupoid onto the holonomy groupoid of a given foliation. In [Hec], see also [Las], one can find an example of a foliation for which the holonomy groupoid is Hausdorff, but the homotopy groupoid is not. The non-Hausdorffness of the homotopy groupoid is an important obstruction in the symplectic integration of Poisson manifolds as well as in their quantization.

A *vanishing cycle* for a foliation $\{L\}$ is a mapping $c : S^1 \times [0, 1]$ such that for any $t \in [0, 1]$ $c_t = c_{|S^1 \times \{t\}}$ is a loop on a leaf of $\{L\}$ and c_0 is not homotopic to the constant loop in the leaf, but for $t > 0$, the loops c_t are homotopic.

If a foliation admits a vanishing cycle, then it admits a vanishing cycle $c : S^1 \times [0, 1] \to M$ for which the curves c_t are tangent to a horizontal subbundle.

In [DH] it is demonstrated that the non-existence of vanishing cycles is equivalent to the Hausdorffness of the homotopy groupoid of the foliation.

R. Wolak [Wol 9] defines a holonomy vanishing cycle for a foliation and, by similar methods, studies its relations with the Hausdorffness of the holonomy groupoid.

A *holonomy vanishing cycle* for a foliation $\{L\}$ is a mapping $c : S^1 \times [0, 1]$ such that for any $t \in [0, 1]$ $c_t = c_{|S^1 \times \{t\}}$ is a loop on a leaf of $\{L\}$ and c_0 has nontrivial holonomy, but for $t > 0$, the loops c_t carry no holonomy. Obviously, a foliation without holonomy has no holonomy vanishing cycles. The graph of the foliation $\{L\}$ is Hausdorff iff $\{L\}$ has no holonomy vanishing cycles [Wol 9].

Associated with any holonomy group (or groupoid) is a family of *infinitesimal holonomy groups* (or groupoids), obtained by taking the j-jets, ($j \leq k$), of the elements of the holonomy group (or groupoid). The most interesting member of this family is the *linear holonomy*, that is, the case where $j = 1$. For any closed curve, this is an element of the general linear group.

For a Riemannian manifold M with a foliation $\{L\}$, the embeddings $h : \mathbb{R}^n \to M$ (see Lemma 1.14) may be chosen orthogonal to a given leaf L at all points of the path α. Then the linear holonomy acts on the orthogonal subspaces $T_{d(t)}L^\perp$ along the path α. For a Riemannian foliation, the holonomy acts by local isometric transformations on (germs of) transversal manifolds and hence is completely determined by its 1-jets at each point, i.e., by linear holonomy – orthogonal transformations of $T_{d(t)}L^\perp$ [Ton 3].

The trajectory of a vector $y \in T_xL^\perp$ under linear holonomy displacement is called an *L-parallel vector field* along curve $\alpha \subset L$ (see Chapter III). Such fields are defined globally on leaves with trivial holonomy. L-parallel vector fields may be characterized with the help of the *Bott connection* in $TL^\perp$

$$\check{\nabla}_x s = P_2[x, Y_s], \ (x \subset TL, \ s \in TL^\perp),$$

where $Y_s \subset TM$ is any vector field projectable to s under the orthoprojector $P_2 : TM \to TL^\perp$. The curvature $\check{R}(x, u)$ is zero for x, $u \in TL$ as a consequence of the Jacobi identity for the bracket of vector fields. This means that $(TL^\perp, \check{\nabla})$, restricted to each leaf, is a flat vector bundle. The parallel transport in $TL^\perp$ along a path in L is a linearized version of holonomy, and hence, it depends only on the homotopy class of a path in a leaf.

An *adapted connection* in $TL^\perp$ is a connection whose restriction along L coincides with the partial Bott connection $\check{\nabla}$. Such a connection exists, for example,

$$\tilde{\nabla}_x s = \begin{cases} P_2[x, Y_s], & x \subset TL \\ P_2 \nabla_x Y_s, & x \subset TL^\perp \end{cases}, \quad s \in TL^\perp$$

with zero curvature $\tilde{R}(x, u)$ for $x,\ u \in TL$. A connection in $TL^\perp$ is *basic* if it is adapted and holonomy invariant. These connections play an important role in gauge theory on a foliated manifold. In fact (Molino, 1971), there is a topological obstruction for the existence of basic connections [KT1], [Ton 3].

We study L-parallel vector fields from a Riemannian geometry point of view in Chapter III with their applications in Chapters IV–VII.

1.3. Ehresmann foliations

Below we consider a foliation $\{L\}$ on a smooth manifold M together with a complementary distribution D. This distribution is transversal, but in the case of a Riemannian manifold may not be orthogonal to TL.

We call a piecewise smooth curve in M (i.e., a piecewise smooth map from $[0, 1]$ to M), whose velocity vector field lies in TL (resp. D), an *L-curve* (resp. *D-curve*).

1.22 Definition. [BlH 2] Every D-curve $\beta : I \to M$ uniquely determines a family of local diffeomorphisms $g_t : V_0 \to V_t$, $(t \in I)$ from one leaf to another (*an element of holonomy along* β) such that 1) V_t is a neighborhood of $\beta(t)$ in the leaf through $\beta(t)$, 2) $g_t(\beta(0)) = \beta(t)$ for all t, 3) for each $x \in V_0$ the curve $g_t(x)$ is tangent to D, 4) g_0 is the identity map of V_0. When the leaves of the foliation have a geometric structure (a measure, a linear connection, or a Riemannian metric) we say that *D preserves the geometry of the leaves* if the elements of holonomy along D-curves are local isomorphisms of the particular geometric structure.

For example, when $\{L\}$ is a totally geodesic, umbilic, or minimal foliation on a Riemannian manifold and $D = TL^\perp$, then an element of holonomy along any horizontal curve is a local isometry, a conformal mapping, or a volume preserving map of the induced metrics on the leaves, respectively.

C. Ehresmann [Ehr 1] defined a *connection in a fiber bundle* as a distribution complementary to fibers with the following completeness condition: every curve in the base admits a horizontal lift into the total space. R. Blumenthal and J. Hebda [BlH 1-6] defined the notion of an *Ehresmann connection* for a foliation on a manifold without boundary as a complementary distribution to the foliation satisfying a certain condition, and proved some results related to it. N. Koike [Ko 1,4] continued these investigations, in particular, for a foliation on a manifold with boundary. The graph of a foliation with an Ehresmann connection is studied by N. Žukova [Žu 4] and R. Wolak [Wol 4].

1.23 Definition. [BlH 2] A piecewise smooth map $\delta : [0,1] \times [0,1] \to M$ such that, for every fixed s_0, the curve $\delta(*, s_0)$ is a D-curve, and for every fixed t_0 the curve $\delta(t_0, *)$ is an L-curve, is called *rectangular* in the foliated manifold M.

Also, the curves $\delta(0,*)$, $\delta(1,*)$, $\delta(*,0)$ and $\delta(*,1)$ are called the *initial L-edge*, the *terminal L-edge*, the *initial D-edge*, and the *terminal D-edge* of δ, respectively. The rectangular whose initial edges are an L-curve α and a D-curve β is called the *rectangular associated to α and β* and is denoted by $\delta_{\alpha,\beta}$. (It is easy to show that two rectangulars with the same initial edges coincide, so the last notion is correct.) For a sufficiently small positive ε, the rectangulars $\delta_{\alpha|_{[0,\varepsilon]},\beta}$ and $\delta_{\alpha,\beta|_{[0,\varepsilon]}}$ exist.

For a manifold with boundary we assume that each boundary component is tangential or transversal to the leaves (we do not treat a case as in Figure 13) and consider only a complementary distribution D to a foliation satisfying the following condition:

D is tangent to each L-transversal boundary component. (*)

Such a distribution can always be constructed [Ko 4].

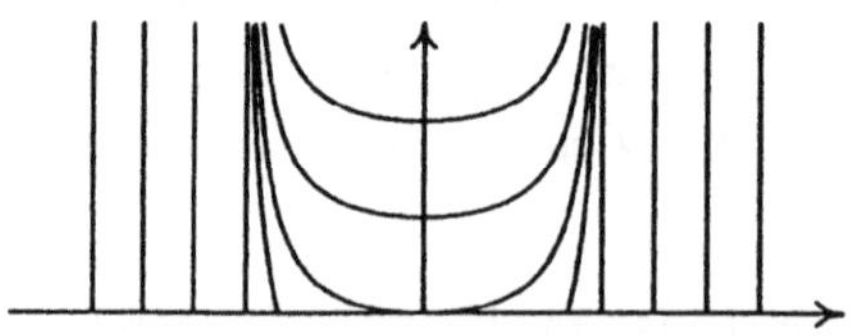

Figure 13. Foliation on a halfplane

1.24 Definition. [BlH 2] A complementary distribution D to a foliation $\{L\}$ is called an *Ehresmann connection* for $\{L\}$, if for every L-curve α and every

D-curve β with the same initial point, i.e., $\alpha(0) = \beta(0)$, the rectangular $\delta_{\alpha,\beta}$ exists. Koike calls $\{L\}$ an *Ehresmann foliation* when such a distribution D exists.

The basic property of an Ehresmann connection is that for any rectangular $\delta_{\alpha,\beta}$ the curve $\delta(1,*)$ depends only on the homotopy class of a curve α, and the curve $\delta(*,1)$ depends only on the homotopy class of a curve β.

There are "boundary restrictions" for an Ehresmann foliation [Ko 4]:

– *An Ehresmann foliation is tangential to all boundary components or transversal to all of them. In the second case each leaf meets all boundary components. In particular, for a* 1*-dimensional Ehresmann foliation that is transversal to all boundary components, M is homeomorphic to $B \times [0, 1]$, $B \times [0, \infty)$ or covered doubly by $B \times [0, 1]$.*

– *If M is compact and $\pi_1(M)$ is finite, then there is no codimension-*1 *Ehresmann foliation on M transversal to the boundary. If M possesses a compact boundary component B with finite fundamental group, then there is neither a codimension-*1 *Ehresmann foliation on M transversal to the boundary nor a codimension-*2 *Ehresmann foliation on M tangential to the boundary.*

For an Ehresmann foliation the following facts are known:

– *Any two leaves are joined by a D-curve.*

– *The universal coverings of any two leaves are diffeomorphic.*

1.25 Theorem (topological product). [BlH 1,2] *Let D be an integrable Ehresmann connection (with leaves $L^\perp$) for a foliation $\{L\}$ on M. Then the universal covering $\tilde{M}$ is topologically the product $\tilde{L} \times \tilde{L^\perp}$ of universal coverings of the leaves L and $L^\perp$.*

1.26 Corollary. [BlH 2] *A compact M with finite fundamental group admits no codimension-*1 *Ehresmann foliation. In particular, the Reeb foliation on S^3 is not geodesible.*

Let $\{L\}$ be a foliation on a Riemannian manifold that is tangential (or transversal) to the boundary. Assume that the leaves are orthogonal to the boundary in the case where $\{L\}$ is transversal to the boundary. Hence the orthogonal complementary distribution $D = TL^\perp$ satisfies the condition (*).

Let α (resp. β) be an L-curve (resp. D-curve). Denote by *Rec* $(\alpha, *)$ the set of all rectangulars whose initial L-edge is α and initial D-edge is a regular curve, and by *Rec* $(*, \beta)$ the set of all rectangulars whose initial D-edge is β and initial L-edge is a regular curve. Define a function $G^\perp_\alpha$ on *Rec* $(\alpha, *)$

by $G_\alpha^\perp(\delta) = \frac{l(\delta_{*,1})}{l(\delta_{*,0})}$ for $\delta \in Rec(\alpha, *)$, and a function $G_\beta^\top$ on $Rec(*, \beta)$ by $G_\beta^\top(\delta) = \frac{l(\delta_{1,*})}{l(\delta_{0,*})}$ for $\delta \in Rec(*, \beta)$, where $l(*)$ is the length of a curve. Note that $G_\beta^\top \equiv 1$ for every D-curve β in the case of a totally geodesic foliation, and $G_\alpha^\perp \equiv 1$ for every L-curve α in the case of a Riemannian foliation (i.e., bundle-like metric).

Sufficient conditions for the normal bundle of the foliation on a Riemannian manifold to be an Ehresmann connection are given below.

1.27 Theorem. [Ko 4] *Let $\{L\}$ be a foliation on a Riemannian manifold M that is tangential (or orthogonal) to the boundary. Assume that one of the following conditions holds:*

a) the induced metrics on the leaves are complete and $\sup G_\beta^\top < \infty$ *for every $TL^\perp$-curve β,*

b) M is complete and $\sup_{s\in[0,1)} G^\perp_{\alpha|[0,s]} < \infty$ *for every L-curve α without a terminal point.*

Then $TL^\perp$ is an Ehresmann connection.

The first part of each of the conditions in Theorem 1.27 is necessary (see Figure 14) as is the second (see the examples of foliations on a plane with Reeb component in [Ko 4]).

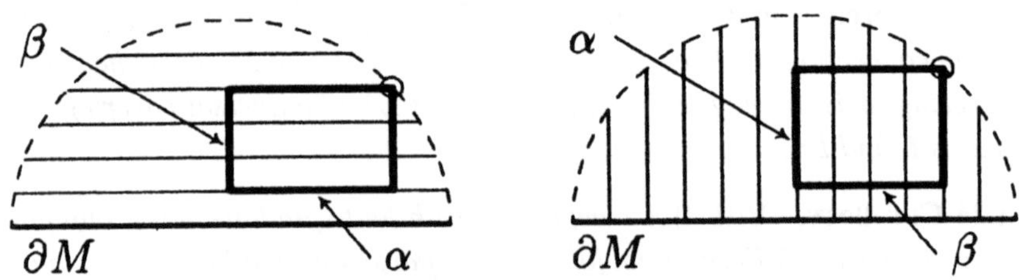

Figure 14. Foliations on half disk

In particular, the orthogonal distribution to a complete totally geodesic foliation and to a Riemannian foliation on a complete Riemannian manifold are Ehresmann connections. However, Ehresmann foliations on M with $\dim M > 2$ are more numerous; they can be neither geodesible nor admit a bundle-like metric, see the example in [BlH 2].

Ehresmann foliations also arise among totally umbilic foliations [BlH 2], and foliations with the following transversal structures: Cartan structure, a G-

structure of finite type, a system of differential equations of arbitrary order (see [Žu 4]), and Lagrange foliations with some assumptions, [Wol 10] foliations on a manifold over local algebra, [Shu] in field theories and [LMM].

For a submersion with compact connected leaves, every distribution complementary to the leaves is an Ehresmann connection. This is not true in the noncompact case:

1.28 Example. [Žu 2] Consider the two sets $X = \{(x, y) \in \mathbb{R}^2 : 0 < x < 2, 0 < y < 2, y < \frac{x}{2} + 1\}$ and $B = \{x \in \mathbb{R}^1 : 0 < x < 2\}$ with induced Euclidean metrics. The map $p : (x, y) \in X \to x \in B$ is a smooth fibration having an Ehresmann connection (the most simple of which is tangent to the complementary foliation by straight lines $\{y = t(\frac{x}{2} + 1) : 0 < t < 1\}$), see Figure 15. But the horizontal (integrable) distribution for p is not an Ehresmann connection. For example, a path $h : I \to B$, given by $h(t) = 1.8 - 1.6t$ has no horizontal lift with initial point $P(1.8, 1.6) \in X$.

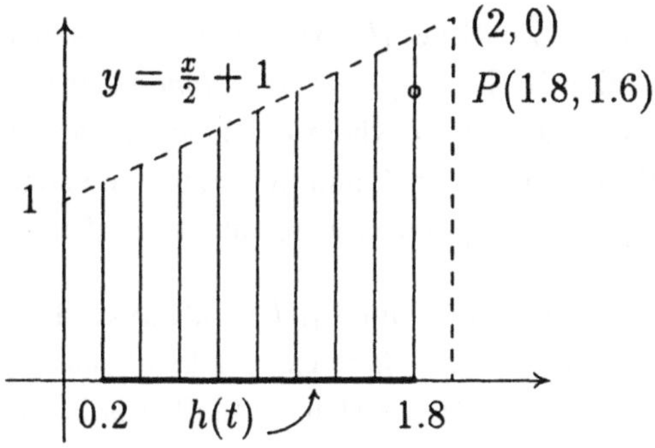

Figure 15. Orthogonal distribution is not an Ehresmann connection

1.29 Theorem. [Rec 3] *If the surjective submersion $\pi : M \to B$ admits an Ehresmann connection D, then it is a fiber bundle.*

1.30 Remark. The condition of Theorem 1.29 is true for two "dual" classes on complete Riemannian manifolds: Riemannian submersions [Her 1] and submersions with totally geodesic fibers [Rec 3] with orthogonal distribution D.

A foliation admitting an Ehresmann connection has no vanishing cycles [Wol 9]. As a corollary, the homotopy groupoid of an Ehresmann foliation is a Hausdorff manifold [Wol 9].

For an Ehresmann foliation, more special holonomy and a graph may be defined in the following way.

1.31 Definition. [BlH 2] Let $\{L\}$ be a foliation on a manifold M with an Ehresmann connection D and let Ω_x be the collection of horizontal curves with initial point x. Then the fundamental group $\pi_1(L, x)$ of leaf L acts on Ω_x in the natural way:

$$\text{for all } \alpha \in \Omega_x, \ [\psi] \in \pi_1(L, x) : \ [\psi]\alpha = \delta(1, *),$$

where δ is the rectangular associated to the curves ψ, α. Let $K_D(L, x) = \{[\psi] \in \pi_1(L, x) : \ [\psi]\alpha = \alpha, \ \forall \alpha \in \Omega_x\}$ be the kernel of this action. The factor-group $\Gamma_D(L, x) = \pi_1(L, x)/K_D(L, x)$ is called the *D-holonomy group* of the leaf L.

$H_D(L, x)$ does not depend on the specific Ehresmann connection and hence is an invariant of the foliation. In some sense $H_D(L, x)$ is a globalization of the germinal holonomy group $\Gamma(L, x)$ and naturally surjects onto it.

1.32 Definition. [Žu 4] Let $\{L\}$ be a foliation on a manifold M with an Ehresmann connection D. The *D-graph (D-holonomy groupoid)* $\Gamma_D(L)$ is the collection of all triples $(x, y, [\alpha])$, where x, y lie in the same leaf L, α is a (piecewise) smooth path from x to y in L and $[\alpha]$ is the equivalence class of α [two paths α, β with the same endpoint x, y are equivalent if $\alpha\beta^{-1} \in K_D(L, x)$].

1.33 Remark. [Žu 4] The graph $\Gamma_D(L)$ is always a Hausdorff ($\dim M + \dim L$)-dimensional manifold, and there exists a local isomorphism $\Gamma_D(L) \to \Gamma(L)$. Moreover, the graph $\Gamma(L)$ is Hausdorff iff it is isomorphic (in the canonical way) to $\Gamma_D(L)$.

1.34 Theorem. [Žu 4], [Wol 4] *Let M be a foliated manifold with an Ehresmann connection D. Then the projections $p_i : \Gamma_D(L) \to M$, $(i = 1, 2)$ are locally trivial fibrations. Their fibers are the holonomy coverings of leaves.*

The graph $\Gamma_D(L)$ is useful for studying stability problems. The stability condition follows very nicely from Theorem 1.34 .

1.35 Corollary (global stability). [BlH 2], [Žu 4], [Wol 4] *Let M be a foliated manifold with an Ehresmann connection D, and L a compact leaf, or a leaf with finite volume; moreover, L has finite D-holonomy. Then all the leaves have finite D-holonomy groups and are compact or with finite volume, respectively.*

1.4. Foliations and curvature

Since 1980, foliations have played an important role not only in topology and analysis, but also in Riemannian geometry; see monographs [Rei 3], [Mol], [Kit 2], and [Ton 1,3]. *Totally geodesic* and *Riemannian foliations* have the simplest external geometry of the leaves, (the tangent or orthogonal distribution has a zero second fundamental form), they are dual in a certain sense, and are investigated in a number of works. In 1979 H. Gluck [Gl 1] raised two basic questions on the correlation between structures of foliations and metrics on manifolds:

FM 1: *The existence of a geodesible metric for a given foliated manifold.*

For a foliation on a 2-dimensional manifold the obstruction for geodesibility is a *Reeb component* (see Section 1.1). Of foliations that are generated by Morse-Smale vector fields without singularities, only suspensions of diffeomorphisms are geodesible [Gl 2]. Note that for $n \geq 4$ there exist non-geodesible foliations of M^n even by closed curves (see [Appendix A, Bes 1] and Section 1.2).

FM 2: *The existence (and classification) of geodesic foliations on a given Riemannian manifold.*

Much is known about geodesic (and Riemannian) foliations on space forms, see Section 4.1. All geodesic foliations on a round 3-sphere were catalogued in [GW], (also see Appendix A), but the same problem is open for S^{2n+1} with $n \geq 2$, for foliations on compact Lie groups, see [Rez 2]; for foliations on some standard hypersurfaces in Euclidean n-spheres, see [Pet]. Later these two problems were extended for minimal, Riemannian, umbilic and other classes of foliations (see [KT 2-4], [Ru 1], and [Hae 3]).

1.36 Definition. A foliation is said to be

1) *tense*, if there exists a Riemannian metric for which the mean curvature form of the leaves is parallel along the leaves [KT 4],

2) *taut*, if there exists a Riemannian metric for which the leaves are minimal submanifolds,

3) *umbilicalizable* (resp. *geodesible*), if there exists a Riemannian metric for which the leaves are totally umbilic [Ca 6] (resp. totally geodesic) submanifolds. (Geodesible $\Rightarrow$ taut $\Rightarrow$ tense),

4) *mean curvature invariant (MCI)* if there exists a Riemannian metric, for which the mean curvature of the leaves is a basic vector field [Wal 6],

5) *Riemannian*, if there exists a Riemannian metric for which the leaves

are locally equidistant submanifolds. (An example is a *homogeneous* foliation obtained by an isometric action of a Lie group.)

Since the curvature of a Riemannian foliation has stronger vanishing properties than that of an arbitrary foliation, its characteristic classes should have special properties. Many of the usual classes vanish, but there are additional classes which can be defined only for Riemannian foliations; see [Rei 3].

The idea of [Ton 2] is to characterize the tautness by cohomological properties of a foliation based on the following theorem (for simplicity let TL and TM be oriented).

1.37 Theorem. [Ru 2], [Sul 2,3] *A Riemannian metric g_L on the leaves (with the volume form ω_L) is induced by a Riemannian metric g on M for which all leaves are minimal, iff ω_L is the restriction of a ν-form χ on M which satisfies the following condition: $d\chi(x_1, \dots, x_{\nu+1}) = 0$, where $\nu\ (= \dim L)$ of the vector fields $\{x_i\}$ are sections of TL.*

For example, a foliation $\mathcal{F}(G, H)$ of a Lie group G by a Lie subgroup H is taut [MV-SC]. A compact orientable foliation of a compact orientable manifold M is taut iff the holonomy groups of all leaves are finite [Rei 3]. The oriented Riemannian foliation is taut iff $H_B^q \neq 0$, where H_B^* represents basic cohomology and $q = \operatorname{codim} L$ [Mas].

The umbilicalizable Riemannian foliation $\{L\}$ on a compact connected manifold is tense, and the following conditions are equivalent for it: $\{L\}$ is taut, $\{L\}$ is geodesible and $\{L\}$ is *cohomologically taut* (i.e., the cohomology of the *basic forms*, which are the differential forms that are locally the pull-backs of forms on the local quotient manifold $\mathbb{R}^{\operatorname{codim} L}$, satisfies Poincaré duality) [Ca 6].

The relations between curvature and topology are fundamental for a Riemannian geometry "in the large" (see [GHL], [BZ], [GKM],and [CE]), where two points of view can roughly be distinguished [GHL]:

– The traditional point of view: By assuming some rather strong properties of the curvature (pinching assumptions, for example) rather precise structure theorems are obtained,

– The coarse point of view: Taking much weaker assumptions, we get for instance, finiteness results for the topological type of the manifolds that are considered.

There are three types of sectional curvature for a foliation: *tangential, mixed* (denoted by K_{mix}) and *transversal*. For a given foliated manifold, a metric with constant tangential, mixed or transversal curvature seems to be *canonical*. Examples are *Hopf fibrations* which have many applications in Riemannian geometry. Their generalization, *Riemannian submersions* with totally geodesic fibers, do not decrease a transversal sectional curvature, have $K_{mix} \geq 0$, and also play an essential role in differential geometry. For example, an open complete manifold with nonnegative sectional curvature admits a Riemannian submersion onto its soul [Per]; given some geometric bounds (on curvature, volume, diameter and second fundamental form) for the base and the fibers, there is a finite number of conjugacy classes of Riemannian submersions between compact Riemannian manifolds [Wal 7].

Over the last few years, the interest of geometers has extended to analogous problems of the existence of adapted metrics and foliations (or distributions) on a manifold with additional curvature restrictions:

FC 1: *Does there exist a complete Riemannian metric g for a manifold M with a fixed foliation $\{L\}$ which induces the given geometrical and curvature properties of a foliation*?

FC 2: *Do there exist foliations $\{L\}$ with given metric and curvature properties for a fixed complete Riemannian manifold (M, g)? If such foliations exist, what is their classification*?

We shall concentrate in Chapter IV (and also in Chapters VI–VII) on mixed curvature such as sectional, Ricci, or scalar curvature (see [Wls 3], [Wal 2,6], [KiT] and [Ze 2]), whose geometric meaning comes from the fact that, for a totally geodesic foliation, certain components of the curvature tensor along any geodesic $\gamma \subset L$ are contained in *the Jacobi equation* $Y'' + R(t)Y = 0$, for the *Jacobi tensor* induced by a foliation, and *the Riccati equation* $B' + B^2 + R(t) = 0$, for a *structural tensor* $B : TL \times TL^{\perp} \to TL^{\perp}$ defined by $B(x, y) = (\nabla_x y)^{\perp}$, $(x \in TL,\ y \in TL^{\perp})$. There is a relation $Y' = B(\gamma', Y)$ between the tensors Y and B, see Section 2.4 and Section 3.1.

In a case of constant curvature, the solutions to the above equations (and thus the relative behavior of geodesics on nearby leaves) are well-known (see Figure 16).

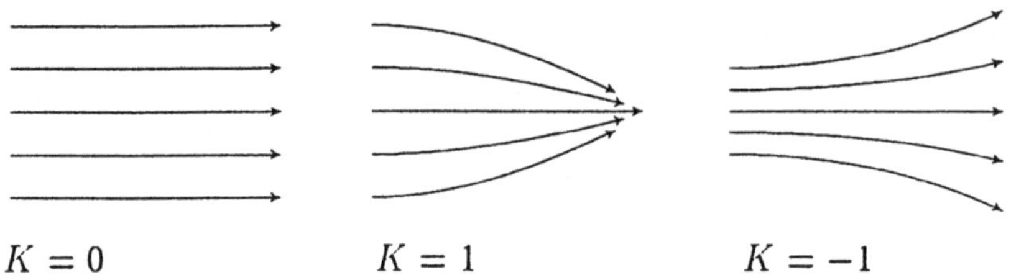

Figure 16. Behavior of geodesics and Jacobi fields for $K_{mix} = const$

For some results on tangential curvature see [Stu], [AF], and [App. C, MS]. For results on transversal curvature see [MRT] and [TV 1]. Integral curvature is discussed in [BLR], [Lan 1], [LL 1,2], [LP], and [Ran 1,2].

From the properties of a Riccati equation with a symmetric solution it follows that a totally geodesic foliation with complete leaves, integrable orthogonal distribution $TL^{\perp}$, and nonnegative mixed Ricci curvature $Ric_1 \geq 0$ splits locally ([BrW], [Ab 3] for $K_{mix} \geq 0$).

It is easy to find the following upper bound for the dimension of leaves

$$\dim L < \operatorname{codim} L \text{ (i.e., } \dim L < \tfrac{1}{2} \dim M)$$

of a compact totally geodesic foliation with $K_{mix} > 0$. The idea is that two compact totally geodesic submanifolds L_1 and L_2 with $\dim L_1 + \dim L_2 \geq \dim M$ in a Riemannian space M of positive (bi) sectional curvature intersect (for example, two great spheres in S^N). In Section 3.4.1, we obtain a generalization for a partial Ricci curvature. Some investigations of foliations with $K_{mix} \leq 0$ are based on ergodic theorems, as in the theory of Riemannian manifolds of nonpositive curvature, (see [BGrS]).

Much is known about foliations with $K_{mix} = const$:

1. *k-nullity foliations* on manifolds with degenerate curvature tensor. These metrics are called *partially hyperbolic, parabolic or elliptic* (see Section 2.3.1). The *k-nullity space* of the curvature tensor at $m \in M$ is given by the formula

$$N_k(m) = \{x \in T_m M : R(x, y)z = k((y, z)x - (x, z)y) \,\forall\, y, z \in T_m M\}.$$

2. totally geodesic *relative nullity foliations* on submanifolds M of space forms $\widehat{M}(k)$ (with constant sectional curvature k). The *relative nullity space* of a second fundamental form h at $m \in M$ is given by the formula

$$\ker h(m) = \{x \in T_m M : h(x, y) = 0 \,\forall\, y \in T_m M\}.$$

Submanifolds $M \subset \widehat{M}$ with a positive relative index of nullity $\mu(M)$ have the structure of developable ruled submanifolds. They were introduced by S. Chern and N. Kuiper in 1956 as *submanifolds with constant nullity* and are also known by the following names: *strongly parabolic* (introduced by A. Borisenko), *with constant rank* (N. Janenko and M. Akivis), *tangentially degenerate* (V. Ryzhkov and M. Akivis), *k-developable* (A. Pogorelov, V. Toponogov and S. Shefel). In particular, when external sectional curvature K_M-$K_{\widehat{M}}$ is nonpositive or zero, *partially hyperbolic or parabolic* immersions are strongly parabolic when codimension is small, see [Bor 1], [Flo 2] and Section 5.1.

A. Borisenko introduced in 1972 (see the survey [BY 2]) a larger class of *parabolic submanifolds*, (see Section 5.1) which have degenerate second quadratic forms A_ξ, i.e., rank $r(M) = \sup\{r(\xi) : \xi \in TM^\perp\} \geq \dim M - \mu(M)$ because $\cap_\xi A_\xi(m) \subset \ker h(m)$. A parabolic submanifold in a space form has a ruled structure, see Section 5.1.2.

Some splitting theorems are known about foliations with $K_{mix} = 0$, see [Tan 1]. The complete strongly parabolic submanifold M in a space form $\widehat{M}(0)$ with the additional condition $Ric_M \geq 0$ is a cylinder with $\mu(M)$-dimensional generator (A. Borisenko, 1985; for $K \geq 0$, Hartman, 1970).

For $K_{mix} = const < 0$ on foliated manifolds with finite volume, see [Ze 1,2] and [CM]. For $K_{mix} = const > 0$, D. Ferus proved the following theorem in 1970 by using the Riccati equation with $R(t) = kE$.

1.38 Theorem. ([Fe 2], *local in transversal directions*) *If a totally geodesic foliation* $\{L\}$ *has* $K_{mix} = k > 0$ *along a complete leaf* L_0, *then* $\dim L < \rho(\operatorname{codim} L)$. (For the definition of $\rho(n)$ see Section 1.1.)

1.39 Corollary. *Let* M^n *be a complete submanifold in a sphere* S^{n+p}.

a) If the index of relative nullity $\mu(M) \geq \rho(n - \mu(M))$, *then* M^n *is a totally geodesic submanifold* [Fe 2].

b) If the extrinsic sectional curvature is nonpositive and $2p < n - \nu(n)$, *then* M^n *is a totally geodesic submanifold* [Flo 2].

1.40 Theorem. [Rov 1] *The estimate* $\dim L < \rho(\operatorname{codim} L)$ *in Theorem* 1.38 *is*
a) **exact***: for all* $\nu < \rho(n)$ *there exists a neighborhood of a* ν*-dimensional great sphere in a* $(n + \nu)$*-dimensional round sphere* $S^{n+\nu}$ *with a foliation by* ν*-dimensional great spheres,* ($\dim L \leq 2$ *for the Kählerian case),*
b) **wrong** *for nonconstant* $K_{mix} > 0$ *: there exists a neighborhood of a (closed) geodesic in any of the projective spaces* KP^n, $(n \geq 2j\ K = \mathbb{C}, \mathbb{H}, \mathbb{C}a)$ *which can be foliated by closed geodesics,*

c) **true** *for ruled submanifolds $M^{n+\nu}$ with ν-dimensional (complete) rulings and $K_{mix} > 0$ in the sphere $S^N(k)$.*

Thus the following questions are natural:

1. When, for a given foliated smooth manifold M, does there exist a complete geodesible metric with nonnegative mixed curvature?

2. What is the structure of such foliations with large dimension? In particular, when do they split?

3. How can we effectively apply the technique of foliations to submanifolds with generators (ruled, parabolic etc.) in a Riemannian space with nonnegative curvature?

The first problem is interesting even locally for a foliation by closed geodesics, i.e., when $\dim L = 1$ and $\operatorname{codim} L = 2m + 1$.

1.41 Problem. *Let $M = S^1 \times B^{2m+1}$ be the product of a circle and an odd-dimensional ball with the product foliation $\{S^1 \times b\}$. For what maximal $\delta \in (\frac{1}{4}, 1)$ does there exist a geodesible metric on M with positive and δ-pinched K_{mix}?*

For $\delta = 0.25$ a locally given geodesic foliation exists even on a tubular neighborhood of closed geodesic in CROSS, see Section 3.1. For $\delta \geq (\frac{m}{m+1})^2$ see Appendix C.

Toponogov conjectured that for a totally geodesic foliation on a compact Riemannian manifold with $K_{mix} > 0$, the inequality $\dim L < \rho(\operatorname{codim} L)$ holds.

Some examples concerning Problem 1.41 with $\delta \to 1$ are given in Appendix C.

CHAPTER II

LOCAL RIEMANNIAN GEOMETRY OF FOLIATIONS

2.1. The main tensors and their invariants

For two arbitrary complementary orthogonal distributions T_1 and T_2 on TM, we define the *structural tensors* $B_1 : T_2 \times T_1 \to T_1$ and $B_2 : T_1 \times T_2 \to T_2$ by the following formulas:

$$B_1(y, x) = P_1(\nabla_x y), \quad B_2(x, y) = P_2(\nabla_y x), \tag{2.1}$$

where $P_i : TM \to T_i$ $(i = 1, 2)$ are orthoprojectors.

In Section 2.1 and Section 2.2, x, u denote elements of T_1 and y, z denote elements of T_2 unless otherwise stated.

2.1 Lemma. *Equation $B_i = 0$ means that a distribution T_i is involutive and tangent to a totally geodesic foliation. Equalities $B_1 = B_2 = 0$ mean that M is locally a Riemannian product $L_1 \times L_2$, where leaves $\{L_1\}$ and $\{L_2\}$ are tangent to T_1 and T_2, respectively.*

The tensors in (2.1) contain all of the information about these two distributions, i.e., all other tensors (see below) may be written in terms of them. For example, the *mean curvature vector fields*, $H_2 \subset T_1$ and $H_1 \subset T_2$, of the distributions T_1 and T_2 respectively, are given by the formulas

$$(H_2, x) = -\text{trace } B_2(x, *), \quad (H_1, y) = -\text{trace } B_1(y, *). \tag{2.2}$$

In the particular case of a totally geodesic foliation $\{L\}$ (i.e., $B_1 = 0$), the tensor corresponding to B_2 is $B = TL \times TL^\perp \to TL^\perp$, which is defined in [Fe 2] with the opposite sign and in [Dom] as follows:

$$B(x, y) = (\nabla_y x)^\perp. \tag{2.3}$$

This tensor (the so-called *co-nullity operator*) is introduced in [Ros 1,2] and [Gr 1] for the nullity foliation of a curvature operator and in [O'N 1] and [O'NS] for a relative nullity foliation on a submanifold in a space form.

For two arbitrary complementary orthogonal distributions T_1 and T_2 on TM, the *second fundamental forms* (symmetric tensors) $h_1 : T_1 \times T_1 \to T_2$, $h_2 : T_2 \times T_2 \to T_1$ and the *integrability tensors* (skew-symmetric) $A_1 : T_1 \times T_1 \to T_2$, $A_2 : T_2 \times T_2 \to T_1$ are defined by the following formulas:

$$h_1(x, u) = P_2(\nabla_x u + \nabla_u x)/2, \quad A_1(x, u) = P_2(\nabla_x u - \nabla_u x)/2,$$

$$h_2(y, z) = P_1(\nabla_y z + \nabla_z y)/2, \quad A_2(y, z) = P_1(\nabla_y z - \nabla_z y)/2. \tag{2.4}$$

These tensors are very important; their geometric meaning is the following: h_i is the second fundamental form at a point $p \in M$ of a submanifold, consisting of geodesics in M, tangent to the subspace $T_i(p)$. In view of the identities

$$A_1(x, u) = \frac{1}{2} P_2[x, u], \qquad A_2(y, z) = \frac{1}{2} P_1[y, z] \tag{2.5}$$

the equality $A_i = 0$ means that the distribution T_i is tangent to the foliation $\{L_i\}$. We also have the formula

$$H_i = \operatorname{trace} h_i, \qquad (i = 1,\ 2). \tag{2.6}$$

2.2 Definition. A distribution T_i on a Riemannian manifold (M, g) is called 1) *geodesic*, 2) *minimal*, or 3) *umbilic*, if its second fundamental form has the property

$$1)\ h_i = 0, \qquad 2)\ H_i = 0, \qquad \text{or} \quad 3)\ h_i = (H_i / \dim T_i) g,$$

respectively.

The geometric meaning of a geodesic distribution is that *every geodesic in M which is tangent to T_i at one point is tangent to T_i at every point.*

Proof. Let $\gamma : I \to M$ be a T_i-geodesic, that is, $\gamma' \subset T_i$ and $\nabla_{\gamma'}\gamma' = h_i(\gamma', \gamma') \perp T_i$. If T_i is a geodesic distribution, i.e., $h_i = 0$, then $\nabla_{\gamma'}\gamma' = 0$ and hence γ is a geodesic in M. □

Denote by B_i^+ and B_i^- $(i = 1,\ 2)$ the symmetric and skew-symmetric components with respect to $T_i^\perp$ of the structural tensors B_i.

2.3 Lemma. B_i^+ $(i = 1,\ 2)$, *may be considered as the self-adjoint Weingarten tensors associated to h_i, and B_i^- $(i = 1,\ 2)$, may be considered as equivalent terms of integrability tensors for distributions T_i in view of the identities*

$$(B_1^+(y, x), u) = -(h_1(x, u), y), \quad (B_2^+(x, y), z) = -(h_2(y, z), x), \tag{2.7}$$

$$(B_1^-(y, x), u) = -(A_1(x, u), y), \quad (B_2^-(x, y), z) = -(A_2(y, z), x). \tag{2.8}$$

Proof. For B_1^+:

$$(h_1(x,u),y) = ((\nabla_x u, y) + (\nabla_u x, y))/2 = -((u, \nabla_x y) + (x, \nabla_u y))/2$$
$$= -((u, B_1(y,x)) + (x, B_1(y,u)))/2 = -(B_1^+(y,x), u). \qquad (2.9)$$

For B_1^-:

$$(A_1(x,u),y) = ((\nabla_x u, y) - (\nabla_u x, y))/2 = -((x, \nabla_u y) - (u, \nabla_x y))/2$$
$$= -((x, B_1(y,u)) - (u, B_1(y,x)))/2 = -(B_1^-(y,x), u). \qquad (2.10)$$

The proof for B_2^+ and B_2^- is analogous. □

Indeed, B_i^+ is a shape operator at a point $p \in M$ of a locally defined transversal submanifold, consisting of geodesics in M which are tangent to the subspace $T_i(p)$. In particular, if $B_2^+(x,*)$ is a homothety map in T_2 (resp., equal to zero), then the distribution T_2 is *umbilic* (resp., *geodesic*). Analogously for $B_1^+(y,*)$. Historically the *configuration tensors, T (vertical)* and *O (horizontal)*, of a Riemannian manifold with two complementary orthogonal distributions T_1 and T_2, were first defined in [Gr 2] and [O'N 3] by formulas different from those in (2.1):

$$T_u v = P_2(\nabla_{P_1 u} P_1 v) + P_1(\nabla_{P_1 u} P_2 v),$$
$$O_u v = P_1(\nabla_{P_2 u} P_2 v) + P_2(\nabla_{P_2 u} P_1 v). \qquad (2.11)$$

It is obvious that $T_y * = O_x * = 0$. A tensor T is determined by its values $T_{x_1} x_2$ or by its values $T_x y$, and similarly for the tensor O:

$$(T_{x_1} x_2, y) = -(T_{x_1} y, x_2), \qquad (O_{y_1} y_2, x) = -(O_{y_1} x, y_2). \qquad (2.12)$$

These classical tensors may be written in terms of our structural tensors B_1 and B_2:

$$T_x y = B_1(y,x), \qquad O_y x = B_2(x,y). \qquad (2.13)$$

In the case of a foliation $\{L\}$ on a Riemannian manifold M, a configuration tensor T is the second fundamental form of the leaves; the identity $T = 0$ means that the leaves are totally geodesic submanifolds. Throughout, vectors and vector fields on TM tangent (resp., orthogonal) to the leaves of a foliation are called *vertical* (resp., *horizontal*). We shall denote by y and z horizontal vectors or *basic* (or projectable, i.e., coming from the base) *vector fields*, by x and x_i vertical vectors, by $\top$, the orthoprojector onto TL, and by $\perp$ the orthoprojector onto $TL^\perp$.

As in the *theory of submanifolds* [Ch 1,2] some *classes of foliations* are given in terms of the second fundamental form of their leaves. Indeed, they correspond to certain classes of distributions (see Definition 2.2) with the additional requirement of integrability.

2.4 Definition. A foliation $\{L\}$ on a Riemannian manifold M is called *totally geodesic, minimal,* or *totally umbilic* if the leaves are totally geodesic, minimal, or totally umbilic submanifolds, respectively.

A foliation is said to be *tense,* if there exists a Riemannian metric, for which the mean curvature of the leaves is parallel along the leaves [KT 4].

Mean curvature invariant (MCI) foliations have a projectable (basic) mean curvature vector field and satisfy the following equation (see [Wal 8] and [Ton 3]):

$$([H,x],y)=0,\ (x\in TL,\ \ y\in TL^{\perp}). \tag{2.14}$$

This means that H is (partially) Bott-parallel (see Section 1.2) along the leaves: $\check{\nabla}_x H = 0$, $(x \in TL)$. Such foliations are invariant under the local flows generated by the mean curvature vector field [Wal 8].

A totally umbilical foliation is called *spherical* if its mean curvature vector field is parallel in the normal bundle along the leaves (see [Rec 2] and [HR]). Using Definition 2.2 some well-known classes of foliations may be defined in terms of their transversal geometry.

2.5 Definition. A foliation $\{L\}$ on a Riemannian manifold M is called *Riemannian* or *conformal* if the orthogonal distribution $TL^{\perp}$ is geodesic or umbilic, respectively.

Some pairs of foliations are dual in a certain sense: totally geodesic and Riemannian foliations are dual (see [Mol]), as are totally umbilic and conformal foliations. Also, a foliation whose orthogonal distribution is minimal (i.e., with holonomy invariant transversal volume form, see Proposition 2.8 below), is dual to a minimal foliation.

2.6 Example. Let $f : M \to N$ be a smooth map of Riemannian manifolds with constant rank $r(f) < \dim M$ (in particular, a submersion when $r(f) = \dim N < \dim M$), $f_* : TM \to TN$ a differential of f, $f^{-1}(TN)$ the induced bundle over M, and $\bar{\nabla}$ a sum of connections of N and M in $f^{-1}(TN) \oplus TM$. Then $TM = T_1 \oplus T_2$, where the distribution $T_1 = \ker f = \{x \in TM :\ f_*x = 0\}$ is always integrable and T_2 is the orthogonal distribution. The symmetric bilinear map $h_f : TM \times TM \to f^{-1}(TN)$ (*the second fundamental form of* f) is defined by the formula

$$h_f(x, y) = \bar{\nabla}_x f_*(y) - f_*(\nabla_x y). \tag{2.15}$$

$H_f = trace\ h_f$ is called the *mean curvature vector field of* f. The map f is called

– *harmonic* if $H_f = 0$; in this case ker f defines a minimal foliation,

– *umbilic* if $h_f(x, y) = H_f \cdot (x, y)$; in this case ker f defines an umbilic foliation and T_2 is tangent to a totally geodesic foliation,

– *geodesic (or projective)* if it preserves the geodesics, i.e., there exists a 1-form ω on M such that $h_f(x, y) = \omega(x) f_* y + \omega(y) f_* x$, [No]; in this case ker f defines a totally geodesic foliation and T_2 is tangent to an umbilic foliation,

– *affine* if it is a geodesic (or projective) mapping and preserves the natural parameterization of geodesics; in this case $h_f(x, y) = 0$ [Ste 7].

The notion of the second fundamental form of a mapping between manifolds endowed with a connection, first constructed by J. Eells [EL] for the study of harmonic mappings, generalizes the second fundamental form of a submanifold isometrically immersed in a Riemannian manifold; it has been used by J. Vilms [Vi] to study totally geodesic mappings and Riemannian submersions, by Z. Har'El [HarE] to study projective mappings, by K. Yano and S. Ishihara [YI] to study harmonic and relatively affine mappings, by T. Nore [No] to study projective and umbilical mappings, and by S. Stepanov [Ste 7] to study such mappings with nonconstant rank.

The vector field y on M generates the local 1-parameter group of diffeomorphisms $\varphi_t : M \to M$. For a given tensor field K the tensor field $\tilde{\varphi} K$ is defined by

$$(\tilde{\varphi} K)_m = \tilde{\varphi}(K_{\varphi^{-1}(m)}), \ m \in M.$$

The *Lie derivative* $\Theta(y)K$ of the tensor field K is defined by the formula [KN]

$$(\Theta(y)K)_m = \lim_{t \to 0} \frac{1}{t} [K_m - (\tilde{\varphi}_t K)_m]. \tag{2.16}$$

An interesting interpretation of the second fundamental form h of a foliation $\{L\}$ follows from the formula

$$\Theta(y) g(x_1, x_2) = -2g(h(x_1, x_2), y), \ (y \in TL^\perp, \ x_1, \ x_2 \in TL) \tag{2.17}$$

i.e., for a vector field y orthogonal to $\{L\}$ the y-component of h is the Lie derivative Θ, with respect to y, of the metric g along the leaves.

2.7 Proposition. [Ton 1] *A totally geodesic (or totally umbilic) foliation $\{L\}$ on (M, g) is characterized by the condition that the induced metric $g_{|L}$ along the leaves is invariant (resp., conformal invariant) under flows of vector fields orthogonal to the foliation, i.e., for $y \in TL^{\perp}$, $\Theta(y)g_{|L} = 0$ (resp., $\Theta(y)g_{|L} = -2(H, y)g_{|L}$).*

Note that Riemannian foliations are characterized by the condition that the induced metric on $TL^{\perp}$ is holonomy invariant: $\Theta(x)g_{|TL^{\perp}} = 0,\ (x \in TL)$, or equivalently that the adapted connection $\tilde{\nabla}$ (see Section 1.2) is metric [Ton 1].

If all leaves L are oriented, then the foliation is said to be *tangentially oriented*. The *characteristic form* χ of a ν-dimensional foliation $\{L\}$ on (M, g) is a ν-form on TM, which gives the value 1 when evaluated on a local oriented orthonormal frame $\{e_j\}$ of $\{L\}$. For arbitrary vectors $u_1, \ldots, u_\nu \in TM$, χ is given by

$$\chi(u_1, \ldots, u_\nu) = \det\{g(u_i, e_j)\}_{ij}. \tag{2.18}$$

From formulas (2.17) and (2.18) we get the following [Ru 1]:

$$\Theta(y)\chi_{|L} = -H(y)\chi_{|L}, \quad (y \in TL^{\perp}). \tag{2.19}$$

The *transversal volume ν-form* $\chi^{\perp}$, which is defined analogously, satisfies the equation dual to (2.19) [Ton 1]:

$$\Theta(x)\chi^{\perp}_{|TL^{\perp}} = -H^{\perp}(x)\chi^{\perp}_{|TL^{\perp}}, \quad (x \in TL). \tag{2.19$'$}$$

Hence, we have

2.8 Proposition. [Ton 1] *a) Minimal foliations are characterized by the condition that the restriction $\chi_{|L}$ of a volume form on the leaves is invariant under flows of vector fields orthogonal to the foliation, i.e., $\Theta(y)\chi_{|L} = 0,\ (y \in TL^{\perp})$.*
b) Transversally oriented foliations with a holonomy invariant transversal volume form $\chi^{\perp}$ are characterized by the condition $H^{\perp} = 0$.

2.2. A Riemannian almost-product structure

Two complementary orthogonal distributions T_1 and T_2 on a Riemannian manifold M (see Section 2.1) may be introduced in an analytical way.

2.9 Definition. [Rei 3] A Riemannian manifold M with a metric $(\, , \,)$ and a tensor field P of type $(1, 1)$ on TM such that

$$P^2 = Id, \qquad (Pu, Pv) = (u, v) \tag{2.20}$$

is called a *Riemannian almost-product structure*. The eigenspaces of P for the eigenvalues $+1$ and -1 are called the *vertical distribution* T_1 and the *horizontal distribution* T_2; they are complementary orthogonal: $TM = T_1 \oplus T_2$.

2.10 Remark. Indeed,

$$P_1 = (Id + P)/2, \qquad P_2 = (Id - P)/2 \tag{2.21}$$

are orthoprojectors from TM onto T_1 and T_2. The tensor P is symmetric and nonsingular, so it may be viewed as a pseudo-Riemannian metric on M for which the vectors in T_1 have positive length and the vectors in T_2 have negative length.

The integrability tensors A_1 and A_2 are given by

$$A_1(u, v) = \frac{1}{2} P_2[P_1 u, P_1 v], \qquad A_2(u, v) = \frac{1}{2} P_1[P_2 u, P_2 v]. \tag{2.22}$$

Evidently, A_1 (resp., A_2) vanishes $\Longleftrightarrow$ T_1 (resp., T_2) is tangent to a foliation.

In view of the formulas in (2.20) and (2.22), the *Nijenhuis tensor* $N = (A_1 + A_2)/8$ may be written in the form

$$N(u, v) = [P, P](u, v) = [u, v] + [Pu, Pv] - P[Pu, v] - P[u, Pv]. \tag{2.23}$$

The second fundamental forms h_1 and h_2 are given by the following formulas:

$$h_1(u, v) = \frac{1}{2} P_2\{P_1 u, P_1 v\}, \qquad h_2(u, v) = \frac{1}{2} P_1\{P_2 u, P_2 v\}, \tag{2.24}$$

where $\{u, v\} = (\nabla_u v + \nabla_v u)/2$ is the *Jordan bracket* of the vector fields u and v. Evidently, h_1 (resp., h_2) vanishes $\Longleftrightarrow$ T_1 (resp., T_2) is a geodesic distribution. Note that the formulas in (2.22) and (2.24) are generalizations of those in (2.4).

In view of the formulas in (2.20) and (2.24), the tensor $L = (h_1 + h_2)/8$ may be written in the form

$$L(u, v) = \{P, P\}(u, v) = \{u, v\} + \{Pu, Pv\} - P\{Pu, v\} - P\{u, Pv\}. \tag{2.25}$$

The vector field $H = 8(H_1 + H_2)$, where $H_1 = trace\, h_1,\ H_2 = trace\, h_2$, may be written in the form

$$H = trace\{P, P\}. \tag{2.26}$$

For each distribution D (T_1 or T_2) we consider the following eight conditions [Gil]:

1) $(\nabla_A P)A = 0$, *GD geodesic distribution,*
2) $\sum_{e_i \in D} (\nabla_{e_i} P)e_i = 0$, *MD minimal distribution,*
3) $(\dim D)(\nabla_A P)B = (A, B) \sum_{e_i \in D} (\nabla_{e_i} P)e_i$, *UD umbilic distribution,*
4) $\emptyset$ *Δ no condition*
5) $\nabla_A P = 0$, *GF totally geodesic foliation,*
6) $(\nabla_A P)B = (\nabla_B P)A,\ \sum_{e_i \in D} (\nabla_{e_i} P)e_i = 0$, *MF minimal foliation,*
7) $(\nabla_A P)B = (\nabla_B P)A$,
   $(\dim D)(\nabla_A P)A = (A, B) \sum_{e_i \in D} (\nabla_{e_i} P)e_i$, *UF totally umbilic foliation,*
8) $(\nabla_A P)B = (\nabla_B P)A$, *F foliation.*

The properties 5–8 are analogous to 1–4 but with *integrability conditions.*

Combining these eight conditions and eliminating the dual cases gives **36** *different classes of almost-product structures*, each of which is characterized by some algebraic condition on ∇P [Nav].

A (GD, Δ) almost-product structure, (i.e., the distribution T_1 is GD and T_2 is Δ; analogous notations are used also for other 35 classes of almost-product structures) is called an *almost foliated metric* in [Vid] and an *anti-foliation* in [Nav]; it is, in other words, a geodesic distribution. (GD, F) and (UD, F) are actually Riemannian and conformal foliations.

Some examples of each of the 36 classes of almost-product structures constructed using hypersurfaces and products of manifolds are given in [Miq]. We consider some of them. There are nine classes of almost-product structures on real (complex) hypersurfaces M in $\mathbb{K}^{n+1}$, ($\mathbb{K} = \mathbb{C}, \mathbb{H}$). Let M be an oriented real hypersurface in $\mathbb{C}^{n+1}$, N the unit normal field on M and J the canonical almost-complex structure on $\mathbb{C}^{n+1}$; see Figure 17 below. The canonical almost-product structure on M is defined by the distributions $V = JN$ and $H = JN^{\perp}$. Observe that $JH = H$. Every such M is in the class (UF, MD).

2.11 Lemma. [Miq] *For $M^{2n+1} \subset \mathbb{C}^{n+1}$, we have the following equivalences (here $x, y \in H$):*

1) M *is* $(UF, GF) \iff h(x, y) = 0,$
2) M *is* $(GF, GD) \iff h(x, Jx) = 0,\ h(x, JN) = 0,$
3) M *is* $(GF, MF) \iff h(x, Jy) = h(Jx, x),\ h(x, Jx) = 0,$
4) M *is* $(UF, GD) \iff h(x, Jx) = 0,$
5) M *is* $(UF, MF) \iff h(x, Jy) = h(Jx, y),$
6) M *is* $(GF, MD) \iff h(x, JN) = 0.$

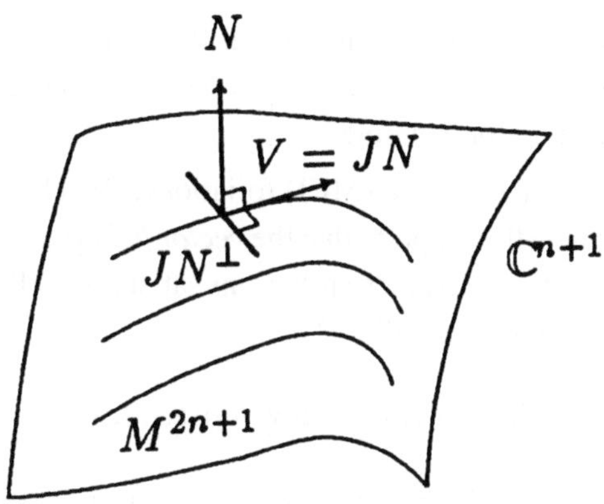

Figure 17. (UF, MD) on hypersurface in complex space

Let M be an oriented real hypersurface in $\mathbb{H}^{n+1}$, N the unit normal field on M and $\{J_1,\ J_2,\ J_3\}$ canonical almost-complex structures on $\mathbb{H}^{n+1}$. These elements define a canonical almost-product structure on M, with V generated by the fields $\{J_1N,\ J_2N,\ J_3N\}$ and $H = V^\perp$. Observe that $J_iH = H$. Every such M is in the class (Δ, MD). There are characterizations in terms of h, similar to those in Lemma 2.11, of all possible such almost-product structures on $M \subset \mathbb{H}^{n+1}$. Analogously, almost-product structures on a complex hypersurface M in $\mathbb{H}^{n+1}$ may be considered.

Conformal transformations of metrics on products of manifolds with distributions lead to some additional examples of almost-product structures [Miq].

The rule for products of manifolds with distributions is the following:

$$
\begin{array}{ccc}
(M,P) & (M',P') & (M\times M', P+P') \\
\hline
(MF,GF) & (GF,MF) & (MF,MF) \\
(MF,GF) & (GF,GD) & (MF,GD) \\
(MF,GF) & (GF,MD) & (MF,MD) \\
(GD,GF) & (GF,GD) & (GD,GD) \\
(GD,GF) & (GF,MD) & (GD,MD) \\
(MD,GF) & (GF,MD) & (MD,MD) \\
\hline
\end{array}
\tag{2.27}
$$

The 36 classes mentioned above are natural from the geometric and algebraic points of view. But there are interesting classes of foliations and distributions outside of this scheme, for example,

1. *Mean curvature invariant* (MCI) foliations, [Wal 8], see formula (2.14).

2. Foliations with the property that the *vector field* $[y,z]^\perp$ *is basic whenever* y *and* z *are basic vector fields;* these foliations satisfy the following infinitesimal condition (see (B.15) in Appendix B):

$$
\begin{aligned}
&((\nabla_z h_2)(y,y) - 2(\nabla_y h_2)(z,y), x) + \\
&(h_2(y,y), B_1(z,x)) - 2(h_2(z,y), B_1(y,x)) = 0.
\end{aligned}
\tag{2.28}
$$

3. *Weak-harmonic distributions* [Roc 2], which generalize the concept of MF.

4. Foliations with restrictions on the extrinsic curvature of the leaves, for instance, foliations of *negative (nonpositive) extrinsic sectional curvature* (see [Bor 7]) and some generalizations for partial Ricci curvature in Section 3.4.1.

2.3. Constructions of geodesic and umbilic foliations

2.3.1. Nullity foliations

Totally geodesic and umbilic foliations come naturally as null-distributions or kernels in the study of manifolds with degenerate curvature-like tensors and certain differential forms. We divide examples of such *nullity foliations* into two classes.

2.12 Examples.

1. A *curvature-like tensor* field R on a Riemannian manifold M is a $(1,3)$-tensor field which satisfies the first and second Bianchi identities. For a curvature-like tensor field R, the *k-nullity space* (the *nullity space* when $k = 0$) at $m \in M$ is defined by the formula

$$N_k(m) = \{x \in T_mM : R(x, y)z = k((y, z)x - (x, z)y) \ \forall y, z \in T_mM\}$$

and its dimension is called the *index of k-nullity* of R at m. Let G be an open (nonempty) subset of M on which the index of k-nullity of R takes its minimum. Then the distribution N_k is involutive and totally geodesic on G, and if M is complete, then the maximal integral manifolds are complete. The nullity space of a Riemannian curvature tensor was first studied in [CK], [Ros 1,2], [Gr 1] and [Mal 1] and later was generalized to any curvature-like tensor in [Gr 4], [CM], [Mal 2] and [Fe 3].

For a hypersurface $M^n \subset \bar{M}^{n+1}(k)$ with *type number* $r(x) \geq 2$ (rank of the second quadratic form) at x, the index of k-nullity $\nu_k(x)$ of the curvature tensor is equal to $n-r(x)$, see [KS]. Since $r(x)$ is equal to 1 or 0 if and only if M^n has constant sectional curvature k at x, the Riemannian manifolds of constant k-nullity $n-2$ are of special interest. It is known that every Riemannian manifold M of nullity $n-2$ is a *semi-symmetric space*, i.e., $R(x, y) \circ R = 0$, $(x, y \in TM)$, where $\circ$ denotes the derivation on the algebra of all tensor fields on M. Because the nullity distribution determines a totally geodesic and locally Euclidean foliation, these spaces are called *foliated semi-symmetric*. (Other semi-symmetric spaces are either locally symmetric spaces, or 2-dimensional surfaces, or *Szabo cones*; see [KS].) Foliated semi-symmetric spaces have been studied more deeply and in a much more explicit form by O. Kowalski and E. Boeckx [Boe 1,2]. The key problem of *classifying the* 3*-dimensional Riemannian manifolds with k-nullity* 1 *with respect to some constant $k \in \mathbb{R}$* was studied in [KS] and [Háj], (for $k = 0$, see [BU 2]). There, a particular system of PDEs is solved and explicit formulas for such metrics are given; using these formulas, interesting hypersurfaces in $\bar{M}^4(k)$ are constructed.

The decomposition of a Riemannian manifold M whose curvature tensor has positive index of nullity is investigated in [Ros 1,2] and [Us 1]. In [Us 1] the last problem is studied locally by Cartan's method for exterior differential forms and by deriving certain PDEs.

The nullity space of the Weyl conformal curvature tensor (which does not satisfy the second Bianchi identity) is studied in [Se 1]: the leaves of certain foliations are totally umbilic and conformally flat. The nullity space of the Bochner curvature tensor (which satisfies the second Bianchi identity) on a Kählerian manifold is studied in [Kas]: the leaves of certain foliations are Kählerian totally geodesic. Contact metric manifolds $(M^{2n+1}\phi, \xi, \eta, g)$ satisfying the condition that a characteristic vector field ξ belongs to the k-nullity distribution, that is, $R(X,Y)\xi = k(\eta(Y)X - \eta(X)Y)$, are studied in [Kou], [Pap]. Sasaki metrics on the unit tangent bundle S_1M (see Section 5.1.3) with an index of $(k > 0)$-nullity are studied in [Yam 2].

2. The *relative nullity space* of the second fundamental form h of a submanifold $M \subset \widehat{M}$ at $m \in M$ is given by

$$\ker h(m) = \{x \in T_mM : \; h(x,y) = 0 \;\text{ for all }\; y \in T_mM\}.$$

Curvature invariant submanifolds of a Riemannian space with positive relative null-index $\mu(M) = \min\{\dim \ker h(m)\}$ have the structure of a ruled developable submanifold, see Chapter V. Relative nullity distributions were first studied in [CL 2] for hypersurfaces in Euclidean space and then in [Fe 1] for submanifolds in spaces of constant curvature; for the complex case see [Ab 1,3,4]. As a generalization, arbitrary *principal curvature functions on submanifolds* in a space form are considered in [Ots], [O'N 2], [Rec 1,2], [HR] and [Bö 1,2], and *spherical foliations* induced by the curvature surfaces of submanifolds are studied. For analogous studies of hypersurfaces in conformally flat manifolds, see [Hw]. The *normal nullity space* is treated in [Mir]. Relative nullity foliations on infinite dimensional submanifolds of a Hilbert space are investigated in [Okr1].

Introducing the notion of L-parallel Jacobi fields, P. Dombrowski developed a method to study the question of the completeness of totally geodesic foliations induced by the kernel of differential forms. Continuing these examinations, H. Reckziegel with S. Hiepko [Rec 2], [HR], and R. Böning [Bö 1] dealt with variations of circles in Riemannian manifolds to achieve results on the completeness of spherical foliations induced by the kernel of certain differential forms.

2.13 Definition. [Dom] Let M be a smooth manifold with a covariant derivative ∇ and ξ a vector bundle over M with a covariant derivative $\hat{\nabla}$. Then for a differential form $\Omega \in A^r(M; \mathrm{Hom}(TM, \xi))$ on M of degree $r \in [0, \dim M]$ with coefficients in a vector bundle $\mathrm{Hom}(TM, \xi)$ we consider the differential forms

$d\Omega \in A^{r+1}(M; \mathrm{Hom}(TM, \xi))$, the *exterior derivative,* and
$i_Y\Omega \in A^{r-1}(M; \mathrm{Hom}(TM, \xi))$, the *inner product.*

We also consider the *kernel of* Ω for $a \in M$:

$$\ker \Omega_a = \begin{cases} \{x \in T_aM : \ \Omega x = 0\}, & \text{for } r = 0 \\ \{x \in T_aM : \ \Omega(y_1, \dots, y_r)x = 0, \ \forall y_i \in T_aM\}, & \text{for } r \geq 1, \end{cases}$$

the *index of nullity* $\nu_\Omega = \min\{\dim \ker \Omega_a : \ a \in M\}$ and

the *domain of regularity* $G_\Omega = \{a \in M : \ \dim \ker \Omega_a = \nu_\Omega\}$.

The r-form $\Omega \in A^r(M; \mathrm{Hom}(TM, \xi))$ on M is called *complex* if M is Kählerian and $\ker \Omega$ is invariant with respect to the given complex structure.

Ω is called a *ξ-valued geodesic r-form on M* with respect to ∇, $\hat{\nabla}$ if there exists $\Theta \in A^1(M; \mathrm{Hom}(\xi, \xi))$ such that

$$\forall a \in M \ \forall x \in \ker \Omega : i_x\Omega_a = 0 \text{ and } i_x(d\Omega)_a = \Theta_a(x) \circ \Omega_a. \qquad (2.29)$$

In particular, $\Omega \in A^r(M; \mathrm{Hom}(TM, \xi))$ is called *kernel-closed* (resp. *closed*), if

$$\forall a \in M \ \forall x \in \ker \Omega : i_x(d\Omega)_a = 0, \ \text{(resp. } d\Omega = 0). \qquad (2.29')$$

2.14 Remark. The covariant derivatives ∇, $\hat{\nabla}$ for M, resp. ξ, induce a covariant derivative $\bar{\nabla}$ for the vector bundle $Hom(TM, \xi)$. For $\omega \in A^r(M, \eta)$ one defines the exterior derivative $d\omega \in A^{r+1}(M, \eta)$ with respect to $(\eta, \bar{\nabla}$ (in our case $\eta = Hom(TM, \xi)$ as follows:

$$d\omega(y_0, \dots, y_r) = \sum_{i=0}^{r} (-1)^i \bar{\nabla}_{y_i}(\omega(y_0, \dots, \hat{y}_i, \dots, y_r))$$
$$+ \sum_{0 \leq j < j \leq r} (-1)^{i+1} \omega([y_i, y_j], y_0, \dots, \hat{y}_j, \dots, y_r), \quad (y_0, \dots, y_r \subset TM),$$

which, for $\eta = \mathbb{R}_M$, reduces to the classical definition.

Let $\{L\}$ be a totally geodesic foliation, $T_1 = TL$ and $\xi = T_2 = TL^\perp$. Then the orthoprojector $P_2 : TM \to T_2$ is a *canonical* ξ-valued geodesic 0-form on M [Dom]. For information on double forms, see [Gr 4].

The analogous concept of a *spherical r-form on M* is introduced in [HR] and is used to construct totally umbilic (or spherical) foliations tangent to $\ker \Omega$.

We can consider the cases of a) geodesic, b) umbilic or c) minimal distributions tangent to $\ker \Omega$, if we replace (2.29) by

$$a)\ \forall a \in M\ \forall x \in \ker\Omega : i_x\Omega_a = 0 \text{ and } i_x(d\Omega)_a x = 0,$$
$$b)\ \forall a \in M\ \forall x_1 \perp x_2 \text{ from } \ker\Omega : i_{x_1}\Omega_a = 0 \text{ and } i_{x_1}(d\Omega)_a x_2 = 0,$$
$$c)\ \forall a \in M\ \forall x \in \ker\Omega : i_{x_1}\Omega_a = 0 \text{ and } \sum_j i_{x_j}(d\Omega)_a x_j = 0,$$

where $\{x_j\}$ is an orthonormal basis of $\ker \Omega_a$.

2.15 Lemma. [Dom] *Let $\Omega \in A^r(M; Hom(TM, \xi))$ be a geodesic r-form on a Riemannian manifold M with positive null-index ν_Ω and let $\gamma \subset G_\Omega$ be the unit speed geodesic tangent to* $\ker \Omega$. *Then for all* $\ker \Omega$*-parallel fields $Y_1, \dots, Y_r$ along γ, the section $\Omega(Y_1, \dots, Y_r)$ of $\gamma^*\xi$ satisfies the following equation:*

$$\hat\nabla_{\gamma'}(\Omega(Y_1, \dots, Y_r)z) = \Theta(\gamma')(\Omega(Y_1, \dots, Y_r)z). \tag{2.30}$$

If Ω is kernel-closed, then $\hat\nabla_{\gamma'}\Omega(Y_1, \dots, Y_r)z = 0$.

The concept of a T-parallel vector field (see Chapter III) allows us to prove the completeness of a nullity foliation tangent to $\ker \Omega$.

2.16 Theorem. [Dom] *Let $\Omega \in A^r(M; Hom(TM, \xi))$ be a geodesic r-form on a (complete) Riemannian manifold M with positive null-index ν_Ω. Then the distribution* $\ker \Omega$ *on G_Ω is tangent to a (geodesically complete) totally geodesic foliation.*

2.3.2. Double-twisted products

One of the most popular generalizations of the product of two (pseudo-) Riemannian manifolds is the *warped product* of such manifolds, introduced by R. Bishop and B. O'Neill [BiN] in 1969. In fact, it appeared in mathematical and physical literature long before that time, for example, polar coordinates of the Euclidean plane $\mathbb{R}^2$ describe a warped product representation of $\mathbb{R}^2 \setminus \{0\}$, and most space-time models are warped products. Warped products allow us to construct new examples of complete Einstein manifolds [Bes 2, II]. In recent investigations of spaces with Ricci curvature bounded below, a class of warped products $(a, b) \times_\lambda M$ for some fixed λ serves as the collection of possible smooth rigid models [ChC 1]. The result by E. Cartan, that a Riemannian manifold M which satisfies the *axiom of s-planes* with fixed $1 < s < \dim M$ is isometric to a space form, was generalized later by geometers in two directions: considerations of more general submanifolds (totally umbilic, extrinsic spheres with parallel second fundamental form) and additional structures (complex, quaternion, etc.), see the survey [LiV].

The warped product of an l-dimensional space form with an arbitrary Riemannian manifold contains a large number of s-dimensional totally geodesic submanifolds for any $s > l$; they are the inverse images of the projection on a base of $(s-l)$-dimensional totally geodesic submanifolds in a given space form. In [Okr 2] this property is described as the synthetic *axiom of (l, s)-planes*, which is also a generalization of Cartan's axiom: *for every point $p \in M$ and any l-dimensional direction $V \subset T_pM$, there exists a totally geodesic s-dimensional submanifold tangent to V.* The conjecture (which is partially proven in [Okr 2]) is that the axiom of (l, s)-planes (for any $1 \leq l < s \leq \dim M - 1$) characterizes the warped products of l-dimensional space forms with an arbitrary Riemannian manifold.

Next, we consider a natural generalization of warped products. For simplicity, denote by $\{L_i\}$ the canonical foliations on a product manifold $M_1 \times M_2$ with natural projections p_i onto M_i, $(i = 1, 2)$. Let P_i be the projection of $T(M_1 \times M_2)$ onto TL_i and also set $P_i^{\perp} = Id - P_i$, $(i = 1, 2)$.

2.17 Definition. [PR], [Ko 2,3] Let (M_i, g_i), $(i = 1, 2)$ be Riemannian manifolds and let $\lambda_i : M_1 \times M_2 \to \mathbb{R}$ be positive differentiable functions. A *double-twisted product* $M_1 \times_{(\lambda_1, \lambda_2)} M_2$ is the differentiable manifold $M_1 \times M_2$ with a Riemannian metric defined by

$$(x, y) = \lambda_1^2 g_1(P_1 x, P_1 y) + \lambda_2^2 g_2(P_2 x, P_2 y)$$

for all vectors x and y tangent to $M_1 \times M_2$. In particular, if the λ_i are independent on the M_i-components, then $M_1 \times_{(\lambda_1, \lambda_2)} M_2$ is called a *double-warped product.*

2.18 Remark. This definition generalizes Bishop's notion $M_1 \times_\lambda M_2$ of an umbilic product, which B. Chen [Ch 2] in 1977 called a *twisted product* and which is the double-twisted product $M_1 \times_{(1,\lambda)} M_2$. If in this case λ depends only on the points of M_1, then $M_1 \times_\lambda M_2$ is called a *warped product.* It also should be mentioned that the conformal change of a Riemannian metric can be interpreted as a twisted product, namely, one where the first factor M_1 consists of only one point. Therefore, formulas and assertions for double-twisted products are applicable in many situations. For information on double-twisted products of more than two factors, see [Ko 3] and [Gau 2].

2.19 Proposition. [Ko 2] *In a double-twisted product $M_1 \times_{(\lambda_1,\lambda_2)} M_2$ the leaves $\{L_i\}$ form a totally umbilic foliation with the mean curvature vector field*

$$H_i = -P_i^{\perp}(grad\,(\log \lambda_i)). \tag{2.31}$$

In the case of a twisted product ($\lambda_1 = 1$) the leaves $\{L_1\}$ are totally geodesic.

Proof. If $y, z \subset TL_2$ are basic vector fields and $x \subset TL_1$, then

$$x(y,z) = x(\lambda_2^2 g_2(y,z)) = 2\frac{x(\lambda_2)}{\lambda_2}(y,z).$$

Since $[x,y] = [x,z] = 0$, then $\nabla_x y = \nabla_y x$ and $\nabla_x z = \nabla_z x$. On the other hand, $x(y,z) = 2(\nabla_y x, z)$. Consequently, the second fundamental form h_2 of L_2 is given by

$$h_2(y,z) = \frac{-P_1(grad\,\lambda_2)}{\lambda_2}(y,z). \tag{2.32}$$

Equation (2.31) now follows from equation (2.32). □

2.20 Lemma. [PR] *The Levi-Civita derivative ∇ and the curvature tensor R of the double-twisted product $M_1 \times_{(\lambda_1,\lambda_2)} M_2$ with $u_i = -grad\,(\log \lambda_i)$ are related to the Levi-Civita derivative $\tilde{\nabla}$ and the curvature tensor $\tilde{R}$ of the usual product of the Riemannian manifolds M_1 and M_2 by*

$$\nabla_x y = \tilde{\nabla}_x y + \sum_i ((P_i x, P_i y)u_i - (x, u_i)P_i y - (y, u_i)P_i x),$$

$$R(x,y) = \tilde{R}(x,y) +$$

$$\sum_i ((\nabla_x u_i - (x,u_i)u_i) \wedge P_i y - (\nabla_y u_i - (y,u_i)u_i) \wedge P_i x) + \sum_{i,j} (u_i, u_j) P_i x \wedge P_j y,$$

where for all $u, v \in T_pM$ the wedge product $u \wedge v$ denotes the linear map $w \to (v,w)u - (u,w)v$.

Double-twisted and warped products can be characterized by the geometry of their canonical foliations.

2.21 Proposition. [PR] *Let g be a (pseudo-) Riemannian metric on a smooth manifold $M_1 \times M_2$ and assume that the canonical foliations $\{L_1\}$ and $\{L_2\}$ intersect perpendicularly everywhere. Then g is a metric of*

a) a double-twisted product $M_1 \times_{(\lambda_1,\lambda_2)} M_2$ iff $\{L_1\}$ and $\{L_2\}$ are totally umbilic foliations,

b) a twisted product $M_1 \times_\lambda M_2$ *iff* $\{L_1\}$ *is a totally geodesic foliation and* $\{L_2\}$ *is a totally umbilic foliation,*

c) a warped product $M_1 \times_\lambda M_2$ *iff* $\{L_1\}$ *is a totally geodesic foliation and* $\{L_2\}$ *is a spherical foliation,*

d) a usual (pseudo-) Riemannian product iff $\{L_1\}$ *and* $\{L_2\}$ *are totally geodesic foliations.*

Recall that a foliation $\{L\}$ *splits* if M decomposes locally as a metric product and its leaves coincide with one of the factors. Note that for a simply connected manifold M, the local splitting is equivalent to a global product structure. On the other hand, the geodesic foliation on a Klein bottle with flat metric splits only locally.

If a complete connected and simply connected Riemannian manifold M has two complementary orthogonal totally geodesic foliations, then these foliations are parallel and de Rham's Decomposition Theorem [KN] asserts that M is isometric to the direct product of the two leaves through a point $m \in M$; for the case where M is not assumed to be simply connected, see [Wan].

Manifolds with two complementary orthogonal totally umbilic foliations are treated by a number of authors (see [BiN], [Bi 1], [Gau 2] and [NR]) and the "twisted and warped decomposition theorems" are obtained.

The case of M with boundary is considered in [CrK]; for the case of M with additional assumptions on the metric, see [Gę 1,2], with an additional complex structure, see [Wat 1], [Okr 3] and [Jo 1]. Finally, for the cases of M with additional symplectic type structures, consult [Chi] and [Wat 2].

Let $L_i(x)$ be a leaf of a foliation $\{L_i\}$ on a Riemannian manifold M through the point $x \in M$ and let $\hat{L}_i(x)$ be the universal covering of $L_i(x)$, identified with the set of all homotopy classes of curves in $\hat{L}_i(x)$ starting at x. Denote by $[a]$ the homotopy class of the curve a in $L_i(x)$ starting at x.

A pair $\{L_1, L_2\}$ of totally umbilic foliations with $TM = TL_1 \oplus TL_2$ is called in [Ko 2,3] a *totally umbilic orthogonal 2-net*. Note that $\{L_1, L_2\}$ is a pair of totally umbilic completely orthogonal foliations if and only if $\{L_1, L_2\}$ is a pair of completely orthogonal conformal foliations.

2.22 Theorem. [Ko 2], [PR] *Let* $\{L_1, L_2\}$ *be a pair of completely orthogonal totally umbilic foliations on a Riemannian manifold* M*. Assume that* $\dim L_1 \geq 3$, *and the induced conformal structures on the leaves* $\{L_1\}$ *are complete (if* $\{L_1\}$ *is a totally geodesic foliation, then we only assume its leaves to be geodesically* *complete). Then for every $x \in M$ there is a Riemannian covering map π of a double-twisted product $\hat{L}_1(x) \times_{(\lambda_1,\lambda_2)} \hat{L}_2(x)$ onto M such that $\pi_*(T\hat{L}_i) = TL_i$, $(i = 1, 2)$, where the λ_i are functions on $\hat{L}_1(x) \times \hat{L}_2(x)$ defined by*

$$\lambda_i([a_1], [a_2]) = \exp\left(-\int_0^1 (T_{a_i}\dot{a}_j, H_i)\, dt\right), \ ([a_1], [a_2]) \in \hat{L}_1(x) \times \hat{L}_2(x), \ i \neq j).$$

Even if M is a simply connected and complete Riemannian manifold, we cannot expect M to be globally isometric to a double-twisted product; if however, one of the foliations is totally geodesic, then M splits globally as a twisted or warped product [PR].

2.4. Curvature identities

2.4.1. Curvature of a Riemannian manifold

Let M^n be a Riemannian manifold with metric tensor g, also denoted by $(,)$, and Levi-Civita connection ∇. We have $\nabla g = 0$. For any vector fields x, y and z tangent to M the *curvature tensor* R of M is given by $R(x, y)z = \nabla_x \nabla_y z - \nabla_y \nabla_x z - \nabla_{[x,y]} z$. (Sometimes we use this tensor with the opposite sign.) If $R = 0$, then M is called *locally flat*. If $e_1, \ldots, e_n$ are local orthonormal vector fields on M, then $Ric\,(x, y) = \sum_{i=1}^n (R(e_i, x)y, e_i)$ defines a symmetric tensor field Ric of type (0,2), called the *Ricci tensor of M*. Using the Ricci tensor, the *scalar curvature* function $s : M \to \mathbb{R}$ is defined by $s = \sum_{i=1}^n Ric\,(e_i, e_i)$. For the curvature tensor R we have the following identities:

$$R(x, y) + R(y, x) = 0, \quad (R(x, y)z, w) = (R(z, w)x, y),$$

$$R(x, y)z + R(y, z)x + R(z, x)y = 0 \quad \textit{(first Bianchi identity)}.$$

For a covariant derivative $\nabla_w R$ of the curvature tensor R the *second Bianchi identity* holds:

$$(\nabla_w R)(x, y)z + (\nabla_x R)(y, z)w + (\nabla_y R)(z, w)x = 0.$$

Let x and y be two linearly independent vectors at a point of M. The *sectional curvature* $K(x, y)$ for the plane section spanned by x and y is defined by

$$K(x, y) = \frac{(R(x, y)y, x)}{(x, x)(y, y) - (x, y)^2}.$$

A curvature tensor R can be completely reconstructed from its sectional curvature values. If the sectional curvature is equal to a constant k (positive, negative, etc.) for all plane sections, then M is called a *space of constant curvature* or a *real space form* (manifold with *positive,* resp., *negative curvature,* etc.). In the first case, we have

$$(R(x, y)z, w) = k\{(x, w)(y, z) - (x, z)(y, w)\}.$$

If for $n > 2$ the sectional curvature depends only on a point of M, then M is a real space form. A Riemannian manifold M is said to be *Einsteinian* if its Ricci tensor satisfies $Ric\,(x, x) = \lambda g(x, x)$ for some function λ on M. For example, any Einsteinian 3-dimensional manifold has constant sectional curvature. A Riemannian manifold M is called a *locally symmetric space* if its curvature tensor is covariant constant, i.e., $\nabla R = 0$. A Riemannian manifold M is said to be *conformally flat* if, for any $m \in M$, there is an open set U containing m and a function $f \in C^\infty(U)$ such that the metric $\exp(2f)g$ is flat. For a Riemannian manifold M^n, $(n > 2)$ the Weyl *conformal curvature tensor* C is defined by

$$\begin{aligned}(C(x, y)z, w) = (R(x, y)z, w) &- \frac{1}{n-2}\{Ric\,(x, w)(y, z) - (x, z)Ric\,(y, w)\}\\ &+ \frac{s}{(n-1)(n-2)}\{(x, w)(y, z) - (x, z)(y, w)\}.\end{aligned}$$

If $n = 3$, then $C \equiv 0$. And if $n \geq 4$, M is *conformally flat* iff $C \equiv 0$.

Let $M = M_1 \times M_2$ be the product of two Riemannian manifolds. If x_1 and x_2 are vector fields on M_1 and M_2, respectively, we define a vector field x on M by setting $x_{(m_1,m_2)} = (x_1, 0)_{m_1} + (0, x_2)_{m_2}$. We will use the short-hand notation $x = x_1 + x_2$. If y is another vector field of type $y_1 + y_2$ then $\nabla_x y = \nabla^1_{x_1} y_1 + \nabla^2_{x_2} y_2$, where ∇, ∇^1, and ∇^2 are the canonical connections of M, M_1 and M_2, respectively. Using similar notation, a curvature tensor is given by

$$(R(x, y)z, t) = (R_1(x_1, y_1)z_1, t_1) + (R_2(x_2, y_2)z_2, t_2),$$

that is, by computing R using decomposed vector fields.

For more detail on curvature of Riemanian manifolds, see [BZ], [GHL], [DoC], [Kli], [KN].

2.4.2. Curvature identities and the Riccati equation

2.23 Definition. [Gr 2] The *horizontal connection* $\tilde\nabla$ and the *horizontal curvature operator* $\tilde R$ of an almost-product structure are given by the following formulas:

$$\tilde{\nabla}_y z = P_2(\nabla_y z), \quad \tilde{R}(y, z)u = \tilde{\nabla}_{P_2[y,z]}u - [\tilde{\nabla}_y, \tilde{\nabla}_z]u, \ (y, z, u \in T_2). \tag{2.33}$$

The *vertical connection* $\hat{\nabla}$ and the *vertical curvature operator* $\hat{R}$ are defined analogously.

The operators $\tilde{R}$ and $\hat{R}$ satisfy all of the curvature symmetries and the Bianchi identities [Gr 2]. The curvature of the vertical and horizontal distributions is related to the structural tensors (as in the theory of submanifolds, see Chapter VI).

2.24 Theorem. [Gr 2] *The following Gauss, Codazzi, and Ricci equations for two complementary orthogonal distributions T_1 and T_2 on TM are satisfied:*

$$(R(y, z)u, w) = (\tilde{R}(y, z)u, w)$$

$$-(O_y u, O_z w) + (O_y w, O_z u) - (O_y z, O_u w) + (O_z y, O_u w), \tag{2.34a}$$

$$(R(y, z)w, x) = ((\nabla_z O)_y w - (\nabla_y O)_z w, x) + (O_y z - O_z y, T_x w), \tag{2.34b}$$

$$(R(y, x_1)z, x_2) = ((\nabla_{x_1} O)_y z + O(O_y x_1, z), x_2) + ((\nabla_y T)_{x_1} x_2 + T(T_{x_1} y, x_2), z), \tag{2.34c}$$

where $y, z, u, w \in T_2$ and $x_1, x_2 \in T_1$.

Proof. We will give only the proof for equation (2.34c):

$$\begin{aligned} P_2 R(x_1, y)x_2 &= P_2 \nabla_{[x_1, y]} x_2 - P_2[\nabla_{x_1}, \nabla_y] x_2 \\ &= T_{P_1[x_1, y]} x_2 + O_{P_1[x_1, y]} x_2 - P_2 \nabla_{x_1}(O_y x_2 \\ &+ P_1 \nabla_y x_2) + P_2 \nabla_{x_1}(T_{x_1} x_2 + P_1 \nabla_{x_1} x_2) \\ &= P_2(\nabla_y T)_{x_1} x_2 - P_2(\nabla_{x_1} O)_y x_2 + T(T_{x_1} y, x_2) - O(O_y x_1, x_2). \quad \square \end{aligned}$$

Gauss and Codazzi equations have dual equations. The Ricci equation and the second Bianchi identity yield a formula for $(R(x_1, x_2)y, z)$. All of these equations taken together solve the problem of computing the curvature of the distributions T_1 and T_2 from the curvature of M as well as the problem of computing the curvature of M from the curvature of the distributions T_1 and T_2. For example, (2.34b) leads to the identity of the integrability tensor A_2 of T_2, (used in Appendix B):

$$(\nabla_y A_2)(z, w) + (\nabla_z A_2)(w, y) + (\nabla_w A_2)(y, z) = 0. \tag{2.35}$$

There are formulas for the covariant derivatives of R corresponding to Theorem 2.24 [Gr 2], but we omit them.

2.25 Lemma. [Rov 18] *The following differential equation holds:*

$$((\nabla_{x_1}B_2)(x_2,y),z)+(B_2(x_2,B_2(x_1,y),z)$$

$$+((\nabla_y B_1)(z,x_1),x_2)+(B_1(z,B_1(y,x_1)),x_2)+(R(y,x_1)x_2,z)=0. \quad (2.36)$$

Proof. For tensors T and B_1 we have

$$((\nabla_y T)_{x_1}x_2,z)=((\nabla_y B_1)(z,x_2,),x_1),\ T(T_{x_1}y,x_2),z)=(B_1(z,B_1(y,x_1)),u)$$

and analogously for the tensors O and B_2. Thus (2.36) follows from (2.34c). □

If $r : I \to M$ is a unit speed T_1-geodesic, i.e., $r' \subset T_1,\ \nabla_{r'}r' \subset T_2$, then the endomorphism field $B_2(t) = B_2(r'(t),*)$ of T_2 along r satisfies the following Riccati differential equation

$$B_2' + B_2^2 + F_2(t) = 0,\ \ (t \in I), \quad (2.37)$$

where $(F_2(t)y,z) = (R(y,r')r',z)+((\nabla_y B_1)(z,r'),r')+(B_1(z,B_1(y,r')),r')$. In the case of totally geodesic foliations, the equation (2.36) was given in [Dom]

$$(\nabla_{x_1}B_2)(x_2,*)+B_2(x_2,B_2(x_1,*))+R(*,x_1)x_2=0, \quad (2.36')$$

and the Riccati equation (2.37) takes the form [Fe 2]

$$B_2' + B_2^2 + R(t) = 0. \quad (2.37')$$

2.26 Remark. In Riemannian geometry, it is sometimes better to use the Riccati inequality instead of the Jacobi equation. In Chapter IV the Riccati equations (2.36) and (2.37) will also be considered in polarized form with respect to one of the given distributions.

2.4.3. Integral formulas for mixed curvature

The first integral formulas for the curvature of a foliated manifold were obtained by G. Reeb.

2.27 Theorem. *Let H be the mean curvature function of a transversally oriented foliation of codimension 1 on a closed oriented Riemannian manifold M. Then $\int_M H\, d\, vol = 0$, where $d\, vol$ is the volume form of M.*

Consequently, if a foliation has a projectable mean curvature function, then, according to Theorem 2.27, one of its leaves has mean curvature equal to zero, so this leaf is minimal [Wal 8]. The following question is studied in [Os 6,7] and [Wal 3]: what are the functions on M that can be the mean curvature of a codimension-1 foliation?

For a given Riemannian manifold M with a codimension-1 foliation, let $K(m)$ be the determinant of the second fundamental form of the leaf through the point $m \in M$. A foliation minimizing the total curvature $\int_M |K|$ is called *tight*. The integral formulas of such foliations are studied, using integral geometry extensively, in [BLR 1,2], [Lan 1,3], [LL 1,2] and [LP]. For example, the following problem is open: *prove (directly!) that S^n does not admit a tight foliation*; for $n = 3$, [Lan 2] obtained the result using Reeb components.

Let $\{x_i\} \subset T_1$ and $\{y_j\} \subset T_2$ be local orthonormal bases and

$$\sum_{i,j} (R(y_j, x_i)x_i, y_j) = s_{mix}. \tag{2.38}$$

Note that

$$\sum_{i,j} (B_2(x_i, B_2(x_i, y_j)), y_j) = ||B_2^+||^2 - ||B_2^-||^2 = ||h_2||^2 - ||A_2||^2, \tag{2.39}$$

and

$$\sum_{i,j} (B_1(y_j, B_1(y_j, x_i)), x_i) = ||B_1^+||^2 - ||B_1^-||^2 = ||h_1||^2 - ||A_1||^2. \tag{2.40}$$

The contraction of equation (2.36)

$$((\nabla_{x_i} B_2)(x_i, y_j), y_j) + (B_2(x_i, B_2(x_i, y_j), y_j) +$$

$$((\nabla_{y_j} B_1)(y_j, x_i), x_i) + (B_1(y_j, B_1(y_j, x_i)), x_i) + (R(y_j, x_i)x_i, y_j) = 0 \tag{2.41}$$

with respect to i and j, in view of the formulas

$$\mathrm{div}_1 H_2 = \mathrm{div}\, H_2 + |H_2|^2, \quad \mathrm{div}_2 H_1 = \mathrm{div}\, H_1 + |H_1|^2, \tag{2.42}$$

gives us the equation

$$\mathrm{div}\,(H_1 + H_2) = s_{mix} + ||h_1||^2 + ||h_2||^2 - ||A_1||^2 - ||A_2||^2 - |H_1|^2 - |H_2|^2. \tag{2.43}$$

An analogous formula is true for more than two orthogonal distributions; for the case of three orthogonal distributions, see [BM]. In the particular case of a Riemannian almost-product manifold, the *mixed scalar curvature* s_{mix} in formula (2.43) takes a special form, for example,

$$\begin{cases} (UD, UD): -\mathrm{div}\,(H_1 + H_2) + ||A_1||^2 + ||A_2||^2 + \frac{\nu^2-1}{\nu^2}|H_1|^2 + \frac{n^2-1}{n^2}|H_2|^2, \\ (GD, GD): -\mathrm{div}\,(H_1 + H_2) + ||A_1||^2 + ||A_2||^2, \\ (GD, MD): \mathrm{div}\,(H_1 + H_2) + ||A_1||^2 + ||A_2||^2 - |h_2|^2, \\ (F, F): \quad -\mathrm{div}\,(H_1 + H_2) - ||h_1||^2 - ||h_2||^2 + |H_1|^2 + |H_2|^2, \\ (MF, MF): -\mathrm{div}\,(H_1 + H_2) - ||h_1||^2 - ||h_2||^2. \end{cases} \tag{2.44}$$

Integration of equation (2.43) along a compact manifold M leads (with the help of Green's Theorem) to *integral formulas*, which have many interesting global corollaries.

2.28 Theorem. *Let T_1 and T_2 be complementary orthogonal distributions on a closed oriented Riemannian manifold M. Then*

$$\int_M \{s_{mix} + ||B_1^+||^2 + ||B_2^+||^2 - ||B_1^-||^2 - ||B_2^-||^2 - |H_1|^2 - |H_2|^2\} = 0. \tag{2.45}$$

For the space form $M(k)$, we have $s_{mix} = k(\dim T_1)(\dim T_2)$. Integral formulas for foliations that are in a form similar to formula (2.45) are introduced in [Ra 2]. In [Roc 1], equation (2.45) was deduced for a Riemannian almost-product manifold in terms of analogously defined differentials ∇P, δP and dP of the tensor P (see (2.20)):

$$\int_M \{4s_{mix} + |\nabla P|^2 - |dP|^2 - |\delta P|^2\} = 0. \tag{2.46}$$

In [Wal 6] equation (2.45) is derived by direct calculation

$$\int_M \{s_{mix} + ||h_1||^2 + ||h_2||^2 - ||A_1||^2 - ||A_2||^2 - |H_1|^2 - |H_2|^2\} = 0. \tag{2.47}$$

In [Ste 7] equation (2.45) takes the form

$$\int_M \{64 s_{mix} + ||L||^2 - ||N||^2 - |H|^2\} = 0, \tag{2.48}$$

where N, L and H are given by formulas (2.23), (2.25) and (2.26) respectively.

Of course,

$$\frac{1}{64}||L||^2 = ||h_1||^2 + ||h_2||^2, \ \frac{1}{64}||N||^2$$
$$= ||A_1||^2 + ||A_2||^2, \ \frac{1}{64}|H|^2 = |H_1|^2 + |H_2|^2. \tag{2.49}$$

This method was applied in [Ste 7] to degenerate maps $f : M \to N$ with nonconstant rank $r(f) < \dim M$, (see Example 2.6). Equation (2.43) could be compared with the formula for a Riemannian foliation [KT 3]

$$|h_1|^2 = |A_2|^2 + (\Delta P_1, P_1) - s_{mix}, \tag{2.50}$$

where ΔP_1 is the Laplacian of the orthoprojector P_1. Also see [GR], [NR] and [Biv].

Equation (2.45) gives us decomposition criteria for a compact Riemannian almost-product manifold using the sign of the mixed scalar curvature s_{mix}.

2.29 Corollary. [Ste 7] *Let T_1 and T_2 be complementary orthogonal umbilic distributions on a closed oriented Riemannian manifold M with $s_{mix} \leq 0$. Then M splits, that is, T_1 and T_2 are tangent to totally geodesic foliations.*

2.30 Corollary. [Ste 7] *Let T_1 and T_2 be complementary orthogonal geodesic and umbilic distributions on a closed oriented Riemannian manifold M with $s_{mix} \geq 0$. If both T_1 and T_2 are integrable, then M splits along them.*

Some global results are related to the integration of equation (2.43) over a compact leaf L_0:

$$\int_{L_0} \{s_{mix} + \mathrm{div}_2 H_1 + ||B_1^+||^2 + ||B_2^+||^2 - ||B_2^-||^2\} = 0. \tag{2.51}$$

2.31 Corollary. *A compact minimal foliation $\{L\}$ on a Riemannian manifold M with integrable normal bundle and mixed scalar curvature $s_{mix} \geq 0$ splits, i.e., locally, $M = L \times L^\perp$ and if the manifold is simply connected, then it is a product globally. (In particular, a minimal foliation $\{L\}$ on a Riemannian manifold M with integrable normal bundle and mixed scalar curvature $s_{mix} > 0$ has no compact leaves* [Wal 6].*)*

Proof. Since $B_2^- = 0$, it follows from equation (2.51) that $B_1^+ = B_2^+ = 0$, i.e., TL and $TL^\perp$ are tangent to totally geodesic foliations. By de Rham's theorem, the foliation splits. □

2.32 Remark. Some authors study transversal infinitesimal automorphisms and basic forms of a foliation. By using Stokes' theorem they state the ***transversal divergence theorems*** for harmonic foliations with holonomy invariant transversal volume form, and for Riemannian foliations [Ton 1,3]. A brief survey of Riemannian foliations is given in Section 2.5 below.

2.5. Riemannian foliations

2.5.1. The holonomy invariant metric in the normal bundle $\bar{\nabla}$

A Riemannian metric $g^\perp$ on the normal bundle $TL^\perp$ of a foliation $\{L\}$ is *holonomy invariant* if

$$\Theta(X)g^\perp = 0, \ (X \subset TL). \tag{2.52}$$

For $s_1, s_2 \subset TL^\perp$ we have by definition

$$(\Theta(X)g^\perp)(s_1, s_2) = X\, g^\perp(s_1, s_2) - g^\perp(\Theta(X)s_1, s_2) - g^\perp(s_1, \Theta(X)s_2).$$

A *Riemannian foliation* is a foliation $\{L\}$ with a holonomy invariant transversal metric $g^\perp$. The study of these foliations was initiated by B. Reinhart [Rei 1,3]. Recent results are in monographs [Mol], [Ton 3], [Kit 2]. In terms of a foliated atlas $\{f_p = P_2 \circ \varphi_p\}$ the requirement is that the local transition functions $\tau_{p,q} := f_q(U_q) \to f_p(U_p)$, relating submersions (for two foliated charts) $\varphi_p : U_p \to \mathbb{R}^\nu \times \mathbb{R}^{m-\nu}$ and $\varphi_q : U_q \to \mathbb{R}^\nu \times \mathbb{R}^{m-\nu}$ with $U_p \cap U_q \neq 0$ (see Section 1.1), are local isometries of suitable Riemannian metrics on $\mathbb{R}^{m-\nu}$. In other words, the leaves of a Riemannian foliation locally are given by the level sets of a Riemannian submersion. A Riemannian metric g on M is *bundle-like* for a foliation $\{L\}$ if the induced metric on the normal bundle is holonomy invariant.

A simple example of a Riemannian foliation is given by the orbits of a non-singular Killing vector field V on (M, g) (for instance, the action of a group S^1 on $S^3 \subset \mathbb{C}^2$ by the isometries $\varphi_{p,q}(t)(z_1, z_2) = (\exp(2\pi ipt)z_1, \exp(2\pi iqt)z_2)$, see Section 1.1). This means that $\Theta(V)g = 0$ or equivalently

$$Vg(y, z) = g([V, y], z) + g(y, [V, z]), \ (y, z \subset TM).$$

The complement $TL^\perp$ is preserved by the flow. For the induced metric $g^\perp$ on $TL^\perp$ we have $\Theta(V)g^\perp = 0$. The holonomy invariance in this case is precisely the invariance under the flow generated by V.

More generally, let a Lie group G act on (M, g) by isometries. If the orbits of the G-action have the same dimension, this gives rise to a Riemannian foliation. The key is that the sections of TL are linear combinations of Killing vector fields arising from the group action, so the previous arguments apply.

We compare condition (2.52) with the condition that the adapted connection (see Section 1.2)

$$\tilde{\nabla}_x s = \begin{cases} [x, Y_s]^\perp, & x \subset TL \\ (\nabla_x Y_s)^\perp, & x \subset TL^\perp \end{cases}, \quad s \in TL^\perp, \tag{2.53}$$

is a metric connection in $TL^\perp$ equipped with the induced metric $g^\perp$, where $Y_s \subset TM$ is such that $(Y_s)^\perp = s$. This condition implies that for $s_1, s_2 \subset TL^\perp$ and now for all $Y \subset TM$,

$$Y g^\perp(s_1, s_2) = g^\perp(\nabla_Y s_1, s_2) + g^\perp(s_1, \nabla_Y s_2), \tag{2.54}$$

and thus condition (2.54) implies condition (2.52) written as $X g^\perp(s_1, s_2) = g^\perp(\nabla_X s_1, s_2) + g^\perp(s_1, \nabla_X s_2)$, $(X \subset TL)$.

2.33 Theorem. [Ton 3] *Let $\{L\}$ be a foliation on (M, g), $g^\perp$ the induced metric in $TL^\perp$ and $\tilde{\nabla}$ the connection on $TL^\perp$ defined by condition* (2.53). *Then the foliation $\{L\}$ is Riemannian and g is bundle-like $\Longleftrightarrow$ $\tilde{\nabla}$ is a metric connection.*

2.34 Theorem. [Ton 3] *Let $g^\perp$ be a holonomy invariant metric in the normal bundle $TL^\perp$. Then there is a unique metric and torsion-free connection in $TL^\perp$.*

The unique metric and torsion-free connection in the normal bundle of a Riemannian foliation $\{L\}$ is the transversal Levi-Civita connection of $\{L\}$. From above it follows that the metric and torsion-free connection in the normal bundle of a Riemannian foliation $\{L\}$ is holonomy invariant.

2.35 Theorem. [Ton 3] *Let $\{L\}$ be a foliation on (M, g). Then the following conditions are equivalent:*
a) $\{L\}$ is Riemannian and g is bundle-like;
b) $g(\nabla_y X, z) + g(y, \nabla_z X) = 0$,
c) $g([X, y], y) = 0, \ |y| = 1$,
d) $g(\nabla_y z + \nabla_z y, X) = 0$,
e) $2g(\nabla_y z, X) = g([y, z], X)$,
where $X \subset TL$ and $y, z \subset TL^\perp$.

Property d) in Theorem 2.35 states that the second fundamental form $h^\perp$ of the normal bundle $TL^\perp$ vanishes (i.e., the operator $B_2(x, *)$ is skew-symmetric for all $x \in TL$). Thus this bundle can be viewed as a *totally geodesic subbundle* of TM. In particular, a geodesic which is orthogonal to a leaf at one point of M is orthogonal to $\{L\}$ at each of its points. Moreover this property characterizes Riemannian foliations. This was one of the initial observations of B. Reinhart [Rei 1] on this subject.

Riemannian foliations can be characterized by the following condition on the transversal mean curvature vector field $H^\perp$ and the second fundamental form $h^\perp$.

2.36 Theorem. [TV 2] *A foliation $\{L\}$ of codimension q on a Riemannian manifold (M, g) is Riemannian if and only if for all unit vectors $y \in TL^\perp$*

$$g_L(H^\perp, h^\perp(y, y)) \geq (q+2) g_L(h^\perp(y, y), h^\perp(y, y)).$$

2.5.2. The basic cohomology

Let $\{L\}$ be a foliation on M. A differential form $\omega \in \mathbf{\Omega}^r(M)$ is *basic* if

$$i(x)\omega = 0, \ \Theta(x)\omega = 0, \ (x \subset TL),$$

where $i(x)$ denotes the interior product with respect to x. In a distinguished chart $(x_1, \ldots, x_\nu; y_1, \ldots, y_{m-\nu})$ of the foliation, this means that

$$\omega = \sum_{\alpha_1 < \cdots < \alpha_r} \omega_{\alpha_1 \cdots \alpha_r} d\, y_{\alpha_1} \wedge \cdots \wedge d\, y_{\alpha_r},$$

where the functions $\omega_{\alpha_1 \cdots \alpha_r}$ are independent of $(x_1, \ldots, x_\nu)$, i.e., $\frac{\partial}{\partial x_i} \omega_{\alpha_1 \cdots \alpha_r} = 0$. The set $\mathbf{\Omega}_B(L) \subset \mathbf{\Omega}(M)$ is a subcomplex, since the exterior derivative preserves basic forms $d_{|\mathbf{\Omega}_B(L)} : \mathbf{\Omega}_B(L) \to \mathbf{\Omega}_B(L)$. We denote $d_{|\mathbf{\Omega}_B(L)}$by d_B. The *basic cohomology* is then by definition $\mathbf{H}_B^\bullet(L) = \mathbf{H}^\bullet(\mathbf{\Omega}_B(L), d_B)$. It plays the role of the de Rham cohomology of the leaf space of a foliated manifold. For the case of a codimension-1 foliation, there are just two groups, $\mathbf{H}_B^0(L)$ and $\mathbf{H}_B^1(L)$.

The *transverse orientability* of a foliation $\{L\}$ is defined by the orientability of $TL^\perp$. For any holonomy invariant transversal metric $g^\perp$, there is a *transversal volume form* $\chi^\perp$ on $TL^\perp$ dual to χ_L. It is holonomy invariant, i.e.,

$$\Theta(X)\chi^\perp = 0, \ (X \subset TL), \tag{2.55}$$

where

$$(\Theta(X)\chi^\perp)(s_1, \ldots, s_{m-\nu}) =$$

$$X\chi^\perp(s_1, \ldots, s_{m-\nu}) - \sum_{i=1}^{q} \chi^\perp(s_1, \ldots, \Theta(X)s_i, \ldots, s_{m-\nu})$$

for $s_1, \ldots, s_{m-\nu} \subset TL^\perp$. Since $i(X)\chi^\perp = 0$ for $X \in TL$, the RHS is indeed well defined. For $m - \nu = 1$ (that is, a codimension-1 foliation) the concept of a transversal and holonomy invariant volume coincides with the concept of a holonomy invariant transversal metric. Since $i(X)\chi^\perp = 0$, $(X \subset TL)$, the holonomy invariance condition $\Theta(X)\chi^\perp = 0$, $(X \subset TL)$, shows that $\chi^\perp \in \mathbf{\Omega}_B^{\operatorname{codim} L}(L)$. It follows that $d\chi^\perp = 0$. It is of interest to examine the cohomology class $[\chi^\perp] \in \mathbf{H}_B^{\operatorname{codim} L}(L)$, which plays the role of the *orientation class* for the leaf space of a foliation. Of particular importance are the conditions implying the nontriviality of this class.

2.37 Theorem. [Ton 3] *Let $\{L\}$ be a transversally oriented minimal foliation with holonomy invariant transversal volume $\chi^\perp$ on a closed oriented manifold M. Then $[\chi^\perp] \neq 0$ in $\mathbf{H}_B^{\operatorname{codim} L}(L)$.*

2.38 Theorem. [Ton 3] *Let $\{L\}$ be a transversally oriented and taut Riemannian foliation on a closed oriented manifold M. Then $[\chi^\perp] \neq 0$ in $\mathbf{H}_B^{\operatorname{codim} L}(L)$.*

For $y \in TL^\perp$, consider the symmetric *Weingarten map* $W(y) : TL \to TL$ given by

$$W(y)x = -(\nabla_x Y)^\top, \quad (x \subset TL, \; Y \subset TL^\perp, \; Y \ni y).$$

Note that $W(y)x = -B_1(y, x)$ and $g_L(W(y)x, u) = -g(h(x, u), y)$, see Section 2.1. The 1-form k given by $k(y) = \operatorname{trace} W(y)$ is the *mean curvature form* of the foliation. The mean curvature vector field $H \subset TL^\perp$ is given by $k(y) = g^\perp(H, y)$ for $y \in TL^\perp$. By construction $i(x)k = 0$ for all $x \in TL$.

2.39 Remark. For an oriented and transversally oriented Riemannian foliation $\{L\}$ on a closed oriented Riemannian manifold M, we have [Mas]:

$$\{L\} \text{ is taut} \iff \mathbf{H}_B^{\operatorname{codim} L}(L) \neq 0.$$

Special cases of this result have been proved by many authors, see [Ton 2]. Furthermore, for every Riemannian foliation there is a well-defined cohomology class $[k_B] \in \mathbf{H}_B^1(L)$ (the basic component of the mean curvature form arising from an orthogonal direct sum decomposition $\mathbf{\Omega} = \mathbf{\Omega}_B(L) \oplus \mathbf{\Omega}_B(L)^\perp$, $k = k_B + k_0$), whose vanishing characterizes the tautness of a foliation. Hence the sufficient conditions for the tautness of a Riemannian foliation of codimension $q \geq 2$ are expressed in terms of the *transversal Ricci operator* $\widetilde{Ric} : TL^\perp \to TL^\perp$, and the *transversal curvature operator* $\tilde{R} : \Lambda^2(TL^\perp)^* \to \Lambda^2(TL^\perp)^*$, in view of the following vanishing results (see [Ton 2]):

1) *if* $\widetilde{Ric} > 0$, *then* $\mathbf{H}^1_B(L) = 0$; 2) *if* $\tilde{R} > 0$, *then* $\mathbf{H}^r_B(L) = 0$ *for* $0 < r < q$.

It is worthwhile to note, in view of O'Neill's formula (2.60 c) below, that this result holds in particular on a Riemannian manifold with positive sectional curvature.

An *infinitesimal automorphism* V (i.e., the flow of V maps leaves into leaves) acts on $s \subset TL^\perp$ by

$$\Theta(V)s = [V, Y_s]^\perp, \tag{2.56}$$

where $Y_s \subset TM$ is such that $(Y_s)^\perp = s$. The RHS of equation (2.56) does not depend on the choice of Y_s. Let $V(L)$ denote the set of all infinitesimal automorphisms of the foliation $\{L\}$. Let $Y \in V(L)$ be an infinitesimal automorphism of $\{L\}$. The *transversal divergence* $div_B Y$ is defined by

$$\Theta(Y)\, \chi^\perp = div_B Y \cdot \chi^\perp.$$

Here we use the fact that $\Theta(Y)\, \chi^\perp \in \mathbf{\Omega}_B^{\operatorname{codim} L}(L) \subset \mathbf{\Lambda}^{\operatorname{codim} L}(TL^\perp)^*$. Observe that $div_B Y \in \mathbf{\Omega}^0_B(L)$ and in fact depends only on $Y^\perp$.

2.40 Theorem. [KTT], [Ton 3] *Let* $\{L\}$ *be a transversally orientable minimal foliation with holonomy invariant transversal volume* $\chi^\perp$ *on a closed Riemannian manifold* (M, g), *and let* Y *be an infinitesimal automorphism of* $\{L\}$. *Then*

$$\int_M div_B(V^\perp)\, d\,vol = 0.$$

Proof. Let χ_L be the characteristic form of foliation and assume M to be oriented by $d\,vol = \chi^\perp \wedge \chi_L$. Then

$$\begin{aligned} div_B(V^\perp) d\,vol &= (div_B(V^\perp) \cdot \chi^\perp) \wedge \chi_L = \Theta(V)\chi^\perp \wedge \chi_L \\ &= (d\,i(V)\chi^\perp) \wedge \chi_L = d(i(V)\chi^\perp \wedge \chi_L) + (-1)^{\operatorname{codim} L} i(V)\chi^\perp \wedge d\chi_L. \end{aligned}$$

By Stokes' theorem it suffices to show that the second term is in fact zero. To prove this we consider the canonical multiplicative filtration [KT 4], [KTT]:

$$\mathbf{F}^r\mathbf{\Omega}^m = \{\omega \in \mathbf{\Omega}^m : i(x_1)\dots i(x_{m-r+1})\,\omega = 0,\ x_j \in TL,\ 1 \le j \le m-r+1\},$$

which vanishes for $r > \operatorname{codim} L$. We have $\chi^\perp \in \mathbf{F}^{\operatorname{codim} L}$ and $i(V)\chi^\perp \in \mathbf{F}^{\operatorname{codim} L - 1}$. Since the leaves are minimal, $d\,\chi_L \in \mathbf{F}^2$. It follows that $i(V)\chi^\perp \wedge d\chi_L$ has a filtration of degree $(\operatorname{codim} L - 1) + 2 = \operatorname{codim} L + 1$, and hence vanishes. □

The value of this integral for the case of a Riemannian (but not necessarily minimal) foliation is given in the following theorem.

2.41 Theorem. [Ton 3] *Let $\{L\}$ be a transversally oriented Riemannian foliation on a closed oriented manifold (M, g). Let $V \subset V(L)$. Then*

$$\int_M div_B(V^\perp)\, d\,vol = \int_M g^\perp(H, V^\perp)\, d\,vol \equiv (H, V^\perp)_\perp, \tag{2.57}$$

the global scalar product of the sections H and $V^\perp$ of $TL^\perp$.

2.5.3. Transversal Killing fields

For an arbitrary foliation $\{L\}$ on (M, g) with induced metric $g^\perp$ on $TL^\perp$ and an infinitesimal automorphism $V \in V(L)$, we have

$$(\Theta(V)g^\perp)(y, z) = (\Theta(V^\top)g^\perp)(y, z) + g^\perp(\nabla_y(V^\perp), z) + g^\perp(y, \nabla_z(V^\perp)), \tag{2.58}$$

where $y, z \subset TL^\perp$ and $V = V^\perp + V^\top$. For a Riemannian foliation, the first term on the RHS of equation (2.58) vanishes, and we obtain the following result:

$$(\Theta(V)g^\perp)(y, z) = g^\perp(\nabla_y(V^\perp), z) + g^\perp(y, \nabla_z(V^\perp)), \ (y, z \subset TL^\perp). \tag{2.59}$$

An infinitesimal automorphism $V \in V(L)$ with the property $\Theta(V)g^\perp = 0$ is called *transversally metric*. If this holds, the $V^\perp$ is called a *transversal Killing field* [Ton 3]. For the point foliation (with $\dim L = 0$), this is the usual definition of a Killing vector field.

Associated to $\tilde{\nabla}$ are the (transversal) *Ricci operator* $\widetilde{Ric} : TL^\perp \to TL^\perp$ and also the *Jacobi operator* $\tilde{J} = \tilde{\Delta} - \widetilde{Ric} : TL^\perp \to TL^\perp$, where $\tilde{\Delta}$ is a *transversal Laplacian*, see [KTT]. The condition $\tilde{J}(\tilde{Y}) = 0$ defines (transversal) *Jacobi fields.*

The transversal divergence theorem (Theorem 2.40) is the key to obtaining the following result:

2.42 Theorem. [KTT] *Let $\{L\}$ be a transversally orientable minimal foliation on a compact Riemannian manifold (M, g) with a bundle-like metric, and let Y be an infinitesimal automorphism of $\{L\}$. Then the following properties are equivalent:*

1) *$Y^\perp$ is a transversal Killing field, that is, $\Theta(Y)g^\perp = 0$;*
2) *$Y^\perp$ is a transversally divergence-free Jacobi field;*
3) *$Y^\perp$ is transversally affine, that is $\Theta(Y)\tilde{\nabla} = 0$.*

Moreover, if $\operatorname{codim} L = 2$*, then the following properties are equivalent to each other:*

4) $Y^\perp$ *is a transversal conformal field, that is,* $\Theta(Y)g^\perp = \lambda \cdot g^\perp$*;*
5) $Y^\perp$ *is a transversal Jacobi field.*

By this theorem, the linear space of transversal Jacobi automorphisms of the (harmonic) Hopf fibration $p : S^3 \to S^2$ is isomorphic to the linear space of infinitesimally conformal fields on S^2; in particular, it is 6-dimensional.

The *operator* $A_Y : TL^\perp \to TL^\perp$ for $Y \subset TL^\perp$ is defined by $A_Y(\tilde{Z}) = -\tilde{\nabla}_Z Y$, where $Z \subset TM$ and $Z^\perp = \tilde{Z}$, [Yor 3]. The following theorem generalizes results for Killing fields on a Riemannian manifold M by B. Kostant and Currás-Bosch.

2.43 Theorem. [Yor 3] *Let* $\{L\}$ *be a minimal foliation on a connected, orientable, complete Riemannian manifold* (M, g) *with a bundle-like metric. Let* Y *be a transverse Killing field with finite global norm. Then for each* $m \in M$*,* $(A_Y)_m$ *belongs to the Lie algebra of the linear holonomy group* $\Psi_{\tilde{\nabla}}(m)$*, where* $\tilde{\nabla}$ *is the transversal Riemannian connection of* $\{L\}$*.*

2.44 Remark. If in Theorem 2.43 the transverse Ricci operator $\widetilde{Ric}$ of $\{L\}$ is nonpositive everywhere and negative for at least one point of M, then every transverse Killing field with finite global norm is trivial [Yor 2]. Note that any Killing field with bounded length (for instance, on a compact manifold) preserves a codimension-1 totally geodesic foliation [Os 3,4].

2.5.4. Transversally symmetric foliations

For the curvature tensor $\tilde{R}$ of the transversal connection $\tilde{\nabla}$, we have that $i(x)\tilde{R} = 0$ and $\Theta(x)\tilde{R} = 0$ for all $x \in TL$. The transversal curvature tensor $\tilde{R}$ is related to the curvature tensor R of M by the formula

$$\tilde{R}(y, z)w = (R(y, z)w)^\perp,$$

where $y, z, w \subset TL^\perp$ are projectable vector fields. This follows from the definition of $\tilde{\nabla}$ together with the fact that for any projectable $u, v \subset TM$, $[u, v]^\perp = [u^\perp, v^\perp]$.

For Riemannian foliations it is true that $O_y z = \frac{1}{2}[y, z]^\perp$ where $y, z \subset TL^\perp$.

2.45 Theorem. [O'N 3] *The sectional curvatures for a Riemannian foliation are given by the formulas*

$$\begin{aligned} &a)\ K(x,u) = K_L(x,u) - |T_x u|^2 - (T_x x, T_u u),\\ &b)\ K(x,y) = ((\nabla_x T)_y y, x) - |T_x y|^2 + |O_x y|^2,\\ &c)\ K(y,z) = \tilde{K}(y,z) - 3|O_y z|^2, \end{aligned} \tag{2.60}$$

where $x, u \in TL$ *and* $y, z \in TL^{\perp}$.

From the third formula (2.60) it follows that the sectional curvature $\tilde{K}$ of the base is not smaller than the transversal sectional curvature.

There are Riemannian foliations whose transversal geometry is locally modeled on a Riemannian symmetric space. A *transversal symmetry* can be characterized by a condition on the canonical Levi-Civita connection on the normal bundle.

2.46 Theorem. [Ton 3] *Let* $\{L\}$ *be a Riemannian foliation on* (M, g) *and let* g *be a bundle-like metric. Then the following conditions are equivalent:*

1) $\{L\}$ *is transversally symmetric;*
2) *the local geodesic symmetries (geodesic reflections) on the model space are isometries;*
3) $\tilde{\nabla}_z \tilde{R}(z, y, z, y) = 0, \ (z, y \in TL^{\perp})$;
4) $\nabla_z R(z, y, z, y) + 2R(z, O_z y, z, y) = -6((\nabla_z O)_z y, O_z y), \ (z, y \in TL^{\perp})$.

All these conditions are purely local, and they are automatically satisfied for a Riemannian foliation of codimension one. For a totally geodesic Riemannian foliation, this characterization can be improved in the analytical case, using the of P. Tondeur and L. Vanhecke: *the reflections in the leaves are isometric if and only if the geodesic reflections on the model space are (local) isometries.*

Recall that in a space of constant curvature, the reflections in totally geodesic submanifolds are isometries. Hence from the results mentioned above we get the following theorem.

2.47 Theorem. [Ton 3] *Let* $\{L\}$ *be a Riemannian foliation on a space* (M, g) *of constant curvature and let* g *be a bundle-like metric. Then* $\{L\}$ *is transversally symmetric if and only if*

$$(O_y z, T(O_y z, y)) = 0, \ (y, z \in TL^{\perp}). \tag{2.61}$$

Note that condition (2.61) is satisfied both in the case of a totally geodesic foliation (i.e., tensor $T = 0$) and in the case of a Riemannian foliation with an integrable normal bundle (i.e., tensor $O = 0$); thus these foliations are necessarily transversally symmetric.

Conversely, the existence of a transversally symmetric foliation has strong implications for the geometry of the foliated space. A typical result is that the transversal symmetry of the foliation, defined by a Killing vector field of unit length on a complete, simply connected (M, g), implies that (M, g) is a naturally reductive space, see [Ton 3]. Riemannian foliations can be compared with transversally symmetric foliations, results with small basic mean curvature, see [Ton 1].

It would be worthwhile to study the geometry of the following classes of Riemannian foliations defined in terms of the eigenvalues and eigenspaces of the *transversal Jacobi operator* $\tilde{R}(\gamma', *)\gamma' : TL^{\perp} \to TL^{\perp}$ along a geodesic γ.

2.48 Definition. [TV 3] A Riemannian foliation $\{L\}$ on (M, g) is a *transversally $\mathfrak{C}$-foliation* if the eigenvalues of $\tilde{R}(\gamma', *)\gamma'$ are constant along each geodesic γ orthogonal to the leaves of the foliation. $\{L\}$ is a *transversally $\mathfrak{P}$-foliation* if the eigenspaces of $\tilde{R}(\gamma', *)\gamma'$ can be spanned by parallel fields of eigenvectors along each geodesic γ orthogonal to the leaves of the foliation.

This definition generalizes the notion of $\mathfrak{C}$–space and $\mathfrak{P}$–space introduced by J. Berndt and L. Vanhecke in 1992.

To give examples of transversally $\mathfrak{C}$– and $\mathfrak{P}$– foliations it suffices to consider warped products $B \times_f F$ over a $\mathfrak{C}$– or $\mathfrak{P}$– space B. These concepts are used for a new characterization of transversally symmetric foliations.

2.49 Theorem. [TV 3] *$\{L\}$ is transversally symmetric iff it is a transversally $\mathfrak{C}$– as well as a transversally $\mathfrak{P}$– foliation.*

The following interesting subclass of transversally symmetric (Riemannian) foliations is introduced and studied in [GGV 1–5].

2.50 Definition. *Locally Killing-transversally symmetric spaces* (*locally KTS-spaces*) are Riemannian manifolds (M, g) equipped with an isometric flow generated by a unit Killing vector field such that the local reflections with respect to the flow lines are isometries.

These foliated spaces are studied in [GGV 5], [GV] by using the extrinsic and intrinsic geometry of geodesic spheres and tubular hypersurfaces around flow lines and geodesics orthogonal to flow lines.

CHAPTER III

T–PARALLEL FIELDS AND MIXED CURVATURE

3.1. Jacobi and Riccati equations

The study of the first and second variations of the length or the energy of curves in a Riemannian manifold leads to the Jacobi equation. Sometimes the simple use of the second variation allows us to establish relationships between curvature properties and the structure of a manifold in general. We'll give some basic material from the *variational theory of geodesics*; for more details see [KN], [BuZ] and [GHL].

For smooth normal vector fields y and z along a geodesic γ (i.e., $y, z \perp \gamma'$), the *index bilinear form* is defined by the well-known formula

$$I(y, z) = (y', z)_{|_0^{l(\gamma)}} - \int_0^{l(\gamma)} (z, y'' + R_{\gamma'} y)\, ds, \tag{3.1}$$

obtained from the second variation of the energy $\mathcal{E}(\gamma) = \frac{1}{2}\int_0^{l(\gamma)} |\gamma'(s)|^2\, ds$ in the direction y in the case where $z = y$. Here $R_{\gamma'} := R(*, \gamma')\gamma' : (\gamma')^{\perp} \to (\gamma')^{\perp}$ is a *Jacobi operator* (see, for example, [BVa]) and $y' = \nabla_{\gamma'} y$, $y'' := \nabla_{\gamma'}\nabla_{\gamma'} y$.

3.1 Definition. A vector field y along a geodesic γ in a Riemannian manifold M is called a *Jacobi vector field* if it satisfies the *Jacobi equation*

$$y'' + R_{\gamma'} y = 0. \tag{3.2}$$

If y and z are Jacobi vector fields along a geodesic γ, then the function $f = (y', z) - (y, z')$ is constant.

3.2 Definition. Let $\gamma : [0, l] \to M$ be a geodesic. The points $m = \gamma(0)$ and $m' = \gamma(l)$ are said to be *conjugate along* γ if there exists a nontrivial Jacobi vector field y along γ such that $y(0) = y(l) = 0$. (A slight modification of this concept leads to the notion of a *focal point*: a Jacobi field y must satisfy the conditions $y(0) \perp \gamma'$, $y(0) = y(l) = 0$). The dimension of the subspace of Jacobi vector fields along γ that vanishes at m and m' is called the *order of conjugacy* of m and m'.

3.3 Lemma. [GHL] *Let $\gamma : [0, l] \to M$ be a minimizing geodesic, y a Jacobi vector field and z a vector field along γ with the same values as y at the ends of γ. Then $I(z, z) \geq I(y, y)$, and equality holds iff $z = y$.*

The tangent and orthogonal components of the Jacobi vector field y along the geodesic γ are also Jacobi vector fields. The Jacobi vector fields orthogonal to the geodesic γ are considered in many situations.

To each rectangular $\delta : Q \to M$ we associate two vector fields: $x = d\delta(\frac{\partial}{\partial t})$ and $y = d\delta(\frac{\partial}{\partial \tau})$. A smooth variation (rectangular) $\delta : Q \to M$ is called a *geodesic variation* if all τ-edges $\delta(\tau, *)$ are geodesics, see Figure 18.

The geometric meaning of a Jacobi vector field is as follows.

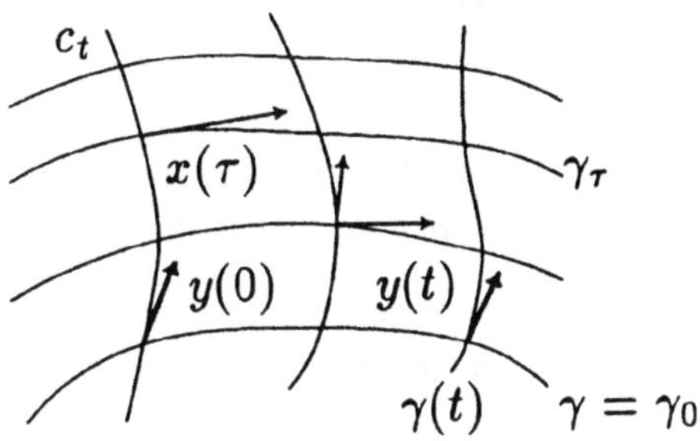

Figure 18. Variations and vector fields

3.4 Lemma. [GHL] *If $\delta : Q \to M$ is a geodesic variation, then the velocity field $y = d\delta(\frac{\partial}{\partial \tau})$ is a Jacobi vector field along every τ-edge. Conversely, every Jacobi vector field can be generated by some geodesic variation.*

The Jacobi equation (3.2) in a space form $M(k)$ can be decomposed and the Jacobi vector fields may be written in the simple form

$$y(t) = \begin{cases} \cos\sqrt{k}t\bar{y} + \sin\sqrt{k}t\bar{z}, & k > 0 \\ t\bar{y} + \bar{z}, & k = 0 \\ \cosh\sqrt{-k}t\bar{y} + \sinh\sqrt{-k}t\bar{z}, & k < 0, \end{cases}$$

where $\bar{y}, \bar{z}$ are parallel vector fields along γ. As a consequence, conjugate and focal points are absent when $k \leq 0$.

There are comparison theorems for the length of a Jacobi vector field when the curvature of one manifold is a majorant for the curvature of another manifold. See Section 4.2 for pinching of curvature. In most cases the model for comparison is a space of constant curvature.

Let $g_{a,b}$ be a solution of the scalar differential equation $g'' + kg = 0$ with initial values $g_{a,b}(0) = a$, $g'_{a,b}(0) = b$, where $k = const$ and either $a > 0$ or $a = 0$ and $b > 0$. This function has the following form:

$$g_{a,b}(t) = \begin{cases} a\cos\sqrt{k}t + \frac{b}{\sqrt{k}}\sin\sqrt{k}t, & k > 0 \\ a + bt, & k = 0 \\ a\cosh\sqrt{-k}t + \frac{b}{\sqrt{-k}}\sinh\sqrt{-k}t, & k < 0. \end{cases}$$

Let $(0, r)$ be a maximal interval where $g_{a,b}(t) > 0$ holds (r may be infinite).

Let $y_{u,v}$ be a Jacobi vector field along a geodesic $\gamma : [0, r] \to M$ with initial values $y_{u,v}(0) = u$ and $y'_{u,v}(0) = v$.

3.5 Lemma. [BuZ] *If the sectional curvature $K(\gamma', *) \leq k$, then*

$$|y_{u,0}(t)| \geq g_{|u|,0}(t), \qquad |y_{0,v}(t)| \geq g_{0,|v|}(t).$$

In particular, if $k > 0$, then points on γ that are conjugate to $\gamma(0)$ are not closer than $\frac{\pi}{\sqrt{k}}$, and points that are focal to $\gamma(0)$ are not closer than $\frac{\pi}{2\sqrt{k}}$.

The proof of Lemma 3.5 is very simple. The case $K_M \geq k$ is more difficult and is studied in Berge's and Rauch's theorems. We'll give some consequences of these theorems.

3.6 Lemma. [BuZ] *Let sectional curvature $K(\gamma', *) \geq k$.*
1) *If there are no points on γ conjugate to $\gamma(0)$, then $|y_{0,v}(t)| \leq g_{0,|v|}(t)$.*
2) *If there are no points on γ focal to $\gamma(0)$, then $|y_{u,0}(t)| \leq g_{|u|,0}(t)$.*
In particular, if $k > 0$, then points on γ conjugate to $\gamma(0)$ are not farther than $\frac{\pi}{\sqrt{k}}$, and points focal to $\gamma(0)$ are not farther than $\frac{\pi}{2\sqrt{k}}$.

Let $m = \gamma(0)$, let $S(m, t)$ be the sphere with center m and radius t, and suppose $\gamma(t_0)$ is not conjugate to $\gamma(0)$ along γ. Then for any Jacobi vector field $y(t)$ along γ with $y(0) = 0$ and for $t = t_0$ we have

$$y' = -A(y), \qquad (y', y) = k_n |y|^2, \tag{3.3}$$

where A is the shape operator of the geodesical sphere $S(m, t)$ at the point $\gamma(t)$ with respect to the normal γ', and k_n is the normal curvature of this sphere in the direction $y(t)$. Note that (y', y) is the nonintegral part in the second variation formulas of length and energy. The operator A satisfies the Riccati equation $A' + A^2 + R_{\gamma'} = 0$. The formulas in (3.3) explain the convexity of geodesical spheres in complete simply connected manifolds with $K \leq 0$. Let $\gamma \subset M$ be a geodesic in a Riemannian manifold which cannot be extended.

3.7 Definition. A smooth (1, 1)-tensor field $Y(t) : \gamma^{\perp} \to \gamma^{\perp}$ is called a *Jacobi tensor* if it satisfies

1) the Jacobi equation

$$Y'' + R_{\gamma'}Y = 0, \tag{3.2'}$$

2) $Y(t)$ is nondegenerate in the sense that $\ker Y(t) \cap \ker Y'(t) = \{0\}$ for all t, (this condition is satisfied precisely when the action of Y on linearly independent parallel sections of $\gamma^{\perp}$ gives rise to linearly independent Jacobi vector fields).

The analysis is clearly valid for any subbundle of $\gamma^{\perp}$ which is invariant under both the covariant derivative and the Jacobi operator $R_{\gamma'}$, in particular, for the horizontal bundle $T_{\gamma}L^{\perp}$ where γ belongs to the leaf of a (totally geodesic) foliation $\{L\}$.

When Y is invertible we can set $B := Y'(t)Y^{-1}(t)$. If we differentiate B covariantly and substitute into equation (3.2′), then we find

$$B' + B^2 + R(t) = 0. \tag{3.4}$$

3.8 Lemma. *Let the matrix function* $Y(t)$, $(\det Y(0) \neq 0)$, *and the matrix function* $B(t)$ *satisfy the equation*

$$Y'(t) = B(t)Y(t). \tag{3.5}$$

Then $Y(t)$ *satisfies the Jacobi equation* (3.2′) *for some matrix function* $R(t)$ *iff* $B(t)$ *satisfies the matrix Riccati equation* (3.5).

3.9 Definition. The *Wronskian* of a pair of Jacobi tensors is defined by the formula $W(Y_1, Y_2) = (Y_1')^*Y_2 - Y_1^*Y_2'$. A Jacobi tensor Y is said to be a *Lagrange tensor* if the Wronskian $W(Y, Y) = (Y')^*Y - Y^*Y'$ is zero, (which, in particular, is true when $Y(t) = 0$ for some t), or equivalently, if the tensor $B := Y'Y^{-1}$ is self-adjoint at all points where Y is invertible.

For example, the following is a version of Meyer's theorem in terms of Jacobi tensors [ES]: *If* $trace\, R(t) \geq c > 0$ *in equation* (3.2) *with* $Y(0) = 0$ *and* $t \geq 0$, *then the first positive value of* t *where* $Y(t)$ *is not invertible satisfies* $t \leq \pi\sqrt{\frac{n}{c}}$ *and* $t = \pi\sqrt{\frac{n}{c}}$ *iff* $R(t) \equiv \frac{c}{n}Id$ (see [Kup] for the version in terms of Fourier coefficients.)

It can be shown that an invertible Jacobi tensor Y is a Lagrange tensor iff the corresponding tensor B is self-adjoint for some t. The important examples are Jacobi tensors, which vanish at some point of a geodesic, and *stable Lagrange tensors* (when they exist) with the condition $Y = \lim_{s\to+\infty} Y_s$, where the Jacobi tensors Y_s satisfy the conditions $Y_s(0) = Id$ and $Y_s(-s) = 0$. In the last case $B(t)$ is the second fundamental form of the (stable) horosphere through $\gamma(t)$ and $trace\, B(t)$ is its mean curvature. Moreover, every Lagrange tensor can be obtained by a normal geodesic variation of γ along some hypersurface [ES].

For the geometry of foliations it is natural to consider a Jacobi (non-Lagrange) invertible tensor Y; for a totally geodesic foliation the corresponding B is the *structural tensor* (2.3). The matrix Riccati equation (3.4) can be polarized with symmetric and skew-symmetric parts:

$$\begin{cases} \text{a) } (B^+)' + (B^+)^2 + (B^-)^2 + R(t) = 0, \\ \text{b) } (B^-)' + B^+B^- + B^-B^+ = 0. \end{cases} \tag{3.6}$$

The *expansion* θ, the *sheaf* tensor σ and the *vorticity* tensor ω are calculated from B as follows:

$$\theta := trace\,(B) = trace\, B^+,\ \sigma := B^+ - a\,\theta\, Id,\ \omega = B^-,$$

where $a = \frac{1}{\dim \gamma^\perp}$ if γ is Riemannian or time-like and $a = \frac{1}{\dim \gamma^\perp - 1}$ if γ is null. Then at points where $Y(t) \neq 0$, we have $\theta = trace\,(Y'Y^{-1}) = \frac{(\det Y)'}{\det Y}$. Note that $\sigma = 0$ exactly for any conformal operator B where $\sigma^2 \geq 0$, but $\omega^2 \leq 0$. For Lagrange tensors in Riemannian or Lorentzian manifolds, the parameter values t for which $\det Y(t) = 0$ are isolated [ES], but this property may fail for (non-Lorentzian) pseudo-Riemannian manifolds at "degenerate" null conjugate points [Hel].

From equation (3.6 a) it follows that θ satisfies the following Riccati equation (which is also called the *Raychaudhuri equation* in Relativity Theory)

$$\theta' + a\theta^2 + \{trace\, R(t) + trace\,(\sigma^2) + trace\,(\omega^2)\} = 0. \tag{3.7}$$

The applications of (3.6) and (3.7) to foliations are given in Chapter IV and Chapter VI.

In particular, if Y is a Lagrange tensor then, since $\omega = 0$, θ satisfies

$$\theta' + a\theta^2 + \{trace\, R(t) + trace\,(\sigma^2)\} = 0. \tag{3.7'}$$

3.10 Lemma. *Let the Riccati matrix equation $B' + aB^2 + R(t) = 0$, where $a = const > 0$ and $R(t)$ is a symmetric continuous $(n \times n)$-matrix, admit a solution $B(t)$ defined for $-\infty < t < \infty$ with a symmetric initial value $B(0)$.*

Then

$$\liminf_{s\to+\infty}\int_{-s}^{s}\left\{trace\,R(t)+trace\,(\sigma^2)\right\}\,dt\le 0, \tag{3.8}$$

and the equality holds iff $trace\,B(t)=trace\,R(t)+trace\,(\sigma^2)\equiv 0$. *Moreover, if* $trace\,R(t)\ge 0$, *then* $B(t)=R(t)\equiv 0$.

For example, Lemma 3.10 allows us to prove the following theorem.

3.11 Theorem. [BrW] *(*[Ab 3] *for* $K_{mix}\ge 0$, [Tan 1] *for* $K_{mix}=0$*) Let* $\{L\}$ *be a totally geodesic foliation with complete leaves, let* $TL^{\perp}$ *be integrable and suppose* $Ric_1(L)\ge 0$. *Then the foliation splits locally.*

3.12 Remark. The application of Lemma 3.10 for manifolds without conjugate points (when a stable Lagrange tensor exists along every geodesic) is given by [EhK]. Indeed, inequality (3.8) may be replaced by the more traditional inequality

$$\liminf_{s\to+\infty}\int_{-s}^{s} trace\,R(t)\,dt\le 0. \tag{3.9}$$

If $trace\,R(t)$ is integrable as in [In 5], then (3.9) may also be replaced by

$$\int_{-\infty}^{+\infty} trace\,R(t)\,dt\le 0. \tag{3.10}$$

Most Riemannian geometers have settled for the less precise inequality (3.9) rather than deriving inequality (3.8) by inadvertently discarding the sheaf tensor $\sigma(B)$ in deriving the Riccati inequality

$$\theta'+a\theta^2+trace\,R(t)\le 0 \tag{3.11}$$

from the trace of equation (3.4) by replacing $trace\,(B^2)$ with $a(trace\,B)^2$, where $a=\frac{1}{n-1}$.

Proof. For $n=1$ this result follows from [Lemma 3.31, EhK], see also [Ab 3]. Since $B^-(0)=0$, from (3.6 b) it then follows that $B^-(t)=0$, i.e., $B(t)$ is symmetric for all t. Then from (3.7′), assuming $f(t)=\{trace\,R(t)+trace\,(\sigma^2)\}$ and $\theta(t)=trace\,B(t)$, we have

$$\theta(s)-\theta(-s)+a\int_{-s}^{s}\theta^2(t)\,dt=-\int_{-s}^{s}f(t)\,dt.$$

Hence, by part b) of Lemma 3.15,

$$\limsup_{s\to+\infty}\left(-\int_{-s}^{s}f(t)\,dt\right)=-\liminf_{s\to+\infty}\int_{-s}^{s}f(t)\,dt\ge 0.$$

If equality holds, then $\theta = 0$, and hence $f = 0$ from (3.7'). From $trace\, R(t) \geq 0$ it follows that $trace\, R(t) = 0$ and $\sigma = 0$, i.e., $B = a\theta = 0$. Finally from the given matrix Riccati equation we obtain $R = 0$. □

We cannot drop the requirement that $B(0)$ is a symmetric matrix in Lemma 3.10, because for every $n > 1$ there exists a symmetric constant $(n \times n)$-matrix $R_n > 0$, such that the Riccati equation (3.4) with $R(t) = R_n$ has a continuous solution $B(t)$ for $t \in \mathbb{R}$. For example, if $n = 2m$ is even, then $R_n = E_n$ and

$$B(t) = \underbrace{\begin{pmatrix} B_2 & \dots & 0 \\ \vdots & \ddots & \vdots \\ 0 & \dots & B_2 \end{pmatrix}}_{m\ blocks}, \text{ where } B_2 = \begin{pmatrix} 0 & 1 \\ -1 & 0 \end{pmatrix}, \tag{3.12}$$

which corresponds to the Hopf fibration $p : S^{2m+1} \to \mathbb{C}P^m$ with the fiber $\{S^1\}$.

If $n = 2m + 1$ is odd, then $R_n = [1, \dots, 1, 4]$ is a diagonal matrix, B_2 is as in (3.12) above and

$$B(t) = \underbrace{\begin{pmatrix} B_2 & \dots & 0 & 0 \\ \vdots & \ddots & \vdots & \vdots \\ 0 & \dots & B_2 & 0 \\ 0 & \dots & 0 & B_3 \end{pmatrix}}_{m\ blocks}, \text{ where } B_3 = \begin{pmatrix} \sin 2t & \cos 2t & -\sin t \\ \cos 2t & -\sin 2t & -\cos t \\ 4\sin t & 4\cos t & 0 \end{pmatrix}, \tag{3.13}$$

which corresponds to a geodesic foliation of some open domain on $\mathbb{C}P^{m+1}$ (see Appendix C and [Rov 1]). In fact, the Jacobi tensor, $Y(t)$, along a geodesic γ, corresponding to the matrices in (3.13), can be induced by an n-parameter family of geodesics. Since this Jacobi tensor is nondegenerate and geodesics in $\mathbb{C}P^{m+1}$ are simple closed curves, this family forms the foliation near γ.

The example (3.13) of the solution of matrix Riccati equation (3.4) (also, see Appendix C) allows us to conclude that the result by G. Cairns (Lemma 3.4 and hence Theorem 3.2 in [Ca 8]) is wrong.

In global Riemannian geometry, it is often better to use the Riccati equation (or inequality) (3.4) rather than the Jacobi equation (3.2'). In an expository article of [Kar], a Riccati equation (or inequality) for the shape operator is employed to study the local differential *geometry of the level sets of a local distance function*. The Lagrange tensors and the Raychaudhuri equation are employed to study *hypersurface geometry and geodesic congruences* [ES], and the *conjugacy problem*, which is closely related to the *boundary rigidity problem* [Cr 1]. The volume comparison and extremal theorems (for balls and horospheres) are also obtained using the Riccati equation [Cr 2], and for the case of semi-Riemannian manifolds, see [AHo]. In [Kup] and [EhK] it is shown how the Riccati inequality technique is related to the Raychaudhuri equation technique of General Relativity.

In Chapters 4–7 techniques using the Riccati-Raychaudhuri equations in Section 4.3 and Section 6.1 and the Jacobi equations in Section 4.2 and Chapters 6-7 are developed and employed to investigate the *geometry of foliations.*

The **scalar Riccati equation** is a useful tool in oscillation theory and in the related comparison theory for the second order linear ODE $y'' + f(t)y = 0$. Making the change of variables $u(t) = y'(t)/y(t)$ for t values for which $y(t) \neq 0$, we obtain the associated nonlinear Riccati equation (first order ODE) $u' + u^2 + f = 0$. For geometric applications it is useful to consider the related ODE

$$u' + au^2 + f = 0, \tag{3.14}$$

where $a > 0$ is a given positive constant, as well as the associated integral equation

$$u(s) - u(s_0) + a \int_{s_0}^{s} u^2(t)\, dt = -\int_{s_0}^{s} f(t)\, dt. \tag{3.14$'$}$$

In the context of proving the existence of conjugate points for geodesics it is often assumed that $f(t) \geq 0$ in (3.14). Because the improper integral of a continuous nonnegative function has a limit, either finite or $+\infty$, the $u(s)$ has a limit as $s \to +\infty$ precisely when $\int_{s_0}^{\infty} f(t)\, dt$ is (conditionally) convergent. Then on an interval on which $u(t)$ is nonvanishing, the inequality

$$\frac{u'}{u^2} + a \leq 0 \tag{3.15}$$

holds and hence integration gives the following direct consequence.

3.13 Lemma. [EhK] *Suppose equation* (3.14) *holds with* $f \geq 0$.

1. *If* $u(t_1) = -\frac{1}{ad} < 0$, *then* $u(t) = -\infty$ *for some* t *in the interval* $(t_1, t_1 + d]$.

2. *If* $u(t_1) = -\frac{1}{ad} > 0$, *then* $u(t) = +\infty$ *for some* t *in the interval* $[t_1 - d, t_1)$.

3.14 Lemma. [EhK] *Suppose* $\int_0^{+\infty} f(t)\, dt$ *and* $\int_{-\infty}^{0} f(t)\, dt$ *converge conditionally and there exists a smooth solution* $u(t)$, $(-\infty < t < \infty)$, *of the Riccati equation* (3.14). *Then*

a) $\lim_{s \to +\infty} u(t) = \lim_{s \to -\infty} u(t) = 0,$

b) $\int_{-\infty}^{+\infty} f(t)\,dt = -a\int_{-\infty}^{+\infty} u^2(t)\,dt \le 0$ *where equality holds iff* $u = f = 0$.

Proof. Since $u(s) = -\int_0^s f(t)\,dt - a\int_0^s u^2(t)\,dt + u(0)$, the hypothesis of this lemma implies that $\lambda := \lim_{s\to+\infty} u(s)$ exists; either λ is finite or $\lambda = -\infty$. But if $\lambda = -\infty$, the order of growth arguments shows that

$$u(s) + a\int_0^s u^2(t)\,dt \tag{3.16}$$

has limit $+\infty$ as $s \to +\infty$, contradicting the finiteness of $\int_0^{+\infty} f(t)\,dt$ in equation (3.14') with $s_0 = 0$. Thus λ is finite. But if $\lambda \ne 0$, then $\int_0^s u^2(t)\,dt = +\infty$, again contradicting the assumed finiteness of (3.16) as $s \to +\infty$. Hence $\lambda = 0$. Similar arguments show that $\lim_{s\to-\infty} u(s) = 0$.

Now for part b): letting $s \to +\infty$ and $s_0 \to -\infty$ in equality (3.14') yields

$$\int_{-\infty}^{+\infty} f(t)\,dt = -a\int_{-\infty}^{+\infty} u^2(t)\,dt \le 0$$

and the equality holds iff $u = 0$. Thus by (3.14), $f = 0$ as well. □

In the absence of any *a priori* integrability assumption on $f(t)$, the following result may be applied.

3.15 Lemma. [Gu] *Let* $u : [0, +\infty) \to \mathbb{R}$ *and* $v : [-\infty, +\infty) \to \mathbb{R}$ *be continuous functions. Then for any* $a > 0$

a) $\limsup_{s\to+\infty}\left(u(s) + a\int_0^s u^2(t)\,dt\right) \ge 0$ *with equality iff* $u = 0$,

b) $\limsup_{s\to+\infty}\left(v(s) - v(-s) + a\int_{-s}^s v^2(t)\,dt\right) \ge 0$ *with equality iff* $v = 0$.

Proof. a) Taking $u = af$, we may assume that $a = 1$. Suppose that

$$\limsup_{s\to+\infty}\left(u(s) + \int_0^s u^2(t)\,dt\right) < -\delta < 0.$$

Then there exists $s_0 > 0$ such that

$$u(s) + \int_0^s u^2(t)\,dt < -\delta, \quad (\forall\, s \ge s_0). \tag{3.17}$$

Set $g(s) = \exp(-\int_0^s(\int_0^t u^2(\lambda)d\lambda)\,dt)$, $(s \ge 0)$. Then

$$g'(s) = -g(s)\int_0^s u^2(t)\,dt \le 0$$

and

$$g''(s) = \left(\int_0^s u^2(t)\,dt + u(s)\right)\left(\int_0^s u^2(t)\,dt - u(s)\right)g(s).$$

From inequality (3.17) it follows that $-u(s) > \delta$ on $[s_0, \infty)$, so $\int_0^s u^2(t)\,dt - u(s) > \delta$. Hence $g''(s) < -\delta^2 g(s) < 0$, $\forall s \in [s_0, \infty)$. But this is a contradiction, because g is positive and nonincreasing. Hence we have proved the inequality in a). Furthermore,

$$\begin{aligned}&\limsup_{s\to+\infty}\left(u(s) + a\int_0^s u^2(t)dt\right) = \\ &\limsup_{s\to+\infty}\left(u(s) + \frac{a}{2}\int_0^s u^2(t)dt\right) + \frac{a}{2}\int_0^\infty u^2(t)dt \geq \frac{a}{2}\int_0^\infty u^2(t)dt \geq 0.\end{aligned} \tag{3.18}$$

Now, if $\limsup\limits_{s\to+\infty}(u(s) + a\int_0^s u^2(t)\,dt) = 0$, then $\int_0^\infty u^2(t)\,dt = 0$ and so u is identically zero.

In order to prove b), we note that for every $s \in \mathbb{R}$

$$\int_{-s}^s v^2(t)\,dt = \int_0^s \left(v^2(t) + v^2(-t)\right)\,dt \geq \frac{1}{2}\int_0^s (v(t) - v(-t))^2\,dt.$$

Then by part a) we have

$$\begin{aligned}&\limsup_{s\to+\infty}\left(v(s) - v(-s) + a\int_{-s}^s v^2(t)dt\right) \geq \\ &\limsup_{s\to+\infty}\left(v(s) - v(-s) + \frac{a}{2}\int_0^s (v(t) - v(-t))^2 dt\right) \geq 0.\end{aligned}$$

Now if $\limsup\limits_{s\to+\infty}\{v(s) - v(-s) + a\int_{-s}^s v^2(t)\,dt\} = 0$, then by part a) we have $v(s) - v(-s) = 0$ identically. Thus $\int_{-\infty}^\infty v^2(t)\,dt = 0$ and so $v = 0$. □

3.2. T-parallel vector fields and the Jacobi equation

For a distribution $T_1 \subset TM$ (and $T_2 = T_1^\perp$), we define a T_1-parallel vector field in analogy with the basic vector field along a leaf for a Riemannian foliation.

3.16 Definition. A vector field $y : \gamma \to T_2(\gamma)$ along the unit speed T_1-*geodesic* $\gamma \subset M$ (i.e., $\gamma' \subset T_1$, $\nabla_{\gamma'}\gamma' \subset T_2$) is called T_1-*parallel* if the following first order ODE is satisfied:

$$P_2(\nabla_{\gamma'} y) = B_2(\gamma', y). \tag{3.19}$$

Let $\{e_i(0)\}$ be an orthonormal basis of T_2 at $m \in \gamma(0)$. We continue it along γ using differential equations $P_2(\nabla_{\gamma'}, e_i) = 0$ and we obtain an orthonormal frame field $\{e_i\}$ of T_2 along γ. Let $\{y_i\}$ with $y_i(0) = e_i(0)$ be the T_1-parallel vector fields along γ. This gives rise to a *T_1-parallel tensor (endomorphism) field* $Y : T_2(\gamma) \to T_2(\gamma)$ along γ given by $y_i = Ye_i$. Clearly $P_2\nabla_{\gamma'}y_i = P_2(\nabla_{\gamma'}Y)e_i$ and

$$P_2(\nabla_{\gamma'}Y) = B_2(\gamma', Y), \ (Y(0) = Id). \tag{3.20}$$

A T_1-parallel vector field y is called *L-parallel* if the distribution T_1 is tangent to a foliation $\{L\}$. Analogously we define a *T_2-parallel vector (tensor) field* $y : \gamma \to T_1(\gamma)$ (resp., $Y : T_1(\gamma) \to T_1(\gamma)$) along any unit speed *$T_2$-geodesic* $\gamma \subset M$.

3.17 Remark. From the homogeneous ODE (3.19) it follows that a T_1-parallel vector field has no zeros or is identically zero. For a foliation $\{L\}$ (i.e., where $TL = T_1$), L-parallel fields coincide with the horizontal basic vector fields y and are defined locally on the leaves (globally along the leaves with trivial holonomy or for a submersion $\pi : M \to B$ with $\ker_* \pi = T_1$) and satisfy the equation

$$P_2(\nabla_x y) = B_2(x, y), \ (x \in TL) \tag{3.19'}$$

in view of the equality [Gr 2]:

$$P_2[y, x] = 0, \ (x \subset TL). \tag{3.21}$$

For the totally geodesic foliation $\{L\}$ case, (i.e., $TL = T_1$), the concept of an *L-parallel Jacobi field* (i.e., induced by foliation) was introduced by [Dom], see Section 3.4 below.

For a structural tensor $B_2(t)$ and a T_1-parallel vector field $y(t)$ (tensor $Y(t)$) along a T_1-geodesic we have equations (2.37) and (3.19), see Lemma 2.25. The "missing" Jacobi type equation for these tensors is given in the following lemma.

3.18 Lemma. [Rov 19] *Let M be a Riemannian manifold with two complementary orthogonal distributions T_1, T_2 and let $y : \gamma \to T_2(\gamma)$, $(y(0) \neq 0)$, be a T_1-parallel vector field along the T_1-geodesic $\gamma \subset M$. Then*

$$P_2\nabla_{\gamma'}P_2\nabla_{\gamma'}y + \Delta_{\gamma'}y = 0, \tag{3.22}$$

where the tensor $\Delta_{\gamma'} : T_2(\gamma) \to T_2(\gamma)$ is given by the formula

$$(\Delta_{\gamma'}y, z) = (R(y, \gamma')\gamma', z) + ((\nabla_y B_1)(z, \gamma'), \gamma') + (B_1(z, B_1(y, \gamma')), \gamma'). \tag{3.23}$$

Proof. In view of (2.36) we have

$$
\begin{aligned}
(P_2\nabla_{\gamma'}P_2\nabla_{\gamma'}y, z) &= (\nabla_{\gamma'}(B_2(\gamma', y), z) \\
&= ((\nabla_{\gamma'}B_2)(\gamma', y), z) + (B_2(P_1\nabla_{\gamma'}\gamma', y), z) + B_2(\gamma', P_2\nabla_{\gamma'}y), z) \\
&= -(R(y, \gamma')\gamma', z) - ((\nabla_y B_1)(z, \gamma'), \gamma') - (B_1(z, B_1(y, \gamma'), \gamma').
\end{aligned}
$$
□

3.19 Remark. For a totally geodesic distribution T_1 (in particular, for the case of a Riemannian foliation $\{L\}$ with $TL = T_2$ where the geodesic γ is orthogonal to the leaves of the foliation) we have

$$(\Delta_{\gamma'}y, z) = (R(y, \gamma')\gamma', z) + (A_1(z, A_1(y, \gamma'), \gamma'). \tag{3.23'}$$

For a totally geodesic foliation $\{L\}$ with $TL = T_1$, (i.e., the geodesic γ is tangent to the leaves of the foliation), $\Delta_{\gamma'}$ coincides with the Jacobi operator

$$\Delta_{\gamma'}y = R(y, \gamma')\gamma'. \tag{3.23''}$$

For these cases (3.23') and (3.23"), the tensor $\Delta_{\gamma'}$ is symmetric.

3.20 Lemma. *For $x \in T_1$ and $y, z \in T_2$ we have*

$$(P_2\nabla_x P_2\nabla_x y, z) = (\nabla_x\nabla_x y, z) + (B_1(y, x), B_1(z, x)).$$

For the case of a Riemannian foliation, see [Ton 3].

Proof.

$$
\begin{aligned}
(P_2\nabla_x P_2\nabla_x y, z) &= (\nabla_x P_2\nabla_x y, z) = \nabla_x(\nabla_x y, z) - (P_2\nabla_x y, \nabla_x z) \\
&= (\nabla_x\nabla_x y, z) + (P_1\nabla_x y, P_1\nabla_x z) = (\nabla_x\nabla_x y, z) + (B_1(y, x), B_1(z, x)).
\end{aligned}
$$

3.21 Definition. ([Ton 3] *for integrable T_2*) The vector field $y \subset T_2(\gamma)$ is called a *T_2-Jacobi field* along the geodesic $\gamma \subset M$ if

$$P_2\{\nabla_{\gamma'}\nabla_{\gamma'}y + R(y, \gamma')\gamma'\} = 0. \tag{3.24}$$

The next corollary follows from Lemma 3.18 and Lemma 3.20.

3.22 Corollary. *Let M be a Riemannian manifold with two complementary orthogonal distributions T_1, T_2 and let $y : \gamma \to T_2(\gamma)$, $(y(0) \neq 0)$, be a T_1-parallel vector field along the T_1-geodesic $\gamma \subset M$. Then for $z \in T_2$,*

$$(\nabla_{\gamma'}\nabla_{\gamma'}y + R(y, \gamma')\gamma', z) = -2(B_1^+(y, \gamma'), B_1(z, \gamma')) - ((\nabla_y B_1^+)(z, \gamma'), \gamma'). \tag{3.25}$$

In particular, for a totally geodesic distribution T_1 and any geodesic $\gamma \subset M$, $(\gamma' \subset T_1)$ we obtain equation (3.24), *i.e., in this case all T_1-parallel vector fields are T_2-Jacobi vector fields.*

We study the relationship between ordinary Jacobi vector fields and T_2-Jacobi fields of a special type along γ.

3.23 Theorem. [Ton 3] *Let $\gamma(t) \subset M$ be a geodesic tangent to a totally geodesic distribution T_1 and suppose that the orthogonal distribution T_2 is integrable. Then the following holds:*
a) an ordinary Jacobi vector field y along γ is orthogonal to T_1 iff it satisfies the initial conditions at $m = \gamma(0)$

$$y(0) = y_0 \in T_2(m), \ \nabla_{\gamma'(0)} y = B_2(\gamma'(0), y_0) + B_1(y_0, \gamma'(0)), \tag{3.26}$$

b) a T_2-Jacobi vector field y along γ is an ordinary Jacobi vector field orthogonal to T_1 iff it satisfies the initial conditions at $m = \gamma(0)$

$$y(0) = y_0 \in T_2(m), \ P_2 \nabla_{\gamma'(0)} y = B_2(\gamma'(0), y_0). \tag{3.27}$$

Proof. a) Let $y \subset T_2(\gamma)$ be an ordinary Jacobi vector field with $y(0) = y_0 \in T_2(m)$. Then

$$\nabla_{\gamma'} y = P_2 \nabla_{\gamma'} y + P_1 \nabla_{\gamma'} y = P_2 \nabla_{\gamma'} y + B_1(y, \gamma').$$

Since y is a Jacobi vector field along γ and $y \perp \gamma'$, we have $\nabla_{\gamma'} y = \nabla_y \gamma'$ and by formulas (2.1) and (3.19),

$$\nabla_{\gamma'} y = B_2(\gamma', y) + B_1(y, \gamma'). \tag{3.28}$$

Thus (3.26) follows.

Conversely let $\bar{y} \subset T_\gamma M$ be an ordinary Jacobi vector field satisfying (3.26). The initial condition $\bar{y}(0) = y_0 \in T_2(m)$ defines in T_2 a Jacobi vector field y along γ by variations of γ through orthogonal geodesics; if $c(s)$ is a curve in the leaf through m with $c(0) = m$, $c'(0) = y_0$, then the geodesic γ_s with $\gamma_s(0) = c(s)$ has the initial velocity given by the unique horizontal lift of $\gamma'(0) \in T_2(m)$ to $T_2(c(s))$. Since $\bar{y}$, y are both Jacobi vector fields and satisfy the same initial conditions (3.26), it follows that $\bar{y} = y$ and $\bar{y}$ is necessarily tangent to T_2.

b) Let $y \subset T_2(\gamma)$ be an ordinary Jacobi vector field with $y(0) = y_0 \in T_2(\gamma(0))$. Then (3.26) holds. The ordinary Jacobi equation implies the T_2-Jacobi equation (3.24). Moreover (3.26) implies (3.27).

Conversely, let $\bar{y} \subset T_\gamma M$ be the T_2-Jacobi vector field satisfying (3.27). Then $\bar{y}$ coincides with the Jacobi vector field $y \subset T_2(\gamma)$ satisfying (3.26).

3.24 Remark. We see that for a Riemannian foliation $\{L\}$ (with $TL = T_1$) the T_2-parallel vector fields may be induced by variations of T_2-geodesics [Ton 3], [Wal 9] (see [Yor 1] for $TL = T_1$ and certain connections).

3.3. L-parallel vector fields and variations of curves

A variations method allows us to generate a Jacobi vector field with prescribed initial values by a one-parameter family of geodesics. In the case of a foliation, the curves of variation are special and belong to leaves.

Let $\{L\}$ be a totally geodesic foliation and γ a geodesic of M tangent to L. Then the foliation selects a $(\dim M - \dim L)$-dimensional linear subspace $J_\gamma(L)$ in the $(2\dim M)$-dimensional space of **all** Jacobi vector fields of M along γ. The subspace is parameterized by the elements of $T_{\gamma(0)}L^\perp$ and the initial values of these Jacobi vector fields are the following

$$y(0) \in T_{\gamma(0)}L^\perp, \ \nabla_{\gamma'(0)}y = B_2(\gamma'(0), y(0)). \tag{3.29}$$

Thus $J_\gamma(L)$ coincides with the space of L-parallel vector fields along γ.

The Jacobi vector fields of $J_\gamma(L)$ are generated by one-parameter families of geodesics of M, each of which is a geodesic of a leaf L. Note that a nontrivial Jacobi vector field y of $J_\gamma(L)$ has no zeros in view of the homogeneous ODE (see (3.19'))

$$\nabla_{\gamma'}y(t) = B_2(\gamma', y(t)).$$

This fact implies numerous global results about totally geodesic foliations. In particular, many authors prove the completeness for the leaves in certain foliations of open submanifolds of complete manifolds where the foliation is given by integral manifolds of the null spaces of certain tensor fields or differential forms, see Section 2.3.1.

For a totally geodesic (umbilic or spherical) foliation, the variational interpretation of L-parallel vector fields is given in [Dom] (resp., in [Rec 2] and [HR]). The case of an arbitrary foliation $\{L\}$ is considered in [Wal 9].

Let $\{L\}$ be a foliation on a Riemannian manifold (M, g), and $T_1 = TL$, $T_2 = TL^\perp$. Let Ω_1 be the space of smooth curves $c : [0, 1] \to M$ tangent to the foliation, i.e., the space of L-*curves*. We equip Ω_1 with the uniform C^1-topology induced by g and the Sasaki metric g_s on $TL_{|M}$. In this way, Ω_1 becomes a metric space with the distance function d_{Ω_1} given by

$$d_{\Omega_1}(c_1, c_2) = \sup_{0\leq t\leq 1} d(c_1(t), c_2(t)) + \sup_{0\leq t\leq 1} d_s(c_1'(t), c_2'(t)), \tag{3.30}$$

where d and d_s are the distance functions on (M, g) and $(TL_{|M}, g_s)$.

3.25 Definition. [Wal 9] A *curve* in Ω_1 is a continuous map $V : [0, 1] \times (a, b) \to M$ such that $V(*, s) \in \Omega_1$ for all $s \in (a, b)$. If $s_0 \in (a, b)$ and $c = V(*, s_0)$, then V is called an L*-variation* of c. The *tangent space* $T_c\Omega_1$, $(c \in \Omega_1)$ is considered to be the space of all variation fields corresponding to all the L-variations of c.

$T_c\Omega_1$ consists of smooth vector fields along c. Obviously, the set $T_c\Omega_1$ is a vector space containing all the fields tangent to TL.

Let $V(t, s) = c_s(t)$ be a regular rectangular (one-parameter variation of L-curves $\{c_s(t)\}$, $(s \geq 0)$) and assume that $\{c_s(t)\}$ are geodesics on the leaves. Let $e_1(s, t)$ and $e_2(s, t)$ be coordinate vector fields on $V(t, s)$, i.e., $e_1 = \frac{\partial}{\partial t}V$ and $e_2 = \frac{\partial}{\partial s}V$. Since these vector fields commute, i.e,

$$[e_1, e_2] = \nabla_{e_1}e_2 - \nabla_{e_2}e_1 = 0,$$

we have

$$R(e_2, e_1)e_1 = \nabla_{e_2}\nabla_{e_1}e_1 - \nabla_{e_1}\nabla_{e_2}e_1.$$

Also

$$\nabla_{e_1}e_1 = P_1\nabla_{e_1}e_1 + h_1(e_1, e_1) = h_1(e_1, e_1), \tag{3.31}$$

because the trajectories of e_1 are L-geodesics. Finally we obtain the equation along $c(t) = V(0, t)$

$$y'' + R(y, c')c' - (\tilde{\nabla}_y h_1)(c', c') - 2h_1(P_1y', c') = 0, \tag{3.32}$$

where

$$y(t) = e_2(0, t),\ y' = \nabla_{c'}y,\ y'' = \nabla_{c'}\nabla_{c'}y,$$
$$(\tilde{\nabla}_y h_1)(c', c') = \nabla_y h_1(c', c') - 2h_1(P_1\nabla_y c', c').$$

A vector field y along the L-geodesic c is called an L*-Jacobi field* if it satisfies equation (3.32). In particular, if such a y is tangent to the leaves, then it is in the usual sense of a Jacobi vector field on a leaf. It is clear that L-Jacobi fields (3.32) form a vector space of dimension $\dim M + \dim L$. If $\{L\}$ is a totally geodesic foliation, then (3.32) is a standard Jacobi equation in M.

The notion of L-Jacobi fields along geodesics on the leaves of a foliation is introduced in [Wal 4,9] during the study of the geodesic flow of a foliation. In [Wal 4] the L-Jacobi fields are used to investigate some ergodic properties of the geodesic flow of a foliation, e.g., to estimate Lyapunov exponents and entropy.

L-Jacobi fields occur as variation fields while varying a leaf geodesic c among leaf geodesics. The second order differential operator

$$J_c y = -y'' - R(y, c')c' + (\tilde{\nabla}_y h_1)(c', c') + 2h_1(P_1 y', c')$$

is similar to the operator $-P_2\nabla_{c'}(P_2\nabla_{c'} y) - \Delta_{c'} y$ (see (3.23)), acting in the space of vector fields along c. Of course, the L-Jacobi operator J_c depends on the curvature of M as well as on the second fundamental form of $\{L\}$. It is shown in [Wal 9], that J_c plays a role in the second variational formula for the arc length $\mathcal{L}$ and the energy $\mathcal{E}$ of the leaf curves, but leaf geodesics appear to be critical for $\mathcal{L}$ and $\mathcal{E}$ only for some so-called *admissible* variations.

3.26 Lemma. [Wal 9] *If y is an L-Jacobi field for a foliation $\{L\}$ along the L-geodesic c, then $(y', c') = const$.*

We cannot claim (as for the Jacobi vector fields on a Riemannian manifold) that y is constantly orthogonal to c if $y(0)$ and $y(0)'$ are.

The geometric meaning of the L-parallel vector field $y(t) \subset T_2$ along an L-curve $c(t)$ is the following:

3.27 Lemma. [Wal 9]

$$y \in T_c\Omega_1 \iff P_2\nabla_{c'}(P_2 y(t)) = B_2(c', P_2 y(t)), \tag{3.33}$$

i.e., $T_c\Omega_1$ contains all L-parallel vector fields along the L-curve c.

In other words, every L-parallel vector field may be induced by the variation of leaf curves.

3.4. Positive mixed curvature

3.4.1. Compact foliations with positive partial mixed curvature

In 1961 T. Frankel [Fr 1] showed that two compact totally geodesic submanifolds M_1 and M_2 in a Riemannian space M with positive sectional curvature (for example, two great spheres in a round sphere) must necessarily intersect if the sum of their dimensions is at least the dimension of M. In [LR] it was proved that M_1 and M_2 in the above statement may be assumed to be complete when the sectional curvature of M is not smaller than a constant $c > 0$. If M is also a Kähler manifold with positive bisectional curvature, then two compact analytic submanifolds M_1 and M_2 must necessarily intersect if the sum of their dimensions is at least $\dim M$ [Fr 1], [GK]. These results were applied by S. Tanno [Tan 2], K. Abe [Ab 1], D. Ferus [Fe 1] and A. Borisenko [Bor 1-3,5] to estimate the dimension of a compact totally geodesic foliation on a Riemannian manifold (or submanifold, see Section 5.1) with positive mixed sectional curvature.

H. Wu [Wu 2] and later Z. Shen, G. Wei, [Shen], [ShW], and K. Kenmotsu with C. Xia [KX 1,2], studied certain curvature functions on a Riemannian manifold, which fill a gap between the sectional curvature and the Ricci curvature. For $q+1$ orthonormal vectors $V = \{x_0, x_1, \dots x_q\}$ the *(partial) q-Ricci curvature* of M is by definition

$$Ric^q(V) = \sum_{i=1}^{q} K(x_0, x_i).$$

The 1-curvature is equivalent to sectional curvature and the $(\dim M - 1)$-curvature is the usual Ricci curvature.

Below we consider manifolds with nonnegative (resp., positive) partial curvature. If Ric^q is nonnegative (positive) and $q' > q$, then $Ric^{q'}$ is also nonnegative (resp., positive), in view of the identity [KX 1,2]

$$\sum_{i=1}^{q'} K(x, z_i) = \frac{q'}{kC_{q'}^k} \sum_{1 \le i_1 < \dots < i_k \le q'} \sum_{j=1}^{k} K(x, z_{i_j}), \tag{3.34}$$

or by induction, using

$$\sum_{i=1}^{q+1} K(x, z_i) = \frac{1}{q} \sum_{i=1}^{q+1} \sum_{j \ne i} K(x, z_j). \tag{3.34'}$$

Thus, by Meyer's theorem, a complete Riemannian manifold with $Ric^q(M) \ge c > 0$ is compact. Moreover, if for M^n we have $Ric^q(M) \ge qc$ (resp., $Ric^q(M) \le qc$), then $Ric^{q+1}(M) \ge (q+1)c$ and $Ric \ge (n-1)c$, (resp., $Ric^{q+1}(M) \le (q+1)c$ and $Ric^{q+1}(M) \le (n-1)c$). Note that if for some $1 < q < \dim M - 1$ the partial curvature $Ric^q(M)$ is constant, then M is a space of constant sectional curvature. For example, since $M^{2n} = S^n(1) \times S^n(1)$

$n > 2$, has Ricci curvature $Ric(M) = Ric^{2n-1}(M) = n-1$ and sectional curvature $K_M = Ric^1(M) \in [0,1]$, its $(n+1)$-th Ricci curvature is positive: $Ric^{n+1}(M) \geq 1$. Hence, the class of Riemannian manifolds with nonnegative (resp., positive) curvature Ric^q is larger than the class of manifolds with nonnegative (resp., positive) sectional curvature. Compact locally symmetric spaces of rank ≥ 2 have positive q (> 1)-Ricci curvature. Slight perturbations of these metrics give nonsymmetric examples.

K. Kenmotsu and C. Xia [KX 1,2] generalized T. Frankel's results above for the case $Ric^q(M) > 0$. We define in Section 3.4.1 analogous functions for the mixed curvature of a foliated Riemannian manifold and estimate the dimension of a compact foliation with additional conditions on the second fundamental form of the leaves and positive partial mixed curvature.

If the radius of a circle of S^2 is "small", it is possible to find a great circle of S^2, which is "far" from it. J. Morvan [Mrv] started from this elementary fact and generalized the result [Fr 1] by giving an upper bound for the distance of two submanifolds of a Riemannian manifold with positive sectional curvature in terms of the length of the second fundamental forms. We generalize in Section 3.4.1 the result by [Mrv] for positive Ric^q and apply it to foliated manifolds.

For a foliated Riemannian manifold M with foliation $\{L\}$ there exist two *mixed q-Ricci curvatures*: $Ric_1^q(y_0, x_1, \dots, x_q)$ when $q \leq \dim L$ and $Ric_2^q(x_0, y_1, \dots, y_q)$ when $q \leq \operatorname{codim} L$, where $x_i \in TL$ and $y_i \in TL^\perp$.

Let $||h_i||$ be the supremum of the norm of the second fundamental forms of a submanifold M_i in a Riemannian manifold M.

3.28 Lemma. *([Mrv] for $q = 1$) Let M_1 and M_2 be compact submanifolds in a Riemannian manifold M with $Ric^q(M) \geq c > 0$ for some $q \leq$ min$\{\dim M_1, \dim M_2\}$. Then the distance $d(M_1, M_2)$ of M_1 and M_2 satisfies the following inequality:*

$$d^2(M_1, M_2) \leq \begin{cases} \frac{q}{c}\{||h_1|| + ||h_2|| + \frac{\pi^2}{4}\}, & if\ q' \leq 0, \\ \frac{q}{c}\{||h_1|| + ||h_2||\} + \frac{(q-q')}{c}\frac{\pi^2}{4}, & if\ 0 < q' < q, \\ \frac{q}{c}\{||h_1|| + ||h_2||\}, & if\ q' \geq q. \end{cases}$$

where $q' = \dim M_1 + \dim M_2 - \dim M + 1$.

For $q = 1$ the first inequality is exact: two orthogonal great circles in $S^3(1)$ have $||h_1|| = ||h_2|| = 0$ and distance $\frac{\pi}{2}$.

We shall prove the version of this result for a foliation. Let $||h_L||$ be the supremum of the norm of the second fundamental forms of the leaves, *diam L* the maximal extrinsic diameter of the leaves and $diam^{\perp} L$ the maximal distance between the leaves of the foliation $\{L\}$.

3.29 Theorem. *Let $\{L^{\nu}\}$ be a compact foliation on a Riemannian manifold $M^{\nu+n}$ with $Ric^{q}(M) \geq c > 0$ for some $q \leq \nu$. Then $diam\, M \leq 2 diam\, L + diam^{\perp} L$, where*

$$(diam^{\perp} L)^2 \leq \begin{cases} \frac{2q}{c}\{\|h_L\| + \frac{\pi^2}{2}\}, & if\ \nu \leq n-1, \\ \frac{2q}{c}\|h_L\| + \frac{(q-\nu+n-1)}{c}\frac{\pi^2}{4}, & if\ n-1 < \nu < n-1+q, \\ \frac{2q}{c}\|h_L\|, & if\ \nu \geq n-1+q. \end{cases}$$

For the Riemannian foliation in Theorem 3.29, the condition on curvature can be weakened to $Ric_1^q \geq c > 0$, since the shortest geodesic between any two leaves (see the proof) is orthogonal to such a foliation.

Proof. Consider two leaves L_1 and L_2 with distance $l = dist\,(L_1, L_2)$, which is attained at the points $m_1 \in L_1$ and $m_2 \in L_2$. An argument using the first variational formula of arclength shows that the shortest geodesic $\gamma(t)$, $(0 \leq t \leq 1)$, with length l between m_1 and m_2 is orthogonal to L_1 and L_2.

Assume the second case: $n - 1 < \nu < n - 1 + q$. Then the parallel displacement of $T_{m_1}L$ along γ will intersect $T_{m_2}L$ in a q'-dimensional subspace V_2, where $\nu - n + 1 \leq q' < q$. We denote the inverse image of V_2 in $T_{m_1}L$ by V_1. Let $T_{m_1}L = V_1 \oplus V_1' \oplus V''$ be the orthogonal decomposition where the parallel image of V_1' is uniquely projected onto $T_{m_2}L$ (denote its orthogonal projection in $T_{m_2}L$ by V_2') and the parallel image of V'' is orthogonal to $T_{m_2}L$. (Note that V'' is zero for small l.) Let the vectors $e_1, \dots, e_{q'}$ form an orthonormal basis of V_1 and extend them to parallel vector fields $\bar{e}_1, \dots, \bar{e}_{q'}$ along γ. Obviously, $\bar{e}_1(m_2), \dots, \bar{e}_{q'}(m_2)$ belong to V_2. Let vectors $a_1, \dots, a_s$ (where $s = \dim V_1'$) form an orthonormal basis of V_1' and vectors $b_1, \dots, b_s$ form an orthonormal basis of V_2', and extend them to parallel vector fields $\bar{a}_1, \dots, \bar{a}_s$ and $\bar{b}_1, \dots, \bar{b}_s$ along γ.

Proposition. *Let V_1 and V_2 be subspaces in $\mathbb{R}^N$, $\dim V_1 = \dim V_2$. Then there exist orthonormal bases $\{a_i\} \subset V_1$ and $\{b_i\} \subset V_2$ (which correspond to extremal values of the angle between the given subspaces) with the property $a_i \perp b_j$, $(i \neq j)$.*

Consider the field of parallel planes $\sigma_i(t)$ along γ, spanned by the vectors $\bar{a}_i(t)$ and $\bar{b}_i(t)$. Assume that $\{a_i\}$ and $\{b_i\}$ correspond to the extremal angles between V_1' and the parallel image of V_2'. Then $\sigma_i(t) \perp \sigma_j(t)$ for $i \neq j$. We

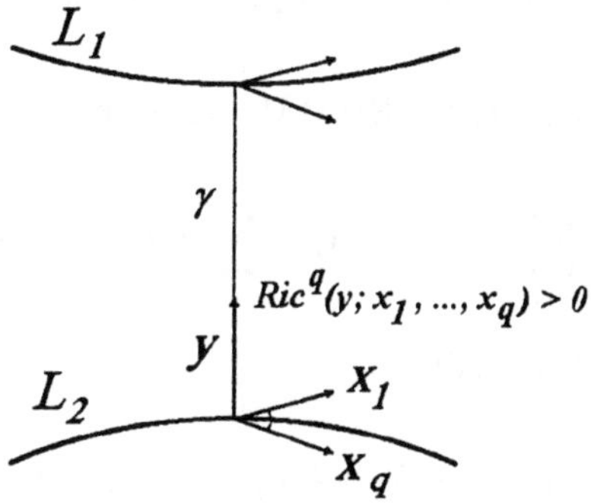

Figure 19.

take the unit vector $\tilde{b}_i(t) \in \sigma_i(t)$ such that $(\bar{a}_i, \tilde{b}_i(t)) = 0$. It is always possible to choose b_i and $\tilde{b}_i(t)$ with the properties $(\bar{a}_i, \bar{b}_i) \geq 0$ and $(\bar{b}_i, \tilde{b}_i(t)) \geq 0$. We introduce the unit vector fields

$$x_i(t) = (\cos\theta_i t)\bar{a}_i + (\sin\theta_i t)\tilde{b}_i(t)$$

along γ, where $\theta_i = \arccos(\bar{a}_i, \bar{b}_i) \in [0, \frac{\pi}{2}]$. Note that $x_i(t)$ is orthogonal to $x_j(t)$ when $i \neq j$. If $q' + s < q$, then we complete the system $\{a_i\}$ with the necessary number of orthogonal vectors from V'' and the system $\{b_i\}$ with the appropriate number of orthogonal vectors from the orthogonal complement to $V_2 \oplus V_2'$ in $T_{m_2}L$. For any such additional pair of vectors, the $x_i(t)$ are defined in the same way, except $\theta_i = \frac{\pi}{2}$. Thus we may assume that $q' + s = q$. By using the second variation of the energy $\mathcal{E}$ of γ along $x_i(t)$ and $\bar{e}_j$ we obtain

$$\mathcal{E}''_{x_i}(0) = (h_L(b_i, b_i), \frac{\gamma'(1)}{l}) - (h_L(a_i, a_i), \frac{\gamma'(0)}{l}) + \theta_i^2 - l^2 \int_0^1 K(\gamma', x_i(t))\, dt \geq 0,$$

$$\mathcal{E}''_{\bar{e}_j}(0) = (h_L(\bar{e}_j, \bar{e}_j), \frac{\gamma'(1)}{l}) - (h_L(e_j, e_j), \frac{\gamma'(0)}{l}) - l^2 \int_0^1 K(\gamma', \bar{e}_j)\, dt \geq 0. \tag{3.35}$$

Since $s = q - q' \leq q - \nu + n - 1$, $\sum_i \theta_i^2 \leq \frac{\pi^2}{4} s$,

$$\sum_{i=1}^{q'} |(h_L(b_i, b_i), \frac{\gamma'(1)}{l}) - (h_L(a_i, a_i), \frac{\gamma'(0)}{l})| \leq 2q' \|h_L\|,$$

$$\sum_{j=1}^{s} |(h_L(\bar{e}_j, \bar{e}_j), \frac{\gamma'(1)}{l}) - (h_L(e_j, e_j), \frac{\gamma'(0)}{l})| \leq 2s||h_L||,$$

and by the given inequality for $Ric^q(M)$ we have

$$\sum_{i=1}^{q'} K(\gamma', x_i(t)) + \sum_{j=1}^{s} K(\gamma', \bar{e}_j) \geq c;$$

then from (3.35) it follows that

$$l^2 c \leq 2q||h_L|| + (q - \nu + n - 1)\frac{\pi^2}{4},$$

which completes the proof of the second formula. The other cases are similar: for the first formula, the subspace V_1 is trivial and for the third formula it is enough to consider only q vectors from V_1. □

By the same method we obtain the following result.

3.30 Corollary. *Let M_1^{n-1}, M_2^l be a complete hypersurface and a submanifold in a Riemannian space $\bar{M}^n$ with the condition $Ric^q_{\bar{M}} \geq c$, where $c = const \geq 0$, $l \geq q$, and let one of M_1^{n-1}, M_2^l be compact. Assume that $\bar{M}^n$ has positive $q-$Ricci curvature either at all points of M_1^{n-1} or at all points of M_2^l. Then the distance $d(M_1, M_2)$ between M_1 and M_2 satisfies the inequality*

$$d^2(M_1, M_2) \cdot \frac{c}{q} < ||h_1|| + |H_2|,$$

where $||h_1||$ denotes the norm of the 2-d quadratic form of M_1, and $|H_2|$ is the mean curvature of M_2.

For $c = 0$ in Corollary 3.30 we obtain a result by [KX 1], that a totally geodesic hypersurface M_1^{n-1} (i.e. $h_1 = 0$) and a minimal submanifold M_2 (i.e. $H_2 = 0$) must intersect.

The next corollary follows from the third case of Theorem 3.29.

3.31 Corollary. *Let $M^{n+\nu}$ be a Riemannian manifold with a compact totally geodesic foliation $\{L^\nu\}$ and let $Ric_1^q(L)$ be positive along some leaf. Then $\nu < n - 1 + q$.*

For $q = 1$ this corollary was proved in [Tan 2] using the ideas of [Fr 1].

Next we consider classes of submanifolds and foliations with additional conditions on the second fundamental form.

3.32 Definition. A submanifold M in a Riemannian manifold $\bar{M}$ is called a *k-saddle* [She], if for every normal $\xi \in TM^{\perp}$ the second quadratic form A_ξ has $\leq k-1$ eigenvalues with the same sign, i.e., its canonical term is

$$a_1 dx_1^2 + \cdots + a_j dx_j^2 - b_1 dx_{j+1}^2 - \cdots - b_i dx_{j+i}^2, \ \max\{i, j\} \leq k-1, \ a_s, \ b_l > 0.$$

A foliation $\{L\}$ on a Riemannian manifold M is called a *k-saddle* if every leaf is a k-saddle submanifold. Totally geodesic submanifolds are 1-saddles.

3.33 Lemma. *Let M_1 and M_2 be compact k_i-saddle submanifolds in a Riemannian manifold M with $Ric^q(M) \geq c > 0$ for some $q \leq min\{\dim M_1 - k_1, \dim M_2 - k_2\} + 1$. Then the distance $d(M_1, M_2)$ of M_1 and M_2 satisfies the inequality*

$$d(M_1, M_2) \leq \begin{cases} \frac{\pi}{2}\sqrt{\frac{q}{c}}, & if \ q' \leq 0, \\ \frac{\pi}{2}\sqrt{\frac{q-q'}{c}}, & if \ 0 < q' < q, \\ 0, \ i.e. \ M_1 \cap M_2 \neq \emptyset, & if \ q' \geq q, \end{cases}$$

where $q' = \dim M_1 + \dim M_2 - \dim M - k_1 - k_2 + 3$.

The third case in Lemma 3.33 means that two compact k-saddle submanifolds M_1 and M_2 in a Riemannian space M with positive $Ric^q(M)$ intersect when $\dim M_1 + \dim M_2 \geq \dim M + 2k - 3 + q$. This statement by A. Borisenko (August, 1995) generalizes [Lemma 8, Bor 2] to k-saddle submanifolds with $q = 1$ and certain results for totally geodesic submanifolds: [KX 1] for $k = 0$ and [Fr 1] for $k = 0$, and $q = 1$. We prove the k-saddle foliation version of Lemma 3.33 using the following lemma.

3.34 Lemma. [Bor 2] *The maximal dimension of a subspace of the cone $\sum_{i=1}^{k} a_i x_i^2 - \sum_{j=k+1}^{r} a_j x_j^2 = 0, \ (a_s > 0)$, is equal to* $\min\{k, r-k\}$.

3.35 Theorem. *Let $M^{n+\nu}$ be a Riemannian manifold with a compact k-saddle foliation $\{L^\nu\}$ and suppose that $Ric_1^q(L)$ is positive for some $q \leq \nu - k + 1$. Then $\nu < q + n + 2k - 3$. Moreover, if $Ric^q(M) \geq c > 0$, then $diam\, M \leq 2\, diam\, L + diam^{\perp} L$, where*

$$diam^{\perp} L \leq \begin{cases} \frac{\pi}{2}\sqrt{\frac{q}{c}}, & if \ \nu \leq n + 2k - 3, \\ \frac{\pi}{2}\sqrt{\frac{q-\nu+n+2k-3}{c}}, & if \ n + 2k - 3 < \nu < q + n + 2k - 3. \end{cases}$$

Proof. Consider two leaves L_1 and L_2 with distance $l = dist\,(L_1, L_2)$, which is reached at the points $m_1 \in L_1$ and $m_2 \in L_2$. The shortest geodesic $\gamma(t)$,

$0 \le t \le 1$, with length l between m_1 and m_2 is orthogonal to L_1 and L_2. Since L_1 and L_2 are k-saddle submanifolds, by Lemma 3.34 the asymptotic cones for the quadratic forms of the normals $\xi_1 = \frac{\gamma'(0)}{l}$ and $\xi_2 = \frac{\gamma'(1)}{l}$ at the endpoints contain subspaces W_1 and W_2 with dimensions $\dim W_i \ge \nu - (k-1)$.

Next we follow the proof of Theorem 3.29 with W_i in the role of $T_{m_i}L$. Assume the second case: $n + 2k - 3 < \nu < q + n + 2k - 3$. Then the parallel displacement of W_1 along γ will intersect W_2 in a q'-dimensional subspace V_2, where $\nu - n - 2k + 3 \le q' < q$. We denote by V_1 the inverse image of V_2 in W_1. Let $W_1 = V_1 \oplus V_1' \oplus V''$ be the orthogonal decomposition, where the parallel image along γ of V_1' is uniquely projected onto W_2 (V_2' is its orthogonal projection in W_2) and the parallel image of V'' is orthogonal to W_2. (Note that V'' is trivial for small l.) Assume that the vectors $e_1, \dots, e_{q'}$ form an orthonormal basis of V_1 and extend them to parallel vector fields $\bar{e}_1, \dots, \bar{e}_{q'}$ along γ. Obviously, $\bar{e}_1(m_2), \dots, \bar{e}_{q'}(m_2)$ belong to V_2. Let the vectors $a_1, \dots, a_s$ (where $s = \dim V_1'$) form an orthonormal basis of V_1' and let $b_1, \dots, b_s$ form an orthonormal basis of V_2'; extend these bases to parallel vector fields $\bar{a}_1, \dots, \bar{a}_s$ and $\bar{b}_1, \dots, \bar{b}_s$ along γ. Consider parallel planes $\sigma_i(t)$ along γ, spanned by the vectors $\bar{a}_i(t)$ and $\bar{b}_i(t)$. Assume that $\{a_i\}$ and $\{b_i\}$ correspond to the extremal angles between V_1' and the parallel image of V_2' (see Proposition 3.30). Then $\sigma_i(t) \perp \sigma_j(t)$ for $i \ne j$.

We choose a unit vector $\tilde{b}_i(t) \in \sigma(t)$ such that $(\bar{a}_i, \tilde{b}_i(t)) = 0$. It is always possible to choose b_i and $\tilde{b}_i(t)$ with the properties $(\bar{a}_i, \bar{b}_i) \ge 0$ and $(\bar{b}_i, \tilde{b}_i(t)) \ge 0$. We introduce the unit vector fields

$$x_i(t) = (\cos\theta_i t)\bar{a}_i + (\sin\theta_i t)\tilde{b}_i(t)$$

along γ, where $\theta_i = \arccos(\bar{a}_i, \bar{b}_i) \in [0, \frac{\pi}{2}]$. Note that $x_i(t)$ is orthogonal to $x_j(t)$ when $i \ne j$.

If $q' + s < q$, then we complete the system $\{a_i\}$ with the necessary number of orthogonal vectors from V'' and we complete $\{b_i\}$ with the appropriate number of orthogonal vectors from the orthogonal complement to $V_2 \oplus V_2'$ in W_2. For any such additional pairs of vectors, the $x_i(t)$ are defined in the same way, except $\theta_i = \frac{\pi}{2}$. Thus we may assume that $q' + s = q$. By using the second variation of the energy $\mathcal{E}$ of γ along $x_i(t)$ and $\bar{e}_j$, in view of

$$\begin{aligned}(h_L(b_i, b_i), \xi_2) &= (h_L(a_i, a_i, \xi_1)) = (h_L(e_j, e_j, \xi_1)) \\ &= (h_L(\bar{e}_j(1), \bar{e}_j(1), \xi_2) = 0,\end{aligned}$$

we obtain

$$\mathcal{E}''_{x_i}(0) = \theta_i^2 - l^2 \int_0^1 K(\gamma', x_i(t))\,dt \geq 0, \quad \mathcal{E}''_{\bar{e}_j}(0) = -l^2 \int_0^1 K(\gamma', \bar{e}_j)\,dt \geq 0. \tag{3.36}$$

Since $s = q - q' \leq q - \nu + n + 2k - 3$, $\sum_i \theta_i^2 \leq \frac{\pi^2}{4} s$, and by the given inequality for $Ric^q(M)$, we have

$$\sum_i K(\gamma', x_i(t)) + \sum_i K(\gamma', \bar{e}_i) \geq c.$$

Then from (3.36) it follows that

$$l^2 c \leq (q - \nu + n + 2k - 3)\frac{\pi^2}{4},$$

which completes the proof of the second case. The first case is similar: the subspace V_1 is trivial. Assuming the opposite of $\nu < q + n + 2k - 3$, we consider for two arbitrary leaves with "small" distance $l > 0$, only q vectors from W_1, and use $Ric_1^q(L) > 0$ to obtain the contradiction: $l = 0$. □

3.36 Definition. ([Bor 8] *for* $s = 1$) A foliation $\{L\}$ on a Riemannian manifold M is said to be of *nonpositive* (resp., *negative*) *extrinsic s-Ricci curvature* if the s-Ricci curvature of its leaves is not more (resp., less) than the s-Ricci curvature of the ambient space along the leaves.

3.37 Lemma. ([Gla] *for* $s = 1$) *A submanifold* $M^n \subset \bar{M}^{n+p}$ *with nonpositive extrinsic s-Ricci curvature is a k-saddle with* $k = p + s$.

Proof. By Lemma 5.48 (in Chapter V) every subspace $V \subset TM$ with dimension greater than $p+s-1$ contains an asymptotic vector. Hence for every unit normal vector $\xi \in TM^\perp$ the second quadratic form A_ξ has not more than $p + s - 1$ nonzero eigenvalues of the same sign. From Definition 3.32 it follows that M^n is a $(p + s)-$saddle submanifold. □

Another generalization of Lemma 3.37 with $s = 1$ was given by [Bor 2] for $\bar{q}(> 2)$-dimensional extrinsic curvature, see Section 5.1. In view of Lemma 3.37 and Theorem 3.35 we have the following corollary.

3.38 Corollary. *a) The distance between two compact submanifolds* M_1 *and* M_2 *with* $Ric_{h_1}^{s_1} \leq 0$, $Ric_{h_2}^{s_2} \leq 0$ *in a Riemannian space* M *with* $Ric^q(M) \geq c > 0$ *for some* $q \leq min\{2\dim M_1 - s_1, 2\dim M_2 - s_2\} - \dim M + 1$ *satisfies the* *inequality*

$$d(M_1, M_2) \leq \begin{cases} \frac{\pi}{2}\sqrt{\frac{q}{c}}, & if\ q' \leq 0, \\ \frac{\pi}{2}\sqrt{\frac{q-q'}{c}}, & if\ 0 < q' < q, \\ 0,\ i.e.\ M_1 \cap M_2 \neq \emptyset, & if\ q' \geq q, \end{cases}$$

where $q' = 2(\dim M_1 + \dim M_2) - 3\dim M + 3 - (s_1 + s_2)$.
b) If $\{L\}$ *is a compact foliation on* M *with nonpositive extrinsic* s*-Ricci curvature and partial mixed Ricci curvature* $Ric_1^q(L) > 0$ *for some* $q \leq \dim L - \mathrm{codim} L - s + 1$, *then* $\dim L < 3\ \mathrm{codim}\, L + q + 2s - 3$.

All of the above conditions include sectional curvature when $s_1 = s_2 = s = q = 1$.

Analogous results are true for the Kählerian case with positive partial bisectional Ricci curvature; see [KX 1,2] for the totally geodesic case.

3.4.2. Ferus's scheme

Theorem 1.38 by D. Ferus contains the most effective estimate of the dimension of a totally geodesic foliation $\{L\}$ with $K_{mix} = k > 0$:

$$\dim L < \rho(\mathrm{codim}\, L). \tag{3.37}$$

From this follow estimates for the index of relative nullity and the k-null index of a curvature tensor of a complete (sub)manifold M.

The proof of Theorem 1.38 is based on the following **scheme**, which was used later by other geometers like K. Abe, M. Magid, S. Tanno, P. Dombrowski, A. Borisenko and the author.

1. B. O'Neill and E. Steil [O'NS] investigated submanifolds with positive index of relative nullity in space forms $\widehat{M}(k)$ and used the following property of the matrix Riccati equation with $K_{\mathrm{mix}} = k = const$

$$B' + B^2 + kE = 0, \tag{3.38}$$

(this is true for the structural tensor $B(t) = B(\gamma'(t), *)$ of a relative nullity foliation $\{L\}$ along geodesic $\gamma \subset L$, see Section 2.3.1): *The eigenvectors of the solution* $B(t)$ *do not depend on* t; *each eigenfunction* $\lambda(t)$ *satisfies the scalar Riccati equation* $\lambda(t)' + \lambda^2(t) + k = 0$, *and hence for* $k > 0$ *a solution can't be extended to the whole real line* $-\infty < t < \infty$.

2. The key to the existence of an eigenvector for the structural tensor $B : TL \times TL^{\perp} \to TL^{\perp}$ of a totally geodesic foliation $\{L\}$ (see Lemma 3.40 below) is the following: *in the case opposite to* (3.37), *for any point* $m \in L$ *there exist nonzero vectors* $x \in T_m L$, $y \in T_m L^{\perp}$ *and a real number* λ *with the property* $B(x, y) = \lambda y$.

The estimate of $\dim L$ in local Theorem 1.38 is exact:

3.39 Theorem. [Rov 1] *For all* $\nu < \rho(n)$ *there exists a domain* U *(a neighborhood of the* ν*-dimensional great sphere) in a* $(n+\nu)$*-dimensional round sphere* $S^{n+\nu}$ *with the foliation* $\{L\}$ *by* ν*-dimensional great spheres.*

Proof. Since $\nu < \rho(n)$, there exist ν orthogonal matrices $\{B_i\}$ of order n which satisfy the identities (see Chapter I)

$$B_i^2 = -E_n, \ B_i B_j = -B_j B_i, \ (i \neq j).$$

Let S_0^{ν} be any great sphere in $S^{n+\nu}$ and let $m \in S_0^{\nu}$. Let $\{e_1, \dots, e_\nu, \partial_1, \dots, \partial_n\}$ be an orthonormal basis of $T_m S^{n+\nu}$ with the first ν vectors from $T_m S_0$. We extend such an orthonormal sub-basis of $T_m S_0$ to the local smooth vector fields $\{e_i\}_{1 \leq i \leq \nu}$ on the orthogonal ball $\exp(T_m S_0^{\perp})$ with the equation at point m

$$B(x, y) = \nabla_y x = \sum_i x_i B_i y, \ \ where \ x = \sum_{i=1}^{\nu} x_i e_i \in T_m S_0.$$

For any $x \in T_m S_0$, the constant extension (under parallel displacement) of operator B from (2.3) along the geodesic γ, $(\gamma'(0) = x)$, is a solution of the Riccati equation (3.37). Thus the n-parameter family of great ν-dimensional spheres, which are tangent to the subspaces $aff\{e_1, \dots, e_\nu\}$, forms a foliation of a small neighborhood of S_0 in $S^{n+\nu}$. □

Note that the second property (see equation (3.38)) is wrong if kE is replaced by a nonconstant symmetric matrix $R(t) > 0$. Moreover, the inequality (3.37) without additional conditions is wrong for totally geodesic foliations with $K_{\text{mixed}} > 0$, see the example in [Rov 1].

The linear operator defined on an even-dimensional vector space $\mathbb{R}^n$, as a rule, has no eigenvectors. A more complicated situation is that of a ν-parameter family of linear operators, in particular, when the linear operators commute with an additional complex structure.

3.40 Lemma. [Fe 2] *Let V_1, V_2 be real vector spaces with dimensions* $\dim V_1 = \nu$, $\dim V_2 = n$ *and let $B : V_1 \times V_2 \to V_2$ be a bilinear operator. If for any nonzero vector $x \in V_1$ the endomorphism $B(x, *) : V_2 \to V_2$ has no real eigenvalues, then $\nu < \rho(n)$.*

Proof. We may assume that V_1 and V_2 are Euclidean spaces. Let $\{e_i\} \subset V_1$ be an orthonormal basis. Then the following ν continuous vector fields $\{w_i\}$

$$w_i(y) = B(e_i, y) - (B(e_i, y), y)y$$

are tangent to the unit sphere $S^{n-1} \subset V_2$.

If these vector fields are linearly dependent at some point $y \in S^{n-1}$, i.e.,

$$\sum_i \lambda_i w_i(y) = 0,$$

then

$$B(\sum_i \lambda_i e_i, y) = [\sum_i \lambda_i (B(e_i, y), y)]y;$$

this is impossible by the hypotheses of the lemma. Thus $\{w_i\}$ are continuous pointwise linearly independent vector fields on the $(n-1)$-dimensional sphere, i.e., $\nu < \rho(n)$. □

3.41 Lemma. [Ab 3] *Let J be a complex structure on the real vector spaces V_1 and V_2 with even dimensions* $\dim V_1 = 2$ *and* $\dim V_2 = n$. *Let $B : V_1 \times V_2 \to V_2$ be a bilinear operator with the property*

$$B(Jx, y) = B(x, Jy) = JB(x, y). \tag{3.39}$$

Then for some nonzero vector $x \in V_1$ the endomorphism $B(x, *) : V_2 \to V_2$ has an eigenvector.

Proof. Since B commutes with the complex structure J, it follows that for any nonzero vector $x \in V_1$ there exists a complex number $\alpha + \beta J$ (nonzero, otherwise the proof is complete) and a nonzero vector $y \in V_2$ with the property $B(x, y) = \alpha y + \beta J y$. In view of property (3.39) and the linearity of B for the nonzero vector $z = \alpha x - \beta J x$ we have

$$B(z, y) = \alpha B(x, y) - \beta J B(x, y) = \alpha(\alpha y + \beta J y) - \beta J(\alpha y + \beta J y) = (\alpha^2 + \beta^2)y.$$

□

Lemmas 3.40 and 3.41 are used as estimates of the dimension of a foliation (distribution) with $K_{mix} = const > 0$ (B plays the role of the structural tensor

(2.1)) and also as estimates of the index of relative nullity of submanifolds in S^N and CP^N. We complete these results with the following lemma.

3.42 Lemma. [Rov 8] *Let J be a complex structure on the real vector spaces V_1 and V_2 with even dimensions* $\dim V_1 = \nu$ *and* $\dim V_2 = n$. *Let $B : V_1 \times V_2 \to V_2$ be a bilinear operator with the property*

$$B(Jx, y) = JB(x, y). \tag{3.40}$$

If for any nonzero vector $x \in V_1$ the endomorphism $B(x, *) : V_2 \to V_2$ has no real eigenvalues, then $\nu = 2$ and n is divisible by 4.

Proof.

1. The cohomology algebra $H^*(CP^m; \mathbb{Z})$ is a truncated polynomial algebra over the ring of integers $\mathbb{Z}$, generated by an element b, whose dimension is two, with the equality $b^{m+1} = 0$. The inclusion $i : CP^n \subset CP^m$ induces the homomorphism $i^* : H^*(CP^m; \mathbb{Z}) \subset H^*(CP^n; \mathbb{Z})$.

The complete Pontryagin class of the tangent bundle $T(CP^m)$ is given by the term

$$p(CP^m) = 1 + C^1_{m+1} b^2 + \cdots + C^i_{m+1} b^{2i} + \ldots.$$

The inclusion $CP^n \subset CP^m$ induces an inclusion of the bundles $T(CP^m)_{|CP^n} \subset T(CP^m)$ and in view of the axiom for characteristic classes [Mln] we have

$$p(T(CP^m)|_{CP^n} = \begin{cases} 1 + C^1_{m+1} b^2 + \ldots, & \text{for } n > 1 \\ 1, & \text{for } n = 1. \end{cases}$$

Finally we note that the normal bundle $N_m(CP^n)$ of the subspace CP^n in CP^m is not trivial for $1 < n < m$, because otherwise we would obtain

$$p(CP^n) = p(T(CP^n) \oplus N_m(CP^n)) = p(T(CP^m)|_{CP^n}),$$

i.e., $C^1_{n+1} = C^1_{m+1}$, which is impossible for $n < m$.

2. Fix a point $a \in CP_0^{\nu/2} \subset CP^{(n+\nu)/2}$. Using property (3.40) we generate an n-parameter family of subspaces $\{L = CP^{\nu/2}\}$ near $L_0 = CP_0^{\nu/2}$, which induces at the point a the given tensor B (see (2.3)). This family of totally geodesic submanifolds generates the sections of the bundle $N_m(CP^{\nu/2})$, which are Jacobi vector fields in $CP^{(n+\nu)/2}$ along the geodesics $\gamma \subset L_0$.

Since the vector bundle $N_m(CP^{\nu/2})$ is nontrivial for $\nu > 2$, the family $\{L\}$ does not form a fibration of a neighborhood of L_0. Thus there exists the induced section Y_1, $(Y_1(a) \neq 0)$, with value zero at some point $b \in L_0$. The restriction of Y_1 along a geodesic $\gamma \subset L_0$, which connects the points a and b, has value zero at b, and since $K_{mix} = 1$, it is equal to the term

$$y(t) = Y_1(a)\cos t + B(\gamma'(a), Y_1(a))\sin t.$$

Consequently, the vector $B(\gamma'(a), Y_1(a))$ is parallel to $Y_1(a)$, i.e., the endomorphism $B(\gamma'(a), *)$ has a real eigenvalue. □

3.43 Remark. Lemma 3.42 can be applied to complex cases; for example, it allows us to prove the following:

1) A complete Kählerian Riemannian manifold M^{2n} with $(k > 0)$-null index $\mu_k(M) > \begin{cases} 2, & n \text{ odd} \\ 0, & n \text{ even} \end{cases}$ is isometric to CP^n.

2) Certain (Riemannian) submersions from CP^7 and HP^3 do not exist; this is assumed in [Es 3].

3.4.3. The length of L-parallel vector fields

The Hopf fibrations of Euclidean spheres in (1.2) and the analogous fibrations of complex projective space (1.2') have the following properties of mixed and tangent curvature:

1. *Mixed sectional curvature is equal to* $k > 0$,
2. *Sectional curvature of the leaves is equal to* k *or* $4k$.

The concept of an L-parallel vector field, and in particular, its norm, helps us to investigate the relationships between these properties of curvature on foliated manifolds.

3.44 Lemma. *Let M be a Riemannian manifold with a totally geodesic foliation $\{L\}$ and suppose that along some complete leaf L_0, the mixed sectional curvature is positive. Then the leaves are locally homogeneous Riemannian manifolds (moreover, locally isometric to each other).*

Proof. Let $m \in L_0$ and $x \in T_m L$ (a unit vector at m). We have the equation (see (2.5) and (2.8))

$$2([Y, Z], x) = (B^-(x, Y), Z). \tag{3.41}$$

Let $Y,\ Z \subset TL_0^\perp$ be local L-parallel Jacobi vector fields near m. The totally geodesic leaves are locally isometric to each other, and the vertical component of the Lie bracket $P_1[Y, Z]$ is a Killing local vector field on L_0 [JW 1].

We shall show that all such vectors $P_1[Y, Z]_m$ linearly generate the whole space $T_m L$. Assume on the contrary that there exists a unit vector $x_0 \in T_m L$ which is orthogonal to every vector of the form $P_1[Y, Z]_m$. Then from (3.41) it follows that the linear operator $B(x_0, *)$ is symmetric. Thus along the unit speed geodesic $\gamma \subset L_0$, $(\gamma'(0) = x_0)$, the Riccati equation (3.4) has a continuous solution with symmetric initial value; this contradicts Lemma 3.10.

So, for any nonzero $x_0 \in T_m L$ there exist vectors $x_i = P_1[y_i, z_i]$, $(1 \le i \le \dim L)$ such that $x_0 = \sum_i P_1[y_i, z_i]$. Note that the sum $X_0 = \sum_i P_1[Y_i, Z_i]$ is also a local Killing vector field on L_0 and hence X_0 generates a one-parameter group of isometries of L_0, which moves the point m in the direction x_0. □

In the case of positive constant mixed sectional curvature, we have more information about the geometry of the leaves.

Note that the function $f = |Y|^2 : \tilde{L} \to \mathbb{R}$ (of an L-parallel Jacobi field on the universal covering of L) has the following expression along any geodesic $\gamma(t) \subset \tilde{L}$:

$$f \circ \gamma(t) = a_1 \cos^2(t) + a_2 \cos(t)\sin(t) + a_3 \sin^2(t),$$

and hence it is a particular case of the following:

3.45 Definition. [Rov 6] The function $f : M \to \mathbb{R}$ on a complete Riemannian manifold M is called *periodic* if there exists a real $T \neq 0$ such that along any unit speed geodesic $\gamma : \mathbb{R} \to M$ we have $f \circ \gamma(t + T) = f \circ \gamma(t)$ for all t.

Since geodesics on CROSS (compact rank one symmetric space, noted by [Bes 1]) are closed with the same length, every continuous function on CROSS is periodic.

3.46 Definition. [Bes 1] A Riemannian manifold M is called a Y_l^m-manifold, if all geodesics through some point $m \in M$ return to this point after the same positive value of the arc-length parameter l.

It is known [Bes 1] that a Y_l^m-manifold is compact, has a finite fundamental group, and its ring of rational cohomologies is generated by one element.

3.47 Lemma. [Rov 6] *Let $f : M \to \mathbb{R}$ be a nonconstant periodic function on a complete connected analytical Riemannian manifold M with* $\dim M \ge 2$. *Then M is a Y_l^m-manifold, where $l = T$ or $l = 2T$ and m is an arbitrary point.*

Lemma 3.47 allows us to solve the problem of reconstructing a metric on a manifold which admits a periodic function.

3.48 Theorem. [Rov 6] *Let M be a complete simply connected homogeneous Riemannian manifold with* $\dim M \geq 2$, *which admits a nonconstant periodic function $f : M \to \mathbb{R}$. Then M is isometric to CROSS.*

Proof. By Lemma 3.47, M is a Y_l^m-manifold for any $m \in M$, where $l = T$ or $l = 2T$. Thus Theorem 3.48 follows from the next theorem.

3.49 Theorem. [Rod] *Let M be a simply connected homogeneous Riemannian manifold and suppose that all the geodesics through some point m come back to this point m for the same positive value of arc-length parameter. Then M is isometric to CROSS.*

From Theorem 3.49 and Lemma 3.44 we get the following theorem.

3.50 Theorem. *Let M be a Riemannian manifold with a totally geodesic foliation $\{L\}$. If the foliation is non-Riemannian along some complete leaf L_0 and*

$$K(x, y) = k > 0, \qquad (x \in TL_0,\ y \in TL_0^{\perp}),$$

then the leaves are locally isometric to CROSS.

CHAPTER IV

RIGIDITY AND SPLITTING OF FOLIATIONS

The known rigidity and splitting results for foliations with nonnegative curvature in mixed directions are not as systematic as those in the theory of Riemannian manifolds of nonnegative curvature; results for foliations usually contain strong additional assumptions, for example, the normal distribution is integrable or totally geodesic, the (co)dimension is equal to one, the curvature is constant, or the result is for a specific class of foliations such as (UF, UF) or (GF, F). In this chapter we begin discussing such results and continue in Chapters VI & VII, which are devoted to a confirmation of the following hypothesis: *If the dimension of a foliation is "large" (for example,* $\dim L \geq \rho(\operatorname{codim} L)$ *or* $\dim L > 2$ *for the Kählerian case) and the mixed sectional curvature is nonnegative, then the situation is locally extremal and we have a metric decomposition of the manifold.*

The results in Chapter IV are local in the sense that a neighborhood of a complete leaf of a foliation is considered. In **Section 4.1** a short survey of foliations on space forms is given. The results of **Section 4.2** are based on the following invariants of foliations introduced by the author: the *turbulence along a leaf (or a geodesic) and the volume of a T-parallel vector field.* We begin with the simple property that for a Jacobi vector field $y(t)$ the function $y'(t) \wedge y(t)$ (which is the *area of a parallelogram*) is constant when $K_{mix} = const$. Then we estimate the derivative of the function above in terms of turbulence and pinching of $K_{mix} \geq 0$. This allows us to state in Section 4.2.1 the special behavior of certain T-parallel (Jacobi) vector fields along the "extremal" geodesic (which starts from the eigenvector of the structural tensor of a foliation). In Section 4.2.2 the development of this method is given based on the volume of a T-parallel vector field.

In **Section 4.3** we consider, in contrast to Section 4.2, the traditional matrix Riccati and Raychaudhuri ODE's in Riemannian geometry. We use, however, a polarized form (in the sense of the symmetric and skew-symmetric components with respect to a certain distribution T_i) of a matrix Riccati equation and study the behavior of structural tensors along T_i-curves under the assumption that a norm of a second fundamental form or an integrability tensor of a distribution is bounded from above by a certain term involving the mixed curvature.

In **Section 4.3.1** we consider a foliation (distribution) with nonnegative mixed Ricci curvature whose orthogonal distribution is "almost" involutive; the integral formula for mixed scalar curvature along a complete leaf; rigidity and splitting results are obtained. **Section 4.3.2** contains facts from ergodic theory used in Section 4.3. In **Section 4.3.3** a foliation (distribution) with nonnegative mixed sectional curvature, whose orthogonal distribution is "almost" totally geodesic, is considered, leading us to a rigidity result. The case when a given distribution is involutive corresponds to a generalization of a totally geodesic Riemannian foliation.

4.1. Foliations on space forms

We consider foliations $\{L\}$ on a real space form $\bar{M}(k)$.

4.1.1. Riemannian submersions and foliations

For $k > 0$, Riemannian submersions from the spheres $S^n(k)$ are Hopf fibrations, except possibly in the case where $\dim M = 15$ and $\dim L = 7$, see [Es 1–3], [Ra 3-4], [GG 1,2]. The class of Riemannian foliations on the spheres $S^n(k)$ is larger than the class of Hopf fibrations (see Appendix A) and for $\dim L \leq 3$ consists only of *homogeneous foliations* (i.e., generated by the isometric action of a Lie group) [GG 2]. There are no homogeneous foliations on the spheres $S^n(k)$ with $\dim L = 7$ [Lu].

For $k = 0$, Riemannian submersions of a complete space form $\bar{M}(0)$ are projections of a direct product onto one of its factors, see [Wls 3], [Chu]. Riemannian foliations of a compact space form $\bar{M}(0)$ split (are locally projections of a direct product onto one of its factors); without compactness, this is false even in $\mathbb{R}^2$: consider, for example, the 1-dimensional foliation generated by glide rotations.

For $k < 0$, there are no Riemannian foliations (and submersions) on a compact space form $\bar{M}(k)$; the same is true for compact locally symmetric spaces M with negative curvature, see [Wls 2]. By contrast, there exist Riemannian submersions from noncompact $\mathbb{H}^n(k)$ with various types of leaves, see [Fe 4], [Bro] and [Gor]. In [Wls 4] Riemannian submersions from $S^n \times \mathbb{R}^k$ are studied.

In [Pak] 1-dimensional Riemannian foliations of Einstein spaces are studied. In [Ber] the concept of a submetry is introduced; this is a generalization of a Riemannian submersion.

4.1 Definition. [Ber] A map $p : M \to N$ of metric spaces is called a *submetry*, if for each point $x \in M$ the p-image of every closed ball with center x is a closed ball of the same radius with center $p(x)$.

4.2 Theorem. [Ber] *Each submetry $p : M \to N$ of complete Euclidean spaces (possibly infinite dimensional) decomposes into $p = i_1 \circ p_1$, where p_1 is an orthogonal projection onto some closed Euclidean subspace M_1 in N and $i_1 : M_1 \to N$ is an isometry.*

Another special case of a Riemannian submersion comes from the notion of a metric fibration as a partition of a metric space by isometric and locally equidistant subsets, see [EVGL], [VGE], [EG], [Gor].

4.3 Definition. [EVGL] Two subsets A and B of a metric space (M, d) are *parallel* $(A \parallel B)$ if, for each point $a \in A$, there exists a unique nearest point $a^* \in B$ and the distance $d(a, a^*)$ is constant (i.e., $d(a, a^*) = d(A, B)$); the same holds for each point $b \in B$. A metric space (M, d), which is the union of its subsets (fibers) $\{L\}$, is called a *metric fibration* if two conditions are satisfied:
1) any two fibers are isometric as subsets of M : $L' \cong L''$,
2) sufficiently close fibers are parallel: $\exists c > 0$ such that $d(L', L'') < c \Rightarrow L' \parallel L''$.

The only metric fibrations of $\mathbb{R}^n$ with connected fibers are the unions of parallel k-planes ($k = 0, 1, \ldots, n$) [VGE]. Hyperbolic geometry is much more interesting from this point of view. Even for the plane $\mathbb{H}^2$, the metric fibration into horocycles (with a common ideal point) is not unique.

4.4 Definition. [EVGL] The union Γ_x of two half horocycles with opposite normals in a common point is called a *broken horocycle*; note that a broken horocycle is not an analytic curve, but only C^1. The trajectory D_x (containing $x \in l$) orthogonal to the family N of all straight lines intersecting the axis l in the angle α is called a *fifth line*.

V. Efremovič called D_x the fifth line because of its importance in hyperbolic geometry (the straight line, the circle, the horocycle and equidistant are the first four lines). Unlike the broken horocycle, the fifth line is an analytic curve.

4.5 Theorem. [VGE] *There are exactly three nontrivial metric fibrations (i.e., different from that of one-pointed fibers and of the whole plane) of $\mathbb{H}^2$ with connected fibers as follows:*

1) *the fibration into horocycles,*
2) *the fibration into broken horocycles,*
3) *the fibration into fifth lines.*

Also in a Lobachevsky space $\mathbb{H}^n$, the fibrations into horospheres (with common ideal point) are not the only metric fibrations, see [Gor] and [EG].

4.1.2. Totally geodesic foliations

For $k > 0$, the totally geodesic foliations on the 3-sphere are classified in [GW]; in the case of S^n with $n \geq 3$, there exist some interesting subclasses such as skew-Hopf fibrations and fibrations associated with curvature tensor (see Appendix A).

For $k < 0$, there are no totally geodesic foliations on a complete space form $\bar{M}(k)$ with a finite volume, see [Ze 1,2]; the same is true on locally symmetric spaces. For $k < 0$, (actually $K_{mix} < 0$) there are no totally geodesic foliations with compact leaves on a space form $\bar{M}(k)$ even locally, see [Rov 18].

Some results for totally geodesic foliations on a hyperbolic space $\mathbb{H}^n(k)$ are given in [Fe 4], [AP], and [Bro].

For $k = 0$, (actually $K_{mix} = 0$) a totally geodesic foliation, given locally on $\bar{M}(0)$ and having compact leaves, splits [Rov 18].

4.6 Remark. In [BaW 1,2] the conformal foliations by geodesics on 3-dimensional space forms $\bar{M}(k)$ are studied. Minimal foliations on space forms $\bar{M}(k)$ are studied for $k > 0$ in [Kam], [LS] and [Got 1,2].

4.2. Area and volume of a T-parallel vector field

4.2.1. The area of a T-parallel vector field

A solution $Y(t) \subset \mathbb{R}^n$ of the matrix Jacobi equation $y'' + R(t)y = 0$ with constant curvature $R(t) = kE > 0$ may be written in the following form, which depends on the initial values $y(0)$ and $y'(0)$:

$$Y(t) = y(0)\cos(\sqrt{k}t) + \frac{y'(0)}{\sqrt{k}}\sin(\sqrt{k}t).$$

If $y(0)$ and $y'(0)$ are linearly independent, then the curve $Y(t)$ is an ellipse in the 2-plane $y(0) \wedge y'(0)$ and the area of the parallelogram $Y(t) \wedge Y'(t)$ is constant. If in addition, $y(0) \perp y'(0)$, then we have the obvious estimate $|Y(t)| \le \max\{|y(0)|, \frac{|y'(0)|}{\sqrt{k}}\}$. We introduce and study an analogous function (called by the *area of a T–parallel vector field*) along geodesics of a Riemannian manifold with foliation or distribution.

The following comparison lemma will be used for estimates of the length of T_i-parallel fields in terms of mixed sectional curvature pinching.

4.7 Lemma. *Let a solution $y(t) \subset \mathbb{R}^n$ of the Jacobi equation*

$$y'' + \Delta(t)y = 0, \quad (0 \le t \le \frac{\pi}{\sqrt{k}}) \tag{4.1}$$

with norm $||\Delta(t) - kE|| \le \varepsilon_1 < \frac{k}{2}$ be written in the form $y(t) = Y(t) + u(t)$, where $Y(t) = y(0)\cos(\sqrt{k}t) + \frac{y'(0)}{\sqrt{k}}\sin(\sqrt{k}t)$. Then

$$|u(t)| \le \frac{\varepsilon_1}{k - 2\varepsilon_1} \int_0^t \sqrt{k}|Y(s)| \sin(\sqrt{k}(t-s))\, ds. \tag{4.2}$$

Proof. Let $z(t) = y'(t)$. Then $z' = -\Delta(t)y$; moreover $\Delta(t) = kE + C(t)$, where the norm of the matrix $C(t)$ satisfies the inequality $||C(t)|| \le \varepsilon_1$. For a norm of the vector $f(t) = C(Y(t))$ we have the estimate $|f(t)| \le \varepsilon_1 |Y(t)|$. If we let $u' = \sqrt{k}v$, then $\sqrt{k}v' = -\Delta(t)u - f(t)$. With the help of the matrices

$$w = \begin{pmatrix} u \\ v \end{pmatrix}, \quad A = \begin{pmatrix} 0 & E \\ -E & 0 \end{pmatrix}, \quad F = -\begin{pmatrix} 0 \\ (Cu + f)/\sqrt{k} \end{pmatrix}, \tag{4.3}$$

we can rewrite the given Jacobi equation as a first order ODE with a constant matrix $\sqrt{k}A$:

$$w' = \sqrt{k}Aw + F, \quad (0 \le t \le \frac{\pi}{\sqrt{k}}). \tag{4.4}$$

Note that $w(0) = 0$ in view of $u(0) = u'(0) = 0$. By using the known formula for the solution of the linear system (4.4), we obtain the integral equation

$$w(t) = \int_0^t \Big(\exp(\sqrt{k}(t-s)A)F(s)\Big)\, ds. \tag{4.5}$$

It is easy to calculate

$$\exp(\sqrt{k}(t-s)A) = \begin{pmatrix} \cos(\sqrt{k}(t-s))E & \sin(\sqrt{k}(t-s))E \\ -\sin(\sqrt{k}(t-s))E & \cos(\sqrt{k}(t-s))E \end{pmatrix}$$

and then

$$\exp(\sqrt{k}(t-s)A)F(s) = -\frac{1}{\sqrt{k}} \begin{pmatrix} \sin(\sqrt{k}(t-s))(Cu + f) \\ \cos(\sqrt{k}(t-s))(Cu + f) \end{pmatrix}.$$

Finally we derive the integral expression for the vector function $u(t)$

$$u(t) = -\int_0^t \frac{\sin(\sqrt{k}(t-s))}{\sqrt{k}}(Cu(s) + f(s))\,ds. \tag{4.6}$$

With the help of (4.6) we estimate the norm $|u|(t) := \sup\{|u(s)| : s \leq t\}$

$$\begin{aligned} |u|(t) \leq & \frac{\varepsilon_1}{k}\int_0^t \sqrt{k}(|Y(s)| + |u(s)|)\sin(\sqrt{k}(t-s))\,ds \leq \\ & \frac{\varepsilon_1}{k}\int_0^t \sqrt{k}|Y(s)|\sin(\sqrt{k}(t-s))\,ds + \frac{2\varepsilon_1}{k}|u|(t), \end{aligned}$$

i.e.,

$$|u|(t)(1 - \frac{2\varepsilon_1}{k}) \leq \frac{\varepsilon_1}{k}\int_0^t \sqrt{k}|Y(s)|\sin(\sqrt{k}(t-s))\,ds, \tag{4.7}$$

from which the inequality (4.2) follows. □

4.8 Definition. For a Riemannian manifold with complementary orthogonal distributions T_1 and T_2, a *turbulence along a T_1-geodesic γ* (a rotation component of a tensor B_2) is defined by

$$\begin{aligned} & a(T_1, \gamma) = \\ & = \sup\{(B_2(x, y), z) : x \in T_1(\gamma),\ y, z \perp T_1(\gamma),\ y \perp z,\ |x| = |y| = |z| = 1\}. \end{aligned}$$

In particular, for a foliation $\{L\}$ on a Riemannian manifold (with $TL = T_1$) a *turbulence along the leaf L_0* is defined by

$$\begin{aligned} & a(L_0) = \\ & = \sup\{(B_2(x, y), z) : x \in TL_0,\ y, z \in TL_0^{\perp},\ y \perp z,\ |x| = |y| = |z| = 1\}, \end{aligned}$$

(see [O'N 4] and [KiT] for Riemannian submersions or foliations).

4.9 Remark. From the equality $a(L) = 0$ for all L, which means $B_2(x, y) = \lambda(x)y$, it follows that the orthogonal distribution $TL^{\perp}$ is tangent to a totally umbilic foliation; moreover, K_{mix} depends only on a point $m \in M$.

In the following theorem the condition for curvature is given for a sheaf of T_1-geodesics from one point.

4.10 Theorem. *(local) Let $M^{n+\nu}$ be a Riemannian manifold with complementary orthogonal distributions T_1 and T_2, let $\dim T_1 = \nu$ and suppose that there exists a point $p \in M$ such that along any T_1-geodesic $\gamma : [0, \frac{\pi}{\sqrt{k}}] \to M$, $(\gamma(0) = p)$ we have*

$$k_2 \geq K(\gamma', y) \geq k_1 > 0,\ (y \in T_2(\gamma)), \tag{4.8}$$

and one of the following inequalities holds:

$$(k_2 - k_1 + 2\varepsilon)\max\{a(T_1,\gamma)^2, k\} \le 0.3k(k_2+\varepsilon), \tag{4.9a}$$

$$\{2(|\nabla \widetilde{B_2^+}|+|\widetilde{B_2^+}|^2(a(T_1,\gamma)+|\widetilde{B_2^+}|))+(k_2-k_1+2\varepsilon)|\widetilde{B_2^+}|\}\max\{a(T_1,\gamma)^4,k^2\}$$
$$\le 0.0082k^2(k_2+\varepsilon)^3, \quad (with\ k_1-\varepsilon \ge 0.583(k_2+\varepsilon)), \tag{4.9b}$$

where $k = \frac{k_1+k_2}{2}$, $\varepsilon < k_1$, $\widetilde{B_2^+}y$ *is the component of* $B_2^+ y$ *that is orthogonal to the vector* y *and*

$$\varepsilon = \sup\{|((\nabla_y B_1)(z,\gamma'),\gamma')| + |(B_1(z, B_1(y,\gamma')),\gamma')|,\ |y| = |z| = 1\}.$$

Then $\nu < \rho(n)$*, and if* T_1 *is invariant with respect to the Kählerian structure on* M*, then* $\nu = 2$ *and* n *is divisible by* 4.

4.11 Remarks. The following are special cases of Theorem 4.10:

1. T_1 is a totally geodesic distribution. Then $\varepsilon = |B_1|^2$;

2. T_1 is a totally geodesic distribution and T_2 is integrable. Then $\varepsilon = |B_1|^2$ (i.e., T_2 is tangent to a Riemannian foliation as in [KiT]);

3. T_1 is tangent to a totally geodesic foliation. Then $\varepsilon = 0$. In this case, which is a generalization of Theorem 1.38, it follows from (4.9a) that $k_1 \ge 0.7k_2$.

In the latter case, Theorem 4.10 does not hold without conditions (4.9a) and (4.9b), but their coefficients 0.3 and 0.0082, respectively, are obtained by the method of proof of the theorem and perhaps can be improved. Condition (4.9a) obviously holds when $k_1 = k_2$. Condition (4.9b) holds when the foliation is conformal (see Definition 2.5).

Proof. First assume that condition (4.9a) holds. We break the proof into steps.

Step 1. In view of Lemma 3.40, it is sufficient to show that the linear operators $B_2(x,*) : T_2(p) \to T_2(p)$, $(x \ne 0)$, which are defined by (2.1), do not have eigenvectors. Suppose on the contrary that there exist unit vectors $x_0 \in T_1(p)$, $y_0 \in T_2(p)$ and a real $\lambda \le 0$ with the property $B_2(x_0,y_0) = \lambda y_0$. Let $\gamma(t) : [0, \frac{\pi}{\sqrt{k}}] \to M$, $(\gamma'(0) = x_0)$ be a T_1-geodesic and let $y(t) : \gamma \to T_2(\gamma)$ be a T_1-parallel vector field along γ containing the vector y_0. By Lemma 3.18 equation (4.1) holds: $y'' + \Delta(t)y = 0$ with $||\Delta(t) - kE|| \le \frac{k_2-k_1}{2} + \varepsilon$, where we denote $y' = P_2\nabla_{\gamma'}y$, $y'' = P_2\nabla_{\gamma'}(P_2\nabla_{\gamma'}y)$.

Idea of Proof of Theorem 4.10*:* The vector field $y(t)$ can be written in the form

$$y(t) = \left(\cos(\sqrt{k}t) + \frac{\lambda}{\sqrt{k}}\sin(\sqrt{k}t)\right)y_0 + u(t), \quad u(0) = u'(0) = 0.$$

(Note that for $k_2 = k_1$ and $\varepsilon = 0$, when T_1 is tangent to a totally geodesic foliation $\{L\}$, it is true that $u(t) = 0$ and hence the L-parallel Jacobi vector field $y(t)$ has a zero at the point $t_0 = \operatorname{arccot}(-\frac{\lambda}{\sqrt{k}})/\sqrt{k}$. This contradiction completes the proof of Theorem 1.38 by D. Ferus).

First we deduce that for (4.8) and $k_1 - \varepsilon \geq 0.583(k_2 + \varepsilon)$, the function

$$|y(t)|, \text{ the length of the } T_2\text{-parallel vector field } y(t),$$

has a local minimum at t_m in the interval $(0, \frac{\pi}{\sqrt{k}})$. The second observation is that the function

$$V(t), \text{ the area of a parallelogram}$$
$$\text{whose sides are the vectors } y(t) \text{ and } y'(t)$$

changes "slowly" along a geodesic γ. (This function is constant for $k_2 = k_1$ in the case where (4.9a) holds and for a conformal foliation in the case where (4.9b) holds.) From these we shall obtain a contradiction because the function $V(t)$ can't increase from zero value $V(0)$ to the "huge" value $V(t_m)$ on a given interval with length t_m bounded from above by $\frac{\pi}{\sqrt{k}}$. Now we continue the proof.

Step 2. We shall prove that the area of a parallelogram $V(t)$ in $T_2(\gamma(t))$, whose sides are the vectors $y(t)$ and $y'(t)$, satisfies the inequality

$$|V(t)'| \leq \left(\frac{k_2 - k_1}{2} + \varepsilon\right) |y(t)|^2. \tag{4.10}$$

In view of (4.1), the derivative of the function $V^2 = y^2(y')^2 - (y, y')^2$ is

$$(V^2)' = -2 \begin{vmatrix} (\Delta(t)y, y') & (\Delta(t)y, y) \\ (y, y') & (y, y) \end{vmatrix}.$$

By using linear combinations of columns in this 2×2-determinant, we obtain

$$(V^2)' = -2(\Delta(t)y, \widetilde{y'})y^2 = -2(\widetilde{\Delta(t)y}, \widetilde{y'})y^2, \tag{4.11}$$

where $\widetilde{\ldots}$ denotes the component orthogonal to the vector $y(t)$. Since

$$(V(t)^2)' = 2V'(t)V(t), \quad V(t) = |y(t)| \cdot |\widetilde{y'}(t)|, \tag{4.12}$$

from (4.11) it follows that $|V(t)'| \leq |\widetilde{\Delta(t)y(t)}| \cdot |y(t)|$. Thus we obtain (4.10) with the help of

$$|\widetilde{\Delta y}(t)| \leq |(\Delta(t) - kE)y(t)| \leq \left(\frac{k_2 - k_1}{2} + \varepsilon\right) |y(t)|.$$

Step 3. Suppose that the first local minimum t_m of a function $|y(t)|$ for $t \geq 0$ belongs to $(0, \frac{\pi}{\sqrt{k}}]$ and let $q = \gamma(t_m)$. We consider (in this step) the opposite parameterization of the geodesic γ with parameter $s = t_m - t$, i.e., $\gamma(0) = q$ and $\gamma(t_m) = p$. Since $\gamma(0)$ is also a local minimum of the function $|y(s)|^2$,

$$y'(0) \perp y(0), \ (y, y)'' \geq 0,$$

and from (4.1) and the equality $\frac{1}{2}(y, y)'' = (y, y'') + (y', y')$ it follows that

$$V(0) = |y'(0)| \cdot |y(0)| \geq \sqrt{k_1 - \varepsilon}|y(0)|^2. \tag{4.13}$$

Moreover $(y'(q), y'(q)) = (B_2(\gamma', y(q)), y'(q))$, i.e.,

$$\frac{|y'(q)|}{|y(q)|} \leq |(B_2\left(\gamma', \frac{y(q)}{|y(q)|}\right), \frac{y'(q)}{|y'(q)|})| \leq a(T_1, \gamma).$$

We can write the vector field $y(s)$ in the form $y(s) = Y_1(s) + u_1(s)$, where $Y_1(s) = y(0)\cos(\sqrt{k}s) + \frac{y'(0)}{\sqrt{k}}\sin(\sqrt{k}s)$. In view of Lemma 4.7 with $\varepsilon_1 = \frac{k_2 - k_1}{2} + \varepsilon$ we have

$$\begin{aligned} 1 &= |y(p)| \leq |Y_1(t_m)| + |u_1(t_m)| \\ &\leq \max\{\frac{a(T_1, \gamma)}{\sqrt{k}}, 1\}|y(q)|\left(1 + (1 - \cos(\sqrt{k}t_m))\frac{k_2 - k_1 + 2\varepsilon}{3k_1 - k_2 - 4\varepsilon}\right), \end{aligned}$$

that is,

$$|y(q)| \geq \frac{3k_1 - k_2 - 4\varepsilon}{k_1 + k_2} / \max\{\frac{a(T_1, \gamma)}{\sqrt{k}}, 1\}. \tag{4.14}$$

Note that from (4.9a) it follows that $k_1 - \varepsilon \geq 0.7(k_2 + \varepsilon)$, but at the same time for $3k_1 - k_2 - 4\varepsilon > 0$ it is sufficient to show that $k_1 - \varepsilon \geq \frac{1}{3}(k_2 + \varepsilon)$. From the eigenvector condition $B_2(x_0, y_0) = \lambda y_0$ it follows that $V(t_m) = 0$. Thus in view of (4.10) and the estimate $|y(s)| \leq 1$ for $0 \leq s \leq t_m$ we obtain the inequality

$$V(0) \leq \int_0^{t_m} |V'(s)| ds \leq \frac{k_2 - k_1 + 2\varepsilon}{2} \int_0^{t_m} |y(s)|^2 ds \leq \left(\frac{k_2 - k_1}{2} + \varepsilon\right) t_m.$$

From this and inequalities (4.13) and (4.14) follows the inequality

$$\left(\frac{\pi}{2} + \tau\right)(1 - \delta)\left(\frac{1 + \delta}{3\delta - 1}\right)^2 \max\{\frac{a(T_1, \gamma)^2}{k}, 1\} \geq \sqrt{2\delta(1 + \delta)} \tag{4.15}$$

where

$$\tau = \sqrt{k}(t_m - t_0), \ \delta = \frac{k_1 - \varepsilon}{k_2 + \varepsilon}, \ \sqrt{k}t_0 = \operatorname{arccot}(-\frac{\lambda}{\sqrt{k}}) \in (0, \frac{\pi}{2}].$$

Step 4. We now go back to the initial parameterization of γ and consider the problem when the function $|y(t)|$ has a local minimum in the interval $(0, \frac{\pi}{\sqrt{k}}]$.

The vector field $y(t)$ can be written in the form $y(t) = Y(t) + u(t)$, where $t \in [t_0, 2t_0]$ and

$$Y(t) = (\cos(\sqrt{k}t) + \frac{\lambda}{\sqrt{k}}\sin(\sqrt{k}t))y_0 = \frac{\sin(\sqrt{k}(t_0 - t))}{\sin(\sqrt{k}t_0)}y_0.$$

In view of Lemma 3.18 and Lemma 4.7 with $\varepsilon_1 = \frac{k_2-k_1}{2} + \varepsilon$, we have

$$|u(t)| \leq \frac{k_2 - k_1 + 2\varepsilon}{3k_1 - k_2 - 4\varepsilon}\frac{1}{\sin(\sqrt{k}t_0)}\int_0^t \sqrt{k}\sin(\sqrt{k}(t-s))\sin(\sqrt{k}|t_0 - s|)\,ds. \tag{4.16}$$

Using the trigonometric identity

$$f = \sin(\sqrt{k}(t-s))\sin(\sqrt{k}(t_0-s)) = \frac{\cos(\sqrt{k}(t-t_0)) - \cos(\sqrt{k}(t+t_0-2s))}{2},$$

and the shorthand notation

$$\tau = \sqrt{k}(t-t_0),\ \tau_0 = \sqrt{k}t_0,\ S = \sin,\ C = \cos,$$

we transform the integral on the right hand side of (4.16):

$$I(t) = \sqrt{k}\left(\int_0^{t_0} f\,ds - \int_{t_0}^t f\,ds\right) = (\tau_0 - \tau)\frac{C(\tau)}{2} + 3\frac{S(\tau)}{4} - \frac{S(\tau + 2\tau_0)}{4}.$$

Since

$$\tau_0,\ \tau \in [0, \frac{\pi}{2}],\ \tau_0 - \tau \geq 0,\ C(\tau) \geq 0,\ S(2\tau_0 + \tau) \geq S(\pi + \tau) = -S(\tau),$$

the function $I(t)$ has its largest upper value at $\tau_0 = \frac{\pi}{2}$:

$$I(t) \leq \left(\frac{\pi}{2} - \tau\right)\frac{C(\tau)}{2} + S(\tau),\quad I(t_0) \leq \frac{\pi}{4}\ (\text{since } \tau = 0).$$

Consequently, (assuming $\delta = \frac{k_1-\varepsilon}{k_2+\varepsilon}$)

$$|u(t)| \leq \left((\frac{\pi}{2} - \tau)\frac{C(\tau)}{2} + S(\tau)\right)\frac{1-\delta}{(3\delta - 1)S(\tau_0)},\quad |u(t_0)| \leq \frac{\pi}{4}\frac{1-\delta}{(3\delta - 1)S(\tau_0)}.$$

Note that $|Y(t)| = \frac{S(\tau)}{S(\tau_0)}$ and $Y(t_0) - 0$. Since $|y|'(0) = \lambda \leq 0$, and in the case of equality $\lambda = 0$ it follows from (4.8) that $|y|''(0) < 0$. Then the function $|y(t)|$ decreases for small values $t \geq 0$. Thus for the property that the function $|y(t)|$

has a local minimum at some $t_m \in (0, t]$ it is sufficient to require $|y(t_0)| \le |y(t)|$, i.e.,

$$|u(t_0)| + |u(t)| \le |Y(t)|.$$

The last inequality (in view of the above estimates) is reduced to

$$\left(\frac{\pi}{4} + (\frac{\pi}{2} - \tau)\frac{C(\tau)}{2} + S(\tau)\right)\frac{1-\delta}{3\delta - 1} \le S(\tau),$$

which is equivalent to

$$\delta \ge 1 - \frac{2S(\tau)}{\frac{\pi}{4} + (\frac{\pi}{2} - \tau)\frac{C(\tau)}{2} + 4S(\tau)}. \tag{4.17}$$

Remember that from (4.9a) it follows that $\delta \ge 0.7$. With the help of a computer we calculate that

$$\frac{2S(\tau)}{\frac{\pi}{4} + (\frac{\pi}{2} - \tau)\frac{C(\tau)}{2} + 4S(\tau)} \ge 0.3$$

for all $\tau \in [0.5, \frac{\pi}{2}]$, i.e., for $\delta \ge 0.7$, the function $|y(t)|$ has a local minimum at $t_m \in [0, \ (\frac{\pi}{2} + 0.5)/\sqrt{k}]$. But for $\delta \in [0.7, 1]$ and $\tau = 0.5$ the inequality

$$0.3\left(\frac{\pi}{2} + \tau\right)\left(\frac{1+\delta}{3\delta - 1}\right)^2 < \sqrt{2\delta(1+\delta)} \tag{4.18}$$

is equivalent to

$$f(\delta) = \left(\frac{3\delta - 1}{1+\delta}\right)^2 \sqrt{2\delta(1+\delta)} > 0.15(\pi + 1). \tag{4.19}$$

Since the function $\frac{3\delta-1}{1+\delta}$ is monotone increasing, the function $f(\delta)$ is also monotone increasing for $\delta > \frac{1}{3}$. It is easy to calculate that $f(0.7) > 0.63 > 0.15(\pi + 1)$. Thus (4.19) and also (4.18) hold for $\delta \ge 0.7$. From (4.18) and (4.15) the inequality follows, which contradicts (4.9a). □

4.12 Remark. We have pointed out the minimal $\delta \approx 0.7$ for which there exists a solution $\tau \in (0, \frac{\pi}{2}]$ for both inequalities (4.17) and (4.18). In other words, Theorem 4.10 holds for distributions with 0.7-pinched mixed sectional curvature and "small" values of ε.

The proof of (4.9b) is broken down into steps using the same plan as for (4.9a).

Step 1. Suppose on the contrary that in view of Lemma 3.40, for any point $m \in M$ there exist unit vectors $x_0 \in T_1(m)$, $y_0 \in T_2(m)$ and a real $\lambda \le 0$ with the property $B_2(x_0, y_0) = \lambda y_0$. Let $\gamma : [0, \frac{\pi}{\sqrt{k}}] \to M$, $(\gamma(0) = m,\ \gamma'(0) = x_0)$ be a T_1-geodesic and let $y : \gamma \to T_2(\gamma)$ be a T_1-parallel vector field along γ, which contains y_0.

Step 2. The function $V(t)$ (the area of the parallelogram in $T_2(\gamma)$ whose sides are the vectors $y(t)$ and $y'(t)$) satisfies the equality (see (4.11))

$$-(V^2)' = 2(\widetilde{\Delta y}, \widetilde{B_2^+ y})y^2 + 2(\widetilde{\Delta y}, B_2^- y)y^2, \tag{4.20}$$

where $\widetilde{\ldots}$ denotes the component orthogonal to the vector $y(t)$ and $B_2 y = B_2(\gamma', y)$. By polarizing the Riccati equation (2.36) (resp., (2.37)) we get its component which is symmetric with respect to $T_2(\gamma)$ (see Lemma 2.3 and equation (3.6)):

$$-\Delta = (B_2^+)' + (B_2^+)^2 + (B_2^-)^2,$$

and transform the right hand side of (4.20) into

$$\begin{aligned} |(\widetilde{\Delta y}, \widetilde{B_2^+ y})| &\le |\widetilde{\Delta y}| \cdot |\widetilde{B_2^+ y}| \le |\widetilde{B_2^+}||y|^2\left(\frac{k_2 - k_1}{2} + \varepsilon\right), \\ |(\widetilde{\Delta y}, B_2^- y)| &\le |(\widetilde{(B_2^+)'}y, B_2^- y) + (\widetilde{(B_2^+)^2}y, B_2^- y) + ((B_2^-)^2 y, B_2^- y)| \le \\ & (|\widetilde{(B_2^+)'}| + |\widetilde{B_2^+}|^2|B_2^-|)y^2 \le (|\widetilde{(B_2^+)'}| + |\widetilde{B_2^+}|^2(a(T_1,\gamma) + |\widetilde{B_2^+}|))y^2. \end{aligned} \tag{4.21}$$

We have used the fact that the inequality

$$|B_2^-| \le a(T_1, \gamma) + |\widetilde{B_2^+}|$$

follows from the identity

$$(B_2^- y, z) = (B_2 y, z) - (B_2^+ y, z).$$

Consequently,

$$|(V^2)'| \le [2(|\widetilde{(B_2^+)'}| + |\widetilde{B_2^+}|^2(a(T_1,\gamma) + |\widetilde{B_2^+}|)) + |\widetilde{B_2^+}|(k_2 - k_1 + 2\varepsilon)]y^4, \tag{4.22}$$

where

$$|\widetilde{B_2^+ y}| \le |B_2^+ y - \frac{\beta_1 + \beta_n}{2} y| \le |y|\frac{\beta_n - \beta_1}{2},$$

and $\beta_1 \le \cdots \le \beta_n$ are eigenvalues of the symmetric linear operator B_2^+. Set

$$F := [2(|\widetilde{(B_2^+)'}| + |\widetilde{B_2^+}|^2(a(T_1,\gamma) + |\widetilde{B_2^+}|)) + |\widetilde{B_2^+}|(k_2 - k_1 + 2\varepsilon)].$$

In particular, for a conformal (and thus for a Riemannian) totally geodesic foliation, we have $V(t) = const$ from (4.22), because from $\beta_n = \beta_1$ it follows that $\widetilde{B_2^+} = 0$.

Step 3. By repeating Step 4 of the proof of Theorem 4.10, we see that to show the existence of a local minimum for the function $|y(t)|$ at $t_m \in [0, \frac{\pi}{\sqrt{k}}]$, it is sufficient to prove the following inequality (see (4.17)):

$$\delta \geq 1 - \frac{2S(\tau)}{\frac{\pi}{4} + (\frac{\pi}{2} - \tau)\frac{C(\tau)}{2} + 4S(\tau)} = f(\tau), \tag{4.23}$$

where $\tau = \sqrt{k}(t - t_0)$, $\delta = \frac{k_1 - \varepsilon}{k_2 + \varepsilon}$. We calculate the minimum $\tau \in [0, \frac{\pi}{2}]$ of the function f. Since the derivative of f is nonpositive, i.e.,

$$f(\tau)' = -\frac{\frac{\pi}{2}C(\tau) + (\frac{\pi}{2} - \tau) + C(\tau)S(\tau)}{[\frac{\pi}{4} + (\frac{\pi}{2} - \tau)\frac{C(\tau)}{2} + 4S(\tau)]^2} \leq 0 \tag{4.24}$$

for $\tau \in [0, \frac{\pi}{2}]$, the function $f(\tau)$ is monotone decreasing; its minimal value is

$$f(\frac{\pi}{2}) = \frac{\pi + 8}{\pi + 16} \approx 0.582.$$

Thus for $\delta \geq 0.583$ the function $|y(t)|$ has its first local minimum in the interval $[0, \frac{\pi}{\sqrt{k}}]$ at a point t_m.

Step 4. As in Step 3 of the proof for the case where (4.9a) holds we derive the following inequalities:

$$V^2(t_m) \geq k_1 |y(t_m)|^4, \tag{4.25}$$

$$|y(t_m)| \geq \frac{3k_1 - k_2 - 4\varepsilon}{k_1 + k_2} / \max\{\frac{a(T_1, \gamma)}{\sqrt{k}}, 1\}. \tag{4.26}$$

In view of (4.21) and the upper estimate $|y(t)| \leq 1$ for $0 \leq t \leq t_m$ we have

$$V^2(0) \leq \int_0^{t_m} |(V^2)'(t)| dt \leq F \int_0^{t_m} |y(t)|^4 dt \leq F t_m. \tag{4.27}$$

From this and inequalities (4.25) and(4.26) follows the inequality opposite to (4.9b):

$$F \max\{a(T_1, \gamma)^4, k^2\} \geq \frac{1}{\pi} k^{3/2} k_1 \frac{(3k_1 - k_2 - 4\varepsilon)^4}{(k_1 + k_2)^4} \geq 0.00827 k^2 (k_2 + \varepsilon)^{3/2}, \tag{4.28}$$

for $\delta \geq 0.583$. This contradiction completes the proof for the case where (4.9b) holds. □

4.13 Definition. A *compact foliation* is a foliation with all of its leaves compact.

4.14 Lemma. *Assume that $M^{n+\nu}$ is a Riemannian manifold with a compact totally geodesic foliation $\{L\}$ and nonpositive mixed sectional curvature. Then M is locally isometric to the product $L \times L^{\perp}$.*

Proof. Note that L-parallel fields are globally defined on any leaf with trivial holonomy. By Lemma 1.16 almost every leaf has trivial holonomy; below L_0 is such a leaf and $Y : L_0 \to TL_0^{\perp}$ is any L-parallel field.

Since the leaves are compact, there exists a point $q \in L_0$, where the function $Y^2 : L_0 \to \mathbb{R}_+$ (square of the length) takes on its maximum. The restriction $y(t)$ of Y along any geodesic $\gamma(t) \subset L_0$ is a Jacobi vector field (see Chapter III). Since the mixed sectional curvature is nonpositive, we obtain

$$\frac{1}{2}((y^2(t))'' = (y, y'') + (y', y') = -K(\gamma', y(t)) + (B(\gamma', y(t))^2 \geq 0. \quad (4.29)$$

The convex function $y^2(t)$ takes on its maximum at $t = 0$ only in the trivial case $y^2(t) = const$. Since q may be connected by a geodesic on L_0 with any point on this leaf, the function Y^2 is also constant. By repeating the same argument for any geodesic $\gamma_1(t) \subset L_0$, $(\gamma_1(0) = p,\ \gamma_1'(0) = x)$, we get

$$-K(x, Y) + (B(x, Y))^2 = 0.$$

This equation holds for any L-parallel vector field Y along L_0, and hence we have

$$K(x, y) = B(x, y) = 0 \qquad (x \in TL_0,\ y \in TL_0^{\perp}). \qquad (4.30)$$

Since (4.30) holds along almost every leaf L_0, the mixed sectional curvature and the structural tensor B are zero on M, and the distribution $TL^{\perp}$ is tangent to the totally geodesic foliation. By de Rham's theorem [KN] M is locally isometric to the product $L \times L^{\perp}$. □

4.15 Remark. The condition Lemma 4.14 places on the mixed sectional curvature of M can be replaced by a weaker inequality with mixed Ricci curvature $Ric_2(y) = \sum_i K(x_i, y)$, where $y \in TL^{\perp}$ and $\{x_i\}$ is a local orthonormal basis of TL. Really, in the case $Ric_2 \leq 0$ the function Y^2 (see the proof of Lemma 4.14) is subharmonic on a certain compact leaf, thus $Y^2 = const$ by the Hopf Lemma [Ch 1].

From Theorem 4.10 and Lemma 4.14 we get the following theorem.

4.16 Theorem. *(decomposition) Assume that $M^{n+\nu}$ is a Riemannian manifold with a compact totally geodesic foliation $\{L^{\nu}\}$. Also assume that*

$$k_2 \geq K(x, y) \geq k_1 \geq 0, \ (x \in TL,\ y \in TL^{\perp}), \qquad (4.31)$$

and that one of the following inequalities holds:

$$(k_2 - k_1)\max\{a(L_0)^2, k\} \le 0.3k_2k, \tag{4.32a}$$

$$\{2(|\nabla\widetilde{B^+}| + |\widetilde{B^+}|^2(a(L_0) + |\widetilde{B^+}|)) + (k_2 - k_1)|\widetilde{B^+}|\}\max\{a(L_0)^4, k^2\} \le 0.0082k^2k_2^3, \quad (with\ k_1 \ge 0.583k_2), \tag{4.32b}$$

where $k = \frac{k_1+k_2}{2}$, *L_0 is some leaf and $\widetilde{B^+}y$ is the component of B^+y orthogonal to the vector y. If $\nu \ge \rho(n)$, ($\nu > 2$ for n divisible by 4 and Kählerian M and $\{L\}$), then $k_1 = k_2 = 0$ and M is locally isometric to the product $L \times L^\perp$.*

4.17 Remark. The assumption in Theorem 4.16 that the leaves are totally geodesic is given for simplicity.

4.2.2. The volume of an L-parallel vector field

Now we shall consider a Riemannian manifold M with a totally geodesic (for simplicity) foliation $\{L\}$ and give the development of the method in Section 4.2.1.

Assume that $x \in T_mL$ is any unit vector, $\nu \in [3, \dim L]$ is an integer, $\gamma \subset L$, $(\gamma'(0) = x)$, is the unit speed geodesic and $Y : L \to TL^\perp$ is an L-parallel (Jacobi) vector field on a neighborhood of γ. We define the following quantities:

$$a_\gamma(Y) = \sup\{|\widetilde{\nabla_u Y}|/|Y| : u \in T_\gamma L,\ |u| = 1\}, \tag{4.33}$$

$$R_\gamma(Y) = \sup\{|\widetilde{R(u, Y)v}|/|Y| :\ u \perp v,\ |u| = |v| = 1\}, \tag{4.34}$$

where $\widetilde{\dots}$ is the component of the given vector that is orthogonal to Y. Obviously, $a_\gamma(Y) \le a(L)$.

4.18 Definition. For any ∇-parallel ν-dimensional subbundle Z of $T_\gamma L$, containing γ', the nonnegative function $V_{\nu+1}(Y)$ is defined as the *$(\nu+1)$-dimensional volume of the parallelepiped* $Y \wedge \nabla_{e_1}Y \wedge \cdots \wedge \nabla_{e_\nu}Y$ in $T_\gamma L^\perp$, where $\{e_i\}_{i\le\nu}$ is an orthonormal basis in Z. (The values $V_{\nu+1}(Y)$ do not depend on the choice of basis in Z. We will call $V_{\nu+1}(Y)$ the $(\nu+1)$*–dimensional volume of* Y.)

4.19 Lemma. [Rov 19] *If the mixed sectional curvature satisfies*

$$k_2 \ge K(u, Y) \ge k_1,\ (u \in Z), \tag{4.35}$$

then the derivative of the $(\nu+1)$*-volume*

$$|\nabla_x V_{\nu+1}(Y)| \le \left((\nu-1)R_\gamma(Y) + \frac{k_2-k_1}{2}\right) a_\gamma(Y)^{\nu-1}|Y|^{\nu+1}. \tag{4.36}$$

4.20 Remark. Note that $V_{\nu+1}(Y) = const$ when M is a real space form. We considered in the proof of Theorem 4.10 a special case of (4.36) for the function $V_2(Y)$, *the area of the parallelogram* $Y \wedge Y'$ in $T_\gamma L^\perp$ (i.e., $\nu = 1$), and obtained the inequality

$$|V_2'| \le \frac{k_2-k_1}{2}|Y|^2. \tag{4.36'}$$

The author will use Lemma 4.19 for the case of $\nu = 2$ and recommends that the reader verify the proof below for this value of ν.

Proof. Let $Y_i = \nabla_{e_i} Y$, $Y_{ij} = \nabla_{e_j} Y_i$, $R_{ij}Y = R(e_i, Y)e_j$ and assume that $e_1 = \gamma'$. The function $V_{\nu+1}^2(Y)$ (the square of the volume) can be written using the Gram determinant

$$V_{\nu+1}^2(Y) = \begin{vmatrix} Y^2 & YY_1 & \cdots & YY_\nu \\ Y_1Y & Y_1^2 & \cdots & Y_1Y_\nu \\ \vdots & \vdots & \ddots & \vdots \\ Y_\nu Y & Y_\nu Y_1 & \cdots & Y_\nu^2 \end{vmatrix}. \tag{4.37}$$

The partial derivative of this determinant with respect to e_1 is equal to the sum of the $\nu+1$ determinants. Each differs from $V_{\nu+1}^2(Y)$ by a certain column

$$\nabla_{e_1}(V_{\nu+1}^2(Y)) =$$

$$\begin{vmatrix} YY_1 & \cdots \\ YY_{11} & \cdots \\ \vdots & \vdots \\ Y_{\nu1}Y & \cdots \end{vmatrix} + \begin{vmatrix} Y^2 & Y_1^2 + YY_{11} & \cdots \\ Y_1Y & 2Y_1Y_{11} & \cdots \\ \vdots & \vdots & \vdots \\ Y_\nu Y & Y_{\nu1}Y_1 + Y_{11}Y_\nu & \cdots \end{vmatrix} + \cdots + \begin{vmatrix} \cdots & Y_1Y_\nu \\ \cdots & Y_{11}Y_\nu + Y_{\nu1}Y_1 \\ \vdots & \vdots \\ \cdots & 2Y_{\nu1}Y_\nu \end{vmatrix} \tag{4.38}$$

(also we subtract the second column from the first column of the first determinant). We collect together the terms in (4.38) which include Y_{i1} and note that the determinant with other terms has two equal rows and so it is equal to zero, thus obtaining

$$\frac{1}{2}\nabla_{e_1}(V_{\nu+1}^2(Y)) = \sum_{i=1}^{\nu}(-1)^i \begin{vmatrix} Y_{i1}Y & Y_{i1}Y_1 & \cdots & Y_{i1}Y_\nu \\ Y^2 & YY_1 & \cdots & YY_\nu \\ \vdots & \vdots & \ddots & \vdots \\ Y_kY & Y_kY_1 & \cdots & Y_kY_\nu \\ \vdots & \vdots & \vdots & \vdots \end{vmatrix}_{k\neq i}. \tag{4.39}$$

We subtract the first column with a certain coefficient in each determinant of (4.39) from the other columns ($\widetilde{\ldots}$ denotes the component of the given vector orthogonal to Y) obtaining

$$\frac{1}{2}\nabla_{e_1}(V_{\nu+1}^2(Y)) = \sum_{i=1}^{\nu}(-1)^i \begin{vmatrix} Y_{i1}Y & \tilde{Y}_{i1}\tilde{Y}_1 & \cdots & \tilde{Y}_{i1}\tilde{Y}_\nu \\ Y^2 & 0 & \cdots & 0 \\ \vdots & \vdots & \ddots & \vdots \\ Y_kY & \tilde{Y}_k\tilde{Y}_1 & \cdots & \tilde{Y}_k\tilde{Y}_\nu \\ \vdots & \vdots & \vdots & \vdots \end{vmatrix}_{k\neq i}. \tag{4.40}$$

We calculate the determinants in (4.40) with respect to the second row (with one nonzero element):

$$\frac{1}{2}\nabla_{e_1}(V_{\nu+1}^2(Y)) = Y^2\sum_{i=1}^{\nu}(-1)^{i+1} \begin{vmatrix} \tilde{Y}_{i1}\tilde{Y}_1 & \cdots & \tilde{Y}_{i1}\tilde{Y}_\nu \\ \vdots & \ddots & \vdots \\ \tilde{Y}_k\tilde{Y}_1 & \cdots & \tilde{Y}_k\tilde{Y}_\nu \\ \vdots & \vdots & \vdots \end{vmatrix}_{k\neq i}. \tag{4.41}$$

We now use the following elementary transformations, which do not change the i-th determinant in (4.41), where $1 \leq i \leq \nu$: subtract from the first row, the row with element $\tilde{Y}_j^2$, with multiple

$$\alpha_j = (\tilde{Y}_{i1}\tilde{Y}_j)/\tilde{Y}_j^2,\ (j \neq i);$$

after this, the first row has only i-th nonzero element $\hat{\tilde{Y}}_{i1}\tilde{Y}_j$, where $\widehat{\ldots}$ is the component of the vector orthogonal to each Y_j except Y_i where $j \neq i$. We calculate the determinants with respect to the first row and obtain

$$\frac{1}{2}\nabla_{e_1}(V_{\nu+1}^2(Y)) = Y^2\sum_{i=1}^{\nu}(-1)^i(\widetilde{Y_{i1}}\widehat{\tilde{Y}_i})\begin{vmatrix} \tilde{Y}_1^2 & \cdots & \tilde{Y}_1\tilde{Y}_k & \cdots & \tilde{Y}_1\tilde{Y}_\nu \\ \vdots & & \vdots & & \vdots \\ \tilde{Y}_j\tilde{Y}_1 & \cdots & \tilde{Y}_j\tilde{Y}_k & \cdots & \tilde{Y}_j\tilde{Y}_\nu \\ \vdots & & \vdots & & \vdots \\ \tilde{Y}_\nu\tilde{Y}_1 & \cdots & \tilde{Y}_\nu\tilde{Y}_k & \cdots & \tilde{Y}_\nu^2 \end{vmatrix}_{j,k\neq i}. \tag{4.42}$$

In view of the equations

$$V_{\nu+1}^2(Y) = Y^2\det\{\tilde{Y}_i\tilde{Y}_j\} \text{ and } \nabla_{e_1}(V_{\nu+1}^2) = 2V_{\nu+1}\nabla_{e_1}V_{\nu+1},$$

we obtain from (4.42)

$$\nabla_{e_1}(V_{\nu+1}(Y)) = V_{\nu+1}(Y)\sum_{i=1}^{\nu}(-1)^i\frac{(\bar{\tilde{Y}}_{i1}\tilde{Y}_i)^2}{|\tilde{Y}_i|}. \tag{4.43}$$

From (4.43) follows the inequality

$$|\nabla_{e_1}(V_{\nu+1}(Y))| \leq a_\gamma(Y)^{\nu-1}|Y|^\nu\sum_{i=1}^{\nu}|\widetilde{Y_{i1}}|. \tag{4.44}$$

By Lemma 3.18 we have $\widetilde{Y_{i1}} = \widetilde{R_{i1}Y}$. Thus in view of the estimate

$$|\widetilde{R(u,Y)u}| \leq |Y|\frac{k_2-k_1}{2} \tag{4.45}$$

we obtain the inequality (4.36) from (4.44). □

We now modify Lemma 4.7.

4.21 Lemma. *Let a solution $y(t) \subset \mathbb{R}^n$ of the Jacobi equation*

$$y'' + R(t)y = 0, \quad (0 \leq t \leq \frac{\pi}{\sqrt{k_1}}, \ 0 < k_1E \leq R(t) \leq k_2E) \tag{4.46}$$

with a symmetric positive matrix $R(t)$ be written in the form $y(t) = Y(t)+u(t)$, where $Y(t) = y(0)\cos(\sqrt{k}t) + \frac{y'(0)}{\sqrt{k}}\sin(\sqrt{k}t)$. Then for $k_1 \geq 0.5k_2$ we have

$$|u(t)| \leq \frac{k_2-k_1}{3.25k_1 - 1.25k_2}\int_0^t \sqrt{k}|Y(s)|\cdot|\sin(\sqrt{k}(t-s))|\,ds. \tag{4.47}$$

Proof. Note that $u(0) = u(0)' = 0$. As in the proof of Lemma 4.7 we obtain

$$u(t) = -\int_0^t \frac{\sin(\sqrt{k}(t-s))}{\sqrt{k}}(R(t) - kE)(u(s) + Y(s))ds. \tag{4.48}$$

From this (in view of $||R(t) - kE|| \leq \frac{k_2 - k_1}{2}$) it follows that

$$\begin{aligned} |u(t)| \leq & \frac{k_2 - k_1}{2k} \int_0^t \sqrt{k}|Y(s)| \cdot |\sin(\sqrt{k}(t-s))|\, ds + \\ & |u|(t)\frac{k_2 - k_1}{2k} \int_0^t \sqrt{k}|\sin(\sqrt{k}(t-s))|\, ds, \end{aligned} \tag{4.49}$$

where $|u|(t) = \max\{|u(s)| : s \leq t\}$. Note that $\int_0^{\frac{\pi}{2\sqrt{k}}} \sqrt{k}\sin(\sqrt{k}(t-s))\, ds = 1$. Define β by

$$\frac{\pi/\sqrt{k_1}}{\pi/(2\sqrt{k})} = 2\sqrt{\frac{k}{k_1}} = 2 + \beta. \tag{4.50}$$

It is obvious that for $\delta = \frac{k_1}{k_2} \geq \frac{2}{7}$ we have $\beta = 2(\sqrt{\frac{k}{k_1}} - 1) \leq 1$. Thus

$$\int_0^{\frac{\pi}{\sqrt{k_1}}} \sqrt{k}|\sin(\sqrt{k}(t-s))|\, ds = 3 - \cos(\frac{\beta\pi}{2}). \tag{4.51}$$

From (4.49) it follows that

$$\begin{aligned} & |u|(t)\left(1 - \frac{k_2 - k_1}{2k}(3 - \cos(\frac{\beta\pi}{2}))\right) \\ & \leq \frac{k_2 - k_1}{2k} \int_0^t \sqrt{k}|Y(s)| \cdot |\sin(\sqrt{k}(t-s))|\, ds \end{aligned}$$

or

$$|u|(t) \leq \frac{k_2 - k_1}{2k - (k_2 - k_1)(3 - \cos(\frac{\beta\pi}{2}))} \int_0^t \sqrt{k}|Y(s)| \cdot |\sin(\sqrt{k}(t-s))|\, ds. \tag{4.52}$$

The denominator on the right-hand side of (4.52),

$$2k - (k_2 - k_1)\left(3 - \cos(\frac{\beta\pi}{2})\right) = 2(2\delta - 1) + (1 - \delta)\cos(\frac{\beta\pi}{2}),$$

is positive for $\delta \geq 0.5$. Note that $\delta_0 = 0.422$ is a solution of the equation

$$2(2\delta - 1) + (1 - \delta)\cos\left(\pi(\sqrt{\frac{1+\delta}{2\delta}} - 1)\right) = 0. \tag{4.53}$$

If we estimate for $\delta = 0.5$ the terms $\beta = 0.45$ and $\cos(\frac{\beta\pi}{2}) = 0.75$ and substitute them into the coefficient before the integral in (4.52), we obtain the inequality (4.47). □

Let B^+ be the part of the tensor B of a totally geodesic foliation which is symmetric with respect to y, (see (2.3) and Lemma 2.3), $k = \frac{k_1+k_2}{2}$ and

$$\delta(R) = \sup\{(R(x,y)u,z) : \\ x,u \in TL,\ y,z \in TL^\perp,\ x \perp u,\ y \perp z,\ |x| = |y| = |u| = |z| = 1\}.$$

4.22 Theorem. *Let M be a Riemannian manifold with a compact totally geodesic foliation $\{L\}$ and suppose the following conditions hold:*

$$k_2 \geq K(x,y) \geq k_1 \geq 0,\ (x \in TL,\ y \in TL^\perp), \tag{4.54}$$

$$(k_2 - k_1 + 2\delta(R))a(L)(diam\, L)\exp(3|B^+|(diam\, L)) \leq 2k_1, \tag{4.55}$$

where $diam\, L$ is the maximal diameter of the leaves. If $\dim L \geq \rho(\operatorname{codim} L)$, *then $k_1 = k_2 = 0$ and M is locally isometric to the product $L \times L^\perp$.*

Proof. Step 1. By Lemma 1.16, almost every leaf has trivial holonomy; below, L_0 is such a leaf. Note that L-parallel fields are globally defined on L_0.

In view of the inequality $\dim L \geq \rho(\operatorname{codim} L)$ and Lemma 3.40, for any point $m \in L_0$ there exist unit vectors $x_0 \in T_m L_0$, $y_0 \in T_m L_0^\perp$ and a real number $\lambda \leq 0$ with the property $B(x_0, y_0) = \lambda y_0$.

Let $Y : L_0 \to TL_0^\perp$, $(Y(m) = y_0)$, be an L-parallel (globally defined) vector field, $m' \in L_0$ a minimum of the length-function $|Y| : L_0 \to R_+$, (m' exists on compact L_0) and $\gamma : [0,l] \to L_0$, $(\gamma(0) = m,\ \gamma(l) = m',\ l \leq \operatorname{diam} L)$, the shortest geodesic between the points m and m'. Let $x_1(t) = \gamma'(t)$ be the velocity vector field and $x_2(t) \in T_{\gamma(t)}L$ a unit vector field, parallel along γ, with the properties:

$x_2(0) \perp x_1(0)$, and *the vectors* $\{x_2(0),\ x_1(0),\ x_0\}$ ***are linearly dependent.***

Let $y(t) = Y(\gamma(t))$ be the restriction of Y along γ and let $y_i(t) = \nabla_{x_i(t)}Y$ be horizontal vector fields along γ.

Step 2. Let $V_3(t)$, $(0 \leq t \leq l)$, be the volume of the parallelepiped $y(t) \wedge y_1(t) \wedge y_2(t)$ in $T_\gamma L^\perp$. In view of Lemma 4.19 (for $\nu = 2$) and the inequality $R_\gamma(Y) \leq \delta(R)$ we obtain

$$|V_3(t)'| \leq \left(\frac{k_2 - k_1}{2} + \delta(R)\right) a(L)|y(t)|^3. \tag{4.56}$$

Let $\{\delta_1, \delta_2\}$ be an orthonormal basis of the plane $x_1(l) \wedge x_2(l)$ with the property that $\nabla_{\delta_1} Y \perp \nabla_{\delta_2} Y$; moreover we have $|\nabla_{\delta_i} Y| \geq \sqrt{k_1}|y(l)|$ (in view of the well-known formula for the second derivative of the length of a Jacobi vector field) and

$$V_3(l) = |y(l)| \cdot |\nabla_{\delta_1} Y| \cdot |\nabla_{\delta_2} Y| \geq k_1 |y(l)|^3. \tag{4.57}$$

(Since $\gamma(l)$ is a minimum point of the function $|Y|$, $y_i = \nabla_{\delta_i} Y$ is orthogonal to the vector $y(l)$). Thus, using $V_3(0) = 0$ we obtain

$$k_1 |y(l)|^3 \leq V_3(l) \leq \int_0^l |V_3(t)'| \, dt \leq a(L) \left(\frac{k_2 - k_1}{2} + \delta(R) \right) \int_0^l |y(t)|^3 \, dt. \tag{4.58}$$

Step 3. Note that from

$$(y^2(t))' = 2(y(t), y_1(t)) = 2(B^+(x_1(t), y(t)), y(t))$$

follows

$$|(y^2(t))'| \leq 2|B^+| \cdot |y(t)|^2$$

and thus

$$|y(t)| \leq |y(l)| \exp(|B^+| l). \tag{4.59}$$

Finally, from (4.58) we obtain the opposite inequality to (4.55).

Step 4. From the extremal situation above it follows that $k_1 = k_2 = 0$ or $a(L) = 0$. In the latter case, the structural tensor B is conformal, in particular, the distribution $TL^\perp$ is integrable. From the inequality $K_{mix} \geq 0$, using the symmetric Riccati equation method (see Lemma 3.10), we obtain the equality $B = 0$ and again $k_1 = k_2 = 0$.

Step 5. $\{L\}$ is thus a compact totally geodesic foliation with the condition $K_{mix} = 0$ so Lemma 4.14 completes the proof. □

4.23 Theorem. *Let M be a Riemannian manifold with a compact totally geodesic foliation $\{L\}$ and suppose the following conditions hold:*

$$k_2 \geq K(x, y) \geq k_1 \geq 0, \ (x \in TL, \ y \in TM), \tag{4.60}$$

$$(k_2 - k_1 + 2\delta(R)) a(L) \max\{a(L)^2, k\}^{3/2} \leq 0.004 k^2 k_2, \tag{4.61}$$

where $k_1 \geq 0.5 k_2$. If $\dim L \geq \rho(\operatorname{codim} L)$, then $k_1 = k_2 = 0$ and M is locally isometric to the product $L \times L^\perp$.

4.24 Remark. The conditions (4.55) and (4.61) automatically hold for the case $\delta(R) = 0$ (for example, when the metric is conformally flat) and $k_1 = k_2$; in contrast to Theorem 4.16 these conditions do not contain an a priori estimate of the mixed curvature pinching.

Proof. Steps 1 & 2 of the proof are the same as those for Theorem 4.22.

Step 3. Assume that $k_1 > 0$. In view of (4.60) the vector field $y(t)$, $(0 \le t \le l \le \frac{\pi}{\sqrt{k_1}})$, may be written in the form $y(t) = Y(t) + u(t)$, where $u(0) = u(0)' = 0$ and

$$Y(t) = y(0)\cos(\sqrt{k}t) + \frac{y'(0)}{\sqrt{k}}\sin(\sqrt{k}t).$$

We obtain the upper estimate for a norm of $y(t)$ with the help of Lemma 4.21:

$$\begin{aligned}|y(t)| \le& |Y(t)| + |u(t)| \le |Y(t)|\left(1 + \frac{k_2 - k_1}{3.25k_1 - 1.25k_2}\left(3 - \cos(\frac{\beta\pi}{2})\right)\right)\\ &\le |Y(t)|\left(1 + \frac{2.25(k_2 - k_1)}{3.25k_1 - 1.25k_2}\right) = |Y(t)|\frac{k_2 + k_1}{3.25k_1 - 1.25k_2}.\end{aligned}$$

Using the inequality $|Y(t)| \le |y(0)|\max\{\frac{a(L)}{\sqrt{k}}, 1\}$ we obtain the estimate

$$|y(t)| \le \frac{k_2 + k_1}{3.25k_1 - 1.25k_2}\max\{\frac{a(L)}{\sqrt{k}}, 1\}|y(0)|. \tag{4.62}$$

Step 4. From inequalities (4.58) and (4.62) it follows that

$$(k_2 - k_1 + 2\delta(R))a(L)\max\{\frac{a(L)}{\sqrt{k}}, 1\}^3 \ge \sqrt{\frac{k_1}{k}}\frac{2}{\pi}\left(\frac{3.25k_1 - 1.25k_2}{k_2 + k_1}\right)^3 k^2 k_1. \tag{4.63}$$

Inequality (4.63) is the opposite of inequality (4.61), because for $k_1 \ge 0.5k_2$ we have

$$\sqrt{\frac{k_1}{k}}\frac{2}{\pi}\left(\frac{3.25k_1 - 1.25k_2}{k_2 + k_1}\right)^3\frac{k_1}{k_2} \ge 0.004.$$

Step 5. From the above extremal situation it follows that $k_1 = k_2 = 0$ (see Step 4 in the proof of Theorem 4.22); this completes the proof in view of Lemma 4.14. □

4.3. Riccati and Raychaudhuri equations

4.3.1. The integral of mixed scalar curvature along a leaf

Using the method of matrix Riccati ODE's, we study the rigidity and splitting of Riemannian manifolds with foliations (distributions) whose mixed Ricci or scalar curvature is nonnegative and whose norms of integrable tensors are bounded above. We obtain in 4.3.1 the integral inequality for mixed scalar curvature along a complete leaf, the rigidity of totally geodesic distributions, and the splitting of foliations with "large" dimension.

In other words, the splitting result of Theorem 3.11 is extended below in two ways: for a totally geodesic distribution (Theorem 4.25) and in the integral sense (Theorem 4.29).

Let M be a Riemannian manifold with complementary orthogonal distributions T_1 and T_2. Let $n = \dim T_2$, $\nu = \dim T_1$ and denote by $\{y_j\}$ a local orthonormal basis of T_2 and by $\{x_i\}$ a local orthonormal basis of T_1 with $x_1 = e$. Let

$$|A_i|(e) = \sup\{|A_i(e, v)| : |v| = 1\}, \ |A_i|(p) = \sup\{|A_i|(e) : e \in T_pM, \ |e| = 1\}.$$

Note that for $e \in T_1(p)$

$$\begin{aligned} \sum_j |B_1^-(y_j, e)|^2 &= \sum_{j,i} (B_1^-(y_j, e), x_i)^2 = \sum_i |A_1(e, x_i)|^2 \le (\nu - 1)|A_1|^2(e), \\ \sum_j |B_2^-(e, y_j)|^2 &= \sum_{i,j} |A_2(y_j, y_i)|^2 \le n(n-1)|A_2|^2(p). \end{aligned} \tag{4.64}$$

The *mixed Ricci and scalar curvature* are defined as follows:

$$\mathrm{Ric}_1(x) = \sum_j K(x, y_j), \ \mathrm{Ric}_2(y) = \sum_i K(x_i, y), \ s_{mix} = \sum_{i,j} (R(y_j, x_i)x_i, y_j).$$

4.25 Theorem. *(Rigidity) Let M be a complete Riemannian manifold equipped with two complementary orthogonal distributions T_1 and T_2* ($\dim T_1 = \nu$, $\dim T_2 = n$), *and suppose that T_1 is totally geodesic.*

a) Then for every T_1-geodesic $\gamma(t) \subset M$

$$\liminf_{s \to +\infty} \int_{-s}^{s} \left\{Ric_1(\gamma') - (\nu - 1)|A_1|^2(\gamma) - n(n-1)|A_2|^2(\gamma)\right\} dt \le 0, \tag{4.65}$$

and equality holds for every γ iff T_2 is a totally geodesic distribution and the skew-symmetric operators B_2 and B_1 have "constant torsion" in the following *sense:*

$$B_2(x, B_2(x, y)) = -|A_2|^2(p)x^2y, \quad B_1(y, B_1(y, x)) = -|A_1|^2(p)y^2x. \tag{4.66}$$

b) If the equality holds in (4.65) *for every* γ *and, in addition,* $\nu \geq \rho(n)$ *(for* $\nu > 2$, *n is divisible by* 4, *when* T_1 *is invariant with respect to the Kählerian structure on* M*), then* T_2 *is also integrable and hence is tangent to a Riemannian totally geodesic foliation; moreover* $K_{mix} \geq 0$.

4.26 Remark. In [So 2] a hyperdistribution T_1, such that property (4.66) is satisfied by B_1, is referred to as having *constant torsion*.

Proof. a) Consider equation (2.36) along any unit speed T_1-geodesic $\gamma : \mathbb{R} \to M$

$$(\nabla_{\gamma'}B_2)(\gamma', y) + B_2(\gamma', B_2(\gamma', y)) + (B_1(*, B_1(y, \gamma')), \gamma') + P_2(R(y, \gamma')\gamma') = 0. \tag{4.67}$$

Here, $((\nabla_y B_1)(*, \gamma'), \gamma') = 0$, because B_1 is skew-symmetric with respect to the totally geodesic T_1. The component of (4.67), symmetric with respect to T_2, is a Riccati equation, see (3.6a),

$$\begin{aligned}(\nabla_{\gamma'}B_2^+)(\gamma', y) &+ B_2^+(\gamma', B_2^+(\gamma', y)) + B_2^-(\gamma', B_2^-(\gamma', y)) \\ &+ (B_1(*, B_1(y, \gamma')), \gamma') + P_2(R(y, \gamma')\gamma') = 0.\end{aligned} \tag{4.68}$$

Consider the Raychaudhuri equation for $\theta(t) = trace\, B_2^+(\gamma')$ (see equation (3.7)):

$$\theta' + \frac{1}{n}\theta^2 + \{Ric_1(\gamma') + trace\,(\sigma^2(\gamma')) - \sum_j |B_2^-(\gamma', y_j)|^2 - \sum_j |B_1(y_j, \gamma')|^2\} = 0. \tag{4.69}$$

From Lemma 3.10 and (4.64) the inequality (4.65) follows and for the equality case

$$\theta = \sigma(\gamma') = B_2^+(\gamma', *) = 0, \quad |B_2^-(\gamma', *)| = |A_2|(\gamma), \; |B_1(*, \gamma')| = |A_1|(\gamma).$$

Applying this to every T_1-geodesic we get, in the case of equality, that $B_2^+ = 0$, i.e., T_2 is a totally geodesic distribution and the equations of (4.66) hold, i.e., the squares of the skew-symmetric operators $B_2(x, *)$ in T_2 and $B_1(y, *)$ in T_1 are homotheties.

b) By part a), the distribution T_2 is totally geodesic and (4.66) holds. Assume on the contrary that T_2 is noninvolutive somewhere. Then by Lemma 2.3 there exists a point $p \in M$ and unit vectors $x_0 \in T_1(p)$ and $y_0 \in T_2(p)$ with the property $B_2(x_0, y_0) \neq 0$. Therefore, $|A_2|(p) \neq 0$ and in view of (4.66) $B_2(x, y) \neq 0$ for each nonzero $x \in T_1(p)$ and $y \in T_2(p)$. Since the skew-symmetric linear operator $B_2(x, *)$, $(x \neq 0)$, has no eigenvectors, following [Fe 2] (see Lemma 3.40) we choose a basis $\{u_i\}$ for $T_1(p)$ and define ν continuous pointwise linearly independent vector fields $\{v_i\}$ on the unit sphere $S^{n-1} \subset T_2(p)$ by the rule $v_i(y) = B_2(u_i, y)$. Thus $\nu < \rho(n)$, which is a contradiction. Assertion b) in the Kählerian case is now an immediate consequence of the property $B_2(Jx, y) = JB_2(x, y)$ and Lemma 3.42. □

4.27 Corollary. *Let $S^{n+\nu}(1)$ be a sphere of curvature one, equipped with two complementary orthogonal distributions T_1 and T_2, $(0 < \dim T_1 = \nu < \dim T_2 = n)$. Assume that T_1 is totally geodesic and that*

$$n|A_2|^2(p) + \frac{\nu - 1}{n - 1}|A_1|^2(p) \leq 1, \qquad (p \in S^{n+\nu}). \tag{4.65'}$$

Then T_1 is tangent to the leaves of the Hopf fibration (with dimension $n = 1, 3, 7$).

4.28 Remark. The special case of Corollary 4.27 when T_1 is tangent to a Riemannian foliation ($A_1 = 0$), is studied in [Ra 2]. Note that for every $\nu < \rho(n)$ there exists a totally geodesic ν-dimensional distribution T_1 on the sphere $S^{\nu+n}$; it is generated by the orthogonal unit vector fields from Theorem 1.9.

Proof. From Theorem 4.25 and inequality (4.65') we deduce that the distribution T_2 is totally geodesic and that (4.66) holds. From $n > \nu$ it follows that a linear operator $B_1(*, x) : T_2(p) \to T_1(p)$ has a nontrivial kernel for every $x \in T_1(p)$, i.e., for every point $p \in M$ there exist unit vectors $x_0 \in T_1(p)$ and $y_0 \in T_2(p)$ with the property $B_1(y_0, x_0) = 0$. From (4.66) we obtain $|A_1|(p) \equiv 0$ and then $B_1 \equiv 0$. Now we may conclude that T_1 is tangent to a Riemannian totally geodesic foliation. Such foliations on spheres are Hopf fibrations [Es 1]. □

Combined with the method in [Gu], where manifolds without conjugate points are studied, we obtain the integral formula for mixed scalar curvature along a leaf.

4.29 Theorem. *(rigidity and splitting) Let $\{L^\nu\}$ be a foliation on a Riemannian manifold $M^{n+\nu}$ with complete minimal leaves. Suppose that the quadratic form*

$$Ric_1(x) + trace\{B_2^-(x, B_2^-(x,*)) - h_1(x, B_1^+(*,x)) - (\nabla_* h_1)(x,x)\}$$

on a unit tangent bundle SL_0 of every leaf L_0 has an integrable positive or negative part. Then

$$\int_{L_0} \{s_{mix} - n(n-1)|A_2|^2\}\, d\,vol \le 0, \tag{4.70}$$

and equality holds for every leaf iff $\{L\}$ is a Riemannian totally geodesic foliation. Moreover, if equality holds in (4.70) for all leaves and $\nu \ge \rho(n)$, (TL is invariant with respect to the Kählerian structure on M, and $\nu > 2$ ensures that n is divisible by 4), then M is locally the product $L \times L^\perp$.

Proof. The proof of Theorem 4.29 is based on some facts from ergodic theory given below. Consider an arbitrary leaf L_0. Using the symmetric part of the structural tensor (2.1) we define a tensor U on SL_0 such that for every $x \in SL_0, U(x) = B_2^+(x,*)$ is a linear operator on $TL_0^\perp$. Then U satisfies the Riccati equation $U' + U^2 + \tilde R(x) = 0$ (see (2.36), (2.37) and (3.6)) where $(\tilde R(x)y, z) = (R(y,x)x, z) + (B_2^-(x, B_2^-(x,y)), z) + (B_1^+(x, B_1^+(y,x)), z) + ((\nabla_y B_1^+)(z,x), x)$, $R(x) = R(*,x)x$ is the Jacobi operator on $TL_0^\perp$, and $B_2^-(x,*)$ is the vorticity tensor, see Section 3.1. The theorem follows from the inequality

$$\int_{SL_0} \widetilde{\mathrm{Ric}}_1(x)\, d\mu \le -\int_{SL_0} \mathrm{trace}\, U^2(x)\, d\mu, \tag{4.71}$$

where
$\widetilde{\mathrm{Ric}}_1(x) = \mathrm{Ric}_1(x) + trace\{B_2^-(x, B_2^-(x,*)) - h_1(x, B_1^+(*,x)) - (\nabla_* h_1)(x,x)\}$.
To see this, note that by Fubini's theorem and Lemma 4.38 below, we obtain

$$\int_{L_0} \tilde s_{mix}\, d\,vol = \frac{n}{\omega_{n-1}} \int_{SL_0} \widetilde{\mathrm{Ric}}_1(x)\, d\mu \le -\frac{n}{w_{n-1}} \int_{SL_0} trace\, U^2(x)\, d\mu \le 0,$$

where $\tilde s_{mix} := trace\, \widetilde{\mathrm{Ric}}_1 = s_{mix} - \|B_2^-\|^2 + \|B_1^+\|^2$ (since $H_1 = 0$ and $s_{mix} = trace\, \mathrm{Ric}_1$, see formula (2.38)) are functions on L_0. Note that

$$\|B_2^-\|^2(p) \le n(n-1)|A_2|^2(p).$$

Furthermore, if $\int_{L_0} \tilde s_{mix}\, d\,vol = 0$, then U^2 is identically zero and so $\tilde R(x) = 0$ on SL_0 and $B_1 = 0$, i.e., L_0 is a totally geodesic submanifold. Since

L_0 is an arbitrary leaf, the distribution $TL^\perp$ and the foliation $\{L\}$ are totally geodesic. Moreover, B_2 has constant torsion in the sense of (4.66). From the given inequality for the dimension of L_0 and Theorem 4.25 b), it follows that the distribution $TL^\perp$ is also integrable. The Local Decomposition Theorem by de Rham completes the proof.

To prove (4.71) let $r_1^+ = (\widetilde{\mathrm{Ric}}_1)^+$ and $r_1^- = (\widetilde{\mathrm{Ric}}_1)^-$ be the positive and negative parts of $\widetilde{\mathrm{Ric}}_1$, respectively. If $\int_{SL_0} r_1^- \, d\mu = +\infty$, then $\int_{SL_0} (\widetilde{\mathrm{Ric}}_1) \, d\mu = -\infty$, because in this case we have $\int_{SL_0} r_1^+ \, d\mu < \infty$. So (4.71) holds.

Hence we may suppose $\int_{SL_0} r_1^- \, d\mu < \infty$. Consider the nonnegative integrable function $f \le trace\,(U^2) + r_1^+$ on SL_0. It suffices to show that

$$\int_{SL_0} f \, d\mu \le \int_{SL_0} r_1^- \, d\mu. \tag{4.72}$$

To prove (4.72) we will verify that the integral of $f - r_1^-$ is nonpositive in each one of the sets D, C^+ and $C^- \setminus C^+$. Note that SL_0 is the disjoint union $SL_0 = D \cup C^+ \cup (C^- \setminus C^+)$. We integrate the Raychaudhuri equation for $u = trace\,U$, which corresponds to the Riccati equation $U' + U^2 + \tilde{R}(x) = 0$; and we obtain

$$u(s) - u(t) + \tfrac{1}{n-1} \int_t^s u^2(\lambda) \, d\lambda + \int_t^s \left\{ trace\,(U^2) + \widetilde{\mathrm{Ric}}_1 \right\} d\lambda = 0,$$

that is,

$$u(s) - u(t) + \frac{1}{n-1} \int_t^s u^2(\lambda) \, d\lambda + \int_t^s (f - r_1^-)(T_\lambda(v)) \, d\lambda = 0 \tag{4.73}$$

for every $v \in SL_0$ and s, $t \in \mathbb{R}$. From Lemma 3.13 we immediately obtain $\int_t^s (f - r_1^-)(T_t(v)) \, dt \le 0$ for almost all $v \in D$. Then by Lemma 4.37

$$\int_D (f - r_1^-) \, d\mu = \int_{\pi(D)} \int_{-\infty}^{\infty} (f - r_1^-)(T_t(v)) \, dt \, d\tilde{\mu} \le 0.$$

In order to compute $\int_{C^+} (f - r_1^-) d\mu$, let $\delta > 0$ and let $f_0 > 0$ be integrable on SL_0. Define $g = f - r_1^- - \delta f_0$. From (4.73) we have

$$\delta \int_0^s f_0(T_t(v)) \, dt - u(0) + u(s) + \frac{1}{n-1} \int_0^s u^2(t) \, dt + \int_0^s g(T_t(v)) \, dt = 0$$

for every $v \in SL_0$ and $s \in \mathbb{R}$. If $v \in C^+$, using Lemma 3.12 a) we conclude that $\liminf_{s\to+\infty} g(T_t(v))\,dt = -\infty$. Then by Theorem 4.34 we obtain

$$\int_{C^+}(f - r_1^- - \delta f_0)\,d\mu = \int_{C^+} g\,d\mu$$

and therefore $\int_{C^+}(f - r_1^-)\,d\mu \le 0$, by letting $\delta \to 0$. In order to compute $\int_{C^+\setminus C^-}(f - r_1^-)\,d\mu$ we derive from (4.73) that

$$\delta\int_{-s}^{0} f_0(T_t(v))\,dt + u(0) - u(s) + \frac{1}{n-1}\int_{-s}^{0} u^2(t)\,dt + \int_{-s}^{0} g(T_t(v))\,dt \le 0$$

for every $v \in SL_0$ and $s \in \mathbb{R}$. Hence

$$\delta\int_{0}^{s} f_0(T_{-t}(v))\,dt + u(0) - u(-s) + \frac{1}{n-1}\int_{0}^{s} u^2(-t)\,dt + \int_{0}^{s} g(T_{-t}(v))\,dt \le 0. \tag{4.74}$$

If $v \in C^-$, then $\liminf_{s\to+\infty} g(T_{-t}(v))\,dt = -\infty$ by (4.74) and Lemma 3.12 a). Thus using Theorem 4.34 (see below) we obtain $\int_{C^-\setminus C^+}(f - r_1^- - \delta f_0)\,d\mu = \int_{C^-\setminus C^+} g\,d\mu$. Hence letting $\delta \to 0$ leads to $\int_{C^-\setminus C^+}(f - r_1^-)\,d\mu \le 0$. □

In the following corollary that generalizes Theorem 3.11 [BrW], the distribution $TL^\perp$ is integrable and the foliation splits without additional assumption for dimensions. Moreover, the integrability of $TL^\perp$ may be assumed along some submanifold which is transversal to TL. For a compact leaf L_0, see Corollary 2.31.

4.30 Corollary. *(local splitting) Let M be a foliated Riemannian manifold with complete minimal leaves $\{L\}$ and let the orthogonal distribution $TL^\perp$ be integrable. Suppose that the function*

$$Ric_1(x) - trace\,\{h_1(x, B_1^+(*, x)) + trace\,(\nabla_* h_1)(x, x)\}$$

on SL_0 of every leaf L_0 has an integrable positive or negative part. Then

$$\int_{L_0} s_{mix}\,d\,vol \le 0 \tag{4.75}$$

and equality holds (for every leaf) iff M is locally the product $L \times L^\perp$.

Proof. By Theorem 4.25 the involutive distribution $TL^\perp$ is totally geodesic. The Local Decomposition Theorem by de Rham completes the proof. □

From Theorem 4.25 and independently from Theorem 4.29 we obtain:

4.31 Corollary. [KiT] *(rigidity) Let $\{L\}$ be a Riemannian foliation on a complete Riemannian manifold M. If for each geodesic γ orthogonal to the leaves we have*

$$Ric_2(\gamma') \geq (n-1)|A_2|^2(\gamma'),$$

then $\{L\}$ is a totally geodesic Riemannian foliation.

4.32 Corollary. *(local splitting) Let M be a foliated Riemannian manifold with complete totally geodesic leaves $\{L\}$ and*

$$Ric_1(x) \geq n(n-1)|A_2|^2(p), \qquad (x \in T_pM).$$

If $\nu \geq \rho(n)$, (TL is invariant with respect to the Kählerian structure on M and $\nu > 2$ ensures that n is divisible by 4), then M is locally the product $L \times L^{\perp}$.

4.33 Remark. The inequality in Corollary 4.31 (where geodesics orthogonal to the leaves are considered) is actually exact: the hypotheses are in particular realized for a foliation $\{L\}$, ($TL = T_2$) of codimension one with $\mathrm{Ric}_1 \geq 0$, see [Heb 3]. Another proof of Corollary 4.32 is given in [Rov 9] using integral curves of a vector field H_2 (also see [BrW] for the case $A_1 = A_2 = 0$), where the inequality is replaced by

$$n(n-1)|A_2|^2(p) \leq \begin{cases} \mathrm{Ric}_1(H_2), & if\ H_2(p) \neq 0 \\ \mathrm{Ric}_1(x), & if\ H_2(p) = 0. \end{cases} \tag{4.76}$$

4.3.2. Necessary facts from ergodic theory

Section 4.3.2 contains facts that were used in Section 4.3.1.

Let L_0 be a leaf of the foliation $\{L\}$, SL_0 a unit tangent bundle, T_t a geodesic flow on SL_0 and μ a Liouville measure on the Borel sets of SL_0 that is σ-finite and invariant under T_t.

4.34 Maximal Ergodic Theorem. [Gu] *Let f be an integrable function on SL_0 and let $D \subset SL_0$ be a T_t-invariant Borel set. Define*

$$E(f) = \{v \in D : \sup_{s>0} \int_0^s f(T_t(v))\, dt > 0\}.$$

Then $\int_{E(f)} f\, d\mu \geq 0$. Set $T = T_1$. If A is a T-invariant Borel set and

$$E[f] = \{v \in A : \sup_{n \geq 0} \sum_{j=0}^{n} f(T^j(v)) > 0\},$$

then $\int_{E[f]} f\, d\mu \geq 0$.

Let f_0 be an integrable function on SL_0. Then the Borel sets

$$D^+ = \{v \in SL_0 : \int_0^\infty f_0(T_t(v))\, dt < \infty\},\ \ C^+ = SL_0 \setminus D^+ \qquad (4.77)$$

are T_t-invariant; they are independent of f_0 in the following sense [Gu]: if f_1 is another integrable function, then the intersection of the set

$$D_1^+ = \{v \in SL_0 : \int_0^\infty f_1(T_t(v))\, dt = \infty\}$$

with D^+ has measure zero. To see this, we note that D_1^+ is a T_f-invariant set. By Theorem 4.34 $\int_{D_1^+}(af_1 - f_0)d\mu \geq 0$ for every $a > 0$. Letting $a \to 0$, we get $\int_{D_1^+} f_0 d\mu = 0$. Since $f_0 > 0$, $\mu(D_1^+) = 0$, as we wished to show. Now if $f_1 > 0$, interchanging f_0 and f_1, we obtain the independence of D^+, up to a subset of measure zero.

The decomposition $SL_0 = D^+ \cup C^+$ is called the *Hopf decomposition* of SL_0 associated to the geodesic flow $\{T_t\}$, $(t \in \mathbb{R})$. The components D^+ and C^+ are called, respectively, the *dissipative* and the *conservative parts* of the decomposition. Denote by $SL_0 = D^- \cup C^-$ the Hopf decomposition of SL_0 associated to the *inverse geodesic flow* $\{T_{-t}\}$, $(t \in \mathbb{R})$, i.e.,

$$D^- = \{v \in SL_0 : \int_{-\infty}^0 f_0(T_t(v))\, dt < \infty\},\ \ C^- = SL_0 \setminus D^-.$$

If $D = D^+ \cap D^-$, then for every integrable function f on SL_0, the Lebesgue integral

$$\int_{-\infty}^\infty f(T_t(v))\, dt \qquad (4.78)$$

exists for almost all $v \in D$. Also, if $g_0 > 0$ is integrable, then the T-invariant sets

$$D[T] = \{v \in SL_0 : \sum_{-\infty}^\infty g_0(T^j(v)) < \infty\},\ \ C(T) = SL_0 \setminus D(T)$$

are independent of g_0, up to a subset of measure zero. Moreover, if $g \geq 0$ is integrable, then

$$\sum_{-\infty}^\infty g(T^j(v)) < \infty \text{ for almost all } v \in D(T). \qquad (4.79)$$

The following lemma establishes a relationship between the sets D and $D(T)$.

4.35 Lemma. [Gu] *If $g \geq 0$ is an integrable function on SL_0, then for almost all $v \in D$, $\sum_{-\infty}^{\infty} g(T^j(v)) < \infty$.*

Proof. Set $g_0(x) = \int_0^1 f_0(T_t(x))\,dt$, for $x \in SL_0$. Then $g_0 > 0$ and

$$\int_{SL_0} g_0\,d\mu = \int_{SL_0}\int_0^1 f_0(T_t(x))\,dt\,d\mu = \int_{SL_0} f_0\,d\mu.$$

Thus g_0 is integrable and by (4.78) we have $\sum_{-\infty}^{\infty} g_0(T^j(v)) = \int_{-\infty}^{\infty} f_0(T_t(v))\,dt < \infty$ for every $v \in D$. If we choose this g_0 in the definition of $D(T)$, then $D \subset D(T)$. In view of (4.79) this completes the proof. □

Next, we consider a measure on the space of orbits. Let $\sim$ be the equivalence relation in SL_0 defined by $v \sim w \iff \exists j \in \mathbb{Z}$ such that $T^j(v) = w$. Let $\tilde{S}L_0$ be the space of orbits and denote by $\pi : \tilde{S}L_0 \to SL_0$ the natural projection $\pi(v) = [v]$. Consider in $\tilde{S}L_0$ the σ-algebra $\tilde{\beta}$ induced by π. Then $\pi(E)$ is measurable for every Borel set E because $\pi^{-1}(\pi(E)) = \bigcup_{-\infty}^{\infty} T^j(E)$.

A Borel set E is called a *wandering set* if $T^j(E) \cap E = \emptyset$ for every $j \geq 1$. For each $\tilde{E} \in \tilde{\beta}$ let $\tilde{\mu}(\tilde{E}) = \sup\{\,\mu(E) : E \in \pi^{-1}(\tilde{E})\,\}$, where E is a wandering set.

4.36 Lemma. [Gu] *$\tilde{\mu}$ is a measure on $\tilde{S}L_0$ with the property that $\tilde{\mu}(\pi(E)) = \mu(E)$ for every wandering set E.*

Given a function f that is integrable on SL_0, the map $x \in D \to \int_{-\infty}^{\infty} f(T_t(x))\,dt$ defines a measurable function on the Borel sets of D by (4.78). Furthermore, this function is constant on each orbit of D. Thus $[x] \in \pi(D) \to \int_{-\infty}^{\infty} f(T_t(x))\,dt$ defines a $\tilde{\mu}$-measurable function on $\pi(D)$.

4.37 Lemma. [Gu] *If f is an integrable function on SL_0, then*

$$\int_D f\,d\mu = \int_{\pi(D)}\int_{-\infty}^{\infty} f(T_t(v))\,dt\,d\tilde{\mu}.$$

Proof. The proof is based on Lemma 4.35, see [Proposition 2.3, Gu]. □

4.38 Lemma. *Let $S^{n-1} \subset \mathbb{R}^n$ be a unit sphere, ω_{n-1} its volume and $A(x)$, ($x \in \mathbb{R}^n$), a quadratic form. Then $\int_{S^{n-1}} A(x)\,dx = \omega_{n-1} \mathit{trace}\,A/n$.*

Proof. The function $f(x) = A(x) - \frac{1}{n}(\mathit{trace}\,A)x^2$ is obviously harmonic in $\mathbb{R}^n$. By the Mean Value Theorem, the integral of $f(x)$ over S^{n-1} is zero. □

4.3.3. Foliations which are close to Riemannian

Let $\{L\}$ be a Riemannian foliation with totally geodesic leaves. Then $B_1 = B_2^+ = 0$ and from (3.6) it follows that

$$\begin{cases} \text{a)}\ \nabla_x B_2^- = 0 \\ \text{b)}\ (B_2^-(x, B_2^-(x, y)) = (R(x, y)y, x), \end{cases} \tag{4.80}$$

and hence $K_{mix} \geq 0$. But if we assume $K_{mix} > 0$, then from (4.80 b) it follows that the skew-symmetric operator B_2^- is nondegenerate and thus has no eigenvectors, i.e., in this case $\nu < \rho(n)$. This simple local property (and its modification for a conformal foliation with complete totally geodesic leaves) is the key in Section 4.3.3.

Let

$$|h_i| = \sup\{|h_i(v, w)| : |v| = |w| = 1,\ v,\ w \in T_i\}.$$

4.39 Theorem. *(rigidity) Let M be a complete Riemannian manifold with a totally geodesic distribution T_1 of dimension $\dim T_1 = \nu$. Suppose that its orthogonal distribution T_2 has dimension $\dim T_2 = n$ and that the following inequality holds:*

$$K_{mix} \geq |h_2|^2 + |A_1|^2. \tag{4.81}$$

If $\nu \geq \rho(n)$, (T_1 is invariant with respect to the Kählerian structure on M and $\nu > 2$ ensures that n is divisible by 4), then the distribution T_2 is totally geodesic.

4.40 Corollary. *Let M be a complete Riemannian manifold with a totally geodesic distribution T_1 of dimension $\dim T_1 = \nu$. Suppose that its orthogonal distribution T_2 has dimension $\dim T_2 = n$ and that there exists a point $p \in M$ such that for every T_1-geodesic $\gamma : \mathbb{R} \to M$ with $\gamma(0) = p$ the following inequality holds:*

$$\liminf_{s\to\infty} \int_{-s}^{s} \{K(\gamma', y_t) - |B_2^+(\gamma', y_t)|^2 + |B_1^-(\gamma', y_t)|^2\} > 0, \tag{4.81'}$$

where $y_t \in T_2(\gamma')$ is parallel along γ. Then $\nu < \rho(n)$, and if T_1 is invariant with respect to the Kählerian structure on M, then $\nu = 2$ and n is divisible by 4.

Special cases of Theorem 4.39 and Corollary 4.40 arise when T_1 is tangent to a totally geodesic foliation.

Proof of Theorem 4.39 and Corollary 4.40. By Lemma 3.40 there exist unit vectors $x_0 \in T_1(p)$ and $y_0 \in T_2(p)$ with the property $B_2^-(x_0, y_0) = 0$. So we can conclude that

$$m = \dim \ker B_2^-(x_0, *) > 0.$$

We shall consider the skew-symmetric with respect to T_2 component of equation (2.36), see (3.6b), along any unit speed T_1-geodesic $\gamma : \mathbb{R} \to M$ with $\gamma(0) = p$

$$(B_2^-)'(t) + B_2^+(t)B_2^-(t) + B_2^-(t)B_2^+(t) = 0, \tag{4.82}$$

where $B_2^+(t) = B_2^+(\gamma'(t), *)$, $B_2^-(t) = B_2^-(\gamma'(t), *)$. The homogeneous ODE (4.82) has the solution

$$B_2^-(t) = F(t)B_2^-(0)G(t), \tag{4.83}$$

where $F(t)$ and $G(t)$ are nondegenerate matrices, in particular, solutions of the following matrix ODE's:

$$G'(t) = G(t)B_2^+(t), \ (G(0) = E), \qquad F'(t) = B_2^+(t)F(t), \ (F(0) = E). \tag{4.84}$$

From (4.83) and (4.84) we conclude that $\ker B_2^-(t)$ is a regular m-dimensional vector subbundle of T_2 along γ. Thus there exist m regular vector fields $\{y_i\} \subset T_2$ along γ with the properties

$$y_i(t) \subset V(t), \ \ (y_i(t), y_j(t)) = \delta_{ij}, \ \ y_i(t)' \perp V(t). \tag{4.85}$$

We shall place $y_i(t)$ in identity (4.82) and in view of

$$(B_2^-(t))'y_i(t) = (B_2^-(t)y_i(t))' - B_2^-(t)y_i(t)' = -B_2^-(t)y_i(t)',$$

we get the equation

$$B_2^-(t)[y_i(t)' - B_2^+(t)y_i(t)] = 0, \ \ (1 \le i \le m). \tag{4.86}$$

From (4.85) and (4.86) it follows that

$$y_i(t)' = B_2^+(t)y_i(t) - \sum_k a_{ik}(t)y_k(t). \tag{4.87}$$

The scalar product of (4.87) and $y_j(t)$ gives

$$a_{ij}(t) = (B_2^+(t)y_i(t), y_j(t)), \ (1 \le i, \ j \le m).$$

Next we consider the symmetric T_2-component of (2.36) along γ,

$$(B_2^+(t))' + (B_2^+(t))^2 + (B_2^-(t))^2 + R(t) + B^-(t) = 0, \tag{4.88}$$

where $R(t) = P_2(R(*, \gamma'(t))\gamma'(t))$ and $B^-(t) = A_1(\gamma', B_1(*, \gamma'))$. Let us consider (4.88) for arguments $y_i(t)$ and form a scalar product with the left hand side of the equality with $y_j(t)$. Then in view of (4.85) and (4.87) we have the Riccati ODE

$$A(t)' + 2A(t)^2 + F(t) = 0, \tag{4.89}$$

for the matrix function $A(t) = \{a_{ij}(t)\}$ with $F(t) = R(t) - D(t) - U(t)$, where

$$R(t) = \{r_{ij}(t) = (R(t)y_i(t), y_j(t))\},$$

$$D(t) = \{d_{ij}(t) = (B_2^+(t)y_i(t),\ B_2^+(t)y_j(t))\},$$

$$U(t) = \{u_{ij}(t) = (B_1(y_i(t), \gamma'(t)),\ B_1(y_j(t), \gamma'(t)))\}$$

are symmetric matrices of order $m \times m$, continuous for $t \in \mathbb{R}$. Note that $D(t) \geq A^2(t)$. The existence of a solution of the matrix equation (4.89) in the case of (4.81′) contradicts Lemma 3.10 and this completes the proof of Corollary 4.40.

In the case (4.81), matrix $F(t)$ is nonnegative definite and by Lemma 3.10 equation (4.89) has a zero solution $A(t) = 0$. Moreover in this extremal situation the subspace $\ker B_2^-(t)$ is invariant under the symmetric operator $B_2^+(t)$ and its restriction is the $|h_2|$-homothety

$$B_2^+(t)y_i(t) = |h_2|y_i(t), \ (1 \leq i \leq m).$$

Finally we obtain

$$|h_2| = (B_2^+ y_i(t), y_i(t)) = \alpha_{ii}(t) = 0,$$

i.e., $h_2 = 0$. Hence the distribution T_2 is totally geodesic. This completes the proof of the theorem. □

4.41 Remark. Corollary 4.40 (and Theorem 4.39) can be generalized for a distribution T_2 that is "close" to conformal if we replace item $B_2^+(\gamma')y$ in (4.81′) by $B_2^+(\gamma')y - \beta(\gamma')y$, where $\beta : T_1 \to R$ is some regular linear functional.

If distribution T_2 is "close" to conformal, then (4.89) is transformed into the following Riccati ODE

$$(A(t))' + A(t)^2 + [F_0(t) + (A(t) - \beta(t)E)^2] = 0, \tag{4.90}$$

where $\beta(t) = \beta(\gamma'(t))$ and the matrix in the square brackets of equation (4.90) is positive semidefinite.

CHAPTER V

SUBMANIFOLDS WITH GENERATORS

5.1. Submanifolds with generators in Riemannian spaces

5.1.1. Basic facts from the theory of submanifolds

In Part 2 we consider the following classes of submanifolds having generators or foliations: ruled submanifolds, (strongly) parabolic submanifolds, k-saddle submanifolds and submanifolds with nonpositive extrinsic curvature. Thus we recall some necessary facts and formulas from the theory of submanifolds [Ch 1-3], [KN], [AkG].

If M is a submanifold of a Riemannian space $\bar{M}$, then

$$\bar{\nabla}_x y = \nabla_x y + h(x, y), \quad (Gauss\ formula), \tag{5.1}$$

$$\bar{\nabla}_x \xi = -A_\xi x + \nabla_x^\perp \xi, \quad (Weingarten\ formula), \tag{5.2}$$

where x and y are vector fields tangent to M, ξ is a vector field orthogonal to M, $h : TM \times TM \to TM^\perp$ is a second fundamental form of embedding which is symmetric with respect to both arguments: $h(x, y) = h(y, x)$, $\nabla^\perp$ is a (metric) normal connection induced in a normal vector bundle $TM^\perp$ and A_ξ is the operator of the second quadratic form for normal vector ξ. Note that $\perp$ indicates the normal component. From the formulas above it follows that

$$(A_\xi x, y) = (h(x, y), \xi). \tag{5.3}$$

Denote by $\bar{R}$, R and $R^\perp$ the curvature tensors of the connections $\bar{\nabla}$, ∇ and $\nabla^\perp$, respectively. Central to the theory of submanifolds are the following *equations of Gauss, Codazzi and Ricci*:

$$(\bar{R}(x, y)z, u) = (R(x, y)z, u) + (h(x, u), h(y, z)) - (h(x, z), h(y, u)), \tag{5.4}$$

$$\bar{R}(x, y)z^\perp = (\bar{\nabla}_x h)(y, z) - (\bar{\nabla}_y h)(x, z), \tag{5.5}$$

$$(\bar{R}(x, y)\xi, \eta) = (R^\perp(x, y)\xi, \eta) - ([A_\xi, A_\eta]x, y) \tag{5.6}$$

where the vectors $x,\ y,\ z,\ u$ are tangent to M, the vectors $\xi,\ \eta$ are orthogonal to M and the derivative $\bar{\nabla} h$ is defined by the rule

$$(\bar{\nabla}_x h)(y, z) = \nabla_x^\perp(h(y, z)) - h(\nabla_x y, z) - h(y, \nabla_x z). \tag{5.7}$$

For a submanifold in a space form $\bar{M}(k)$ (and some submanifolds in CROSS – *compact symmetric spaces of rank one*) we have

$$\bar{R}(x, y)z^{\perp} = 0, \quad (x, y, z \in TM), \tag{5.8}$$

which is equivalent to

$$(\bar{\nabla}_x h)(y, z) = (\bar{\nabla}_y h)(x, z) \quad (x, y, z \in TM). \tag{5.8$'$}$$

A submanifold with the property (5.8) is called *invariant under a curvature tensor of the ambient space* or, in short, *a curvature invariant submanifold.*

A submanifold $M \subset \bar{M}$ has a *flat normal connection* if $R^{\perp} = 0$. A vector sub-bundle μ of $TM^{\perp}$ is said to be *parallel* if $\nabla^{\perp}_x \xi \in \mu$ for any $x \in TM$ and any vector field ξ in μ.

5.1 Definition. For an n-dimensional submanifold $M \subset \bar{M}$, the *mean curvature vector* field $H \subset TM$ is defined by the formula $H = \frac{1}{n} \sum_i h(e_i, e_i)$, where $\{e_i\}$ is a local orthonormal basis in TM. A submanifold with one of the conditions

$$h = 0, \; h(x, y) = (x, y)H, \; H = 0$$

is called *totally geodesic, totally umbilic,* or *minimal,* respectively. A totally umbilic submanifold with parallel nonzero mean curvature vector, i.e., $\nabla^{\perp} H = 0$, is called an *extrinsic sphere.* If there exists a function λ on the submanifold M such that

$$(h(x, y), H) = \lambda(x, y)$$

for any vector fields x and y on M, then M is called a *pseudo-umbilical submanifold.* A submanifold M in a Kähler Riemannian space $\bar{M}$ (i.e., $\bar{\nabla} J = 0$) with the condition $J(TM) \subset TM$, (resp., $J(TM) \perp TM$) is called *Kähler* (resp., *totally real*).

A totally geodesic submanifold is always curvature-invariant; for recent results, see [Ts].

It is clear that every minimal submanifold is pseudo-umbilical with $\lambda \equiv 0$ and that for a pseudo-umbilical submanifold $\lambda = (H, H)$. A totally umbilical submanifold is totally geodesic iff it is minimal. In a quaternionic Kähler Riemannian space, three types of submanifolds can be considered: quaternionic Kähler, totally complex and totally real. Note that quaternionic submanifolds in $\mathbb{H}P^m$ are actually domains of totally geodesic quaternionic Kähler subspaces, see [Bor 2].

5.2 Definition. [Ch 1] A vector bundle E with a metric over a Riemannian manifold M is called a *Euclidean vector bundle.* If such a bundle also has a second fundamental form, i.e., a smooth section h in $Hom\,(TM \oplus TM, E)$ satisfying the symmetry condition $h(y, z) = h(z, y)$ for all $y, z \in TM$, then it is called a *Riemannian vector bundle.*

5.3 Theorem. [Ch 1] *a) Let M^n be a simply connected Riemannian manifold with a Riemannian m-dimensional vector bundle E, which satisfies equations* (5.4), (5.6) *and* (5.8). *Then M^n can be isometrically embedded into a space form $\bar{M}^{n+m}(k)$ with normal bundle E. b) Let $f_1, f_2 : M \to \bar{M}(k)$ be two isometric embeddings with normal bundles E_1, E_2 and suppose there exists an isometry $\varphi : M \to M$, such that φ can covered by a bundle map $\bar{\varphi} : E_1 \to E_2$ which preserves the metrics and the second fundamental forms of the bundles E_1 and E_2. Then there exists a rigid motion $g : \bar{M}(k) \to \bar{M}(k)$ with the property $g \circ f_1 = f_2 \circ \varphi$.*

5.1.2. Ruled and (strongly) parabolic submanifolds

5.4 Definition. *A ruled submanifold* is defined to be a C^2-smooth submanifold $M^{\nu+n}$, $(\nu, n > 0)$, of a Riemannian space $\bar{M}$ together with a C^1-foliation $\{L\}$ with ν-dimensional complete leaves (*generators or rulings*) that are totally geodesic in $\bar{M}$. If $\bar{M}$ is a Kählerian manifold and M and $\{L\}$ are invariant, then M is called a *complex ruled submanifold.* A ruled submanifold with stationary normal space $TM^{\perp}$ along the generators is called *developable.* If a developable ruled submanifold is locally isometric to the product $L \times L^{\perp}$, then it is called *cylindrical.*

Ruled submanifolds in real space forms are studied in [BD], [BDJ], [Bai], [BK], [Er], [Koc], [Th 1-4],[Lum 1], [CDVV], [Ed], [AB], [Di], [Au 1,2], [HV], [Kü], [MR 1,2], [Rov 5,7] and [Sab]; for $\bar{M} = CP^n$ or $\bar{M} = HP^n$, see [Kimu 1,2], [ALS], [Fuj], [Mar], [Mae], [MaS], [KK], [Pe], [Suh], [Pyo], [Zh], and [FW], for Minkowski space, see [Cho] and [DWV].

From (5.1) it follows that the vectors that are tangent to the rulings of a ruled submanifold are asymptotic:

$$h(x, x) = 0, \quad (x \in TL). \tag{5.9}$$

For a developable ruled submanifold we have

$$h(x, y) = 0, \quad (x \in TL, \; y \in TM). \tag{5.10}$$

Hence the rulings are tangent to relative nullity subspaces defined below.

5.5 Definition. The *index of relative nullity at* $m \in M$ of a submanifold $M \subset \bar{M}$ is the dimension $\mu(m)$ of the *relative nullity space*

$$N(h, m) = \{y \in T_m M : h(y, *) = 0\}.$$

The *rank of the second quadratic forms at* $m \in M$ of a submanifold $M \subset \bar{M}$ is the number $r(m) = \max\{r(\xi) : \xi \in T_m M^{\perp}\}$, where $r(\xi) = rank\, A_{\xi}$ is the rank of the second quadratic form A_{ξ} for normal ξ. A submanifold $M \subset \bar{M}$ is called a) *strongly parabolic* if the index of relative nullity $\mu(M) := \min\{\mu(m) : m \in M\}$ is positive [CK], ($\bar{G} := \{m \in M : \mu(m) = \mu(M)\}$ is a *regularity domain*); b) *parabolic* if the maximal rank of the second quadratic forms $r(M) = \max\{r(m) : m \in M\}$ is smaller than $\dim M$ [Bor 2].

Since for any point $m \in M$ we have $\ker h = \bigcap_{\xi} \ker A_{\xi}$ (we can choose a finite number of codim M normals which form a basis in $TM^{\perp}$), it follows that

$$\mu(M) \leq \dim M - r(M). \tag{5.11}$$

For a totally geodesic submanifold the equalities $\mu(M) = \dim M$ and $r(M) = 0$ hold. For a hypersurface the two notions in (5.11) are equivalent: $\mu(M) = \dim M - r(M)$. Note that a k-saddle submanifold $M \subset \bar{M}$ (see Definition 3.32) is parabolic with $r(M) \leq 2(k-1)$. For a submanifold $M \subset \bar{M}$ we have [CK]

$$\mu(M) \leq \nu(M) \leq \mu(M) + \operatorname{codim} M, \tag{5.12}$$

where $\nu(M)$ (see Section 2.1) is the index of nullity of the curvature tensor.

Strongly parabolic submanifolds have generators (a ruled structure) under some conditions on a curvature tensor of the ambient space, see Section 2.3.1.

5.6 Theorem. [Mal 2] *Let* $M \subset \bar{M}$ *be a (complete) submanifold satisfying condition* (5.8) *and* $\mu(M) > 0$. *Then the regularity domain* $\bar{G} \subset M$ *is foliated by (complete) totally geodesic submanifolds (generators) in* $\bar{M}$ *and the normal space of* M *is stationary along the leaves.*

Thus a curvature invariant submanifold with constant index of relative nullity has a developable ruled structure. Properties (5.8) and (5.13) below are true for all submanifolds in real space forms and for certain cases of submanifolds in CROSS. In [Okr 1] infinite dimensional strongly parabolic submanifolds in a Hilbert Riemannian space are studied and a version of Theorem 5.6 is given. The index of relative nullity can be estimated when the partial Ricci curvature (see Section 3.4) is positive. From Theorem 5.6 and Corollary 3.31 we get the following corollary.

5.7 Corollary. ([Bor 2] *for* $q = 1$) *Let* M *be a complete curvature invariant submanifold in a Riemannian space* $\bar{M}$. *Suppose that* M *is not totally geodesic. If* $Ric^q(M)$ *is positive and* M *is compact, then* $\mu(M) < \frac{1}{2}(\dim M + q - 1)$. *If* $Ric^q(\bar{M})_{|M}$ *is positive and* $\bar{M}$ *is compact, then* $\mu(M) < \frac{1}{2}(\dim \bar{M} + q - 1)$.

In [Mrz] the *normal nullity space* at a point $m \in M$,

$$T_1(m) = \{x \in T_m(M) \colon R^{\perp}(x, y) = 0 \ \forall\, y \in T_m M\},$$

of a submanifold M in a Riemannian space $\bar{M}$ is defined by the curvature tensor $R^{\perp}$ of the normal connection, and the tests for the involutiveness of a normal nullity distribution are given (one of the sufficient conditions is that $\bar{M}$ be a real space form).

5.1.3. A Sasaki metric on the tangent and normal bundles of a submanifold

A metric on the tangent bundle of a Riemannian manifold induced by parallel displacement of tangent vectors was constructed by S. Sasaki, see [Sas 1,2].

Let (x^i) be local coordinates on a Riemannian manifold M with a metric tensor $ds^2 = g_{ij}dx^i dx^j$. Then for the induced local coordinates $(x^i,\ \xi^i)$ on the tangent bundle TM, a *Sasaki metric* is

$$d\sigma^2 = g_{ij}dx^i dx^j + g_{ij}D\xi^i D\xi^j,$$

where $D\xi^i = d\xi^i + \Gamma^i_{jk}\xi^j dx^k$ is the covariant derivative of the tangent vector field. In other words, if $d\theta$ is the angle between the tangent vector (ξ^i) at the initial point (x^i) and the vector $(\xi^i + d\xi^i)$ at the point $(x^i + dx^i)$ after parallel displacement to the initial point along the shortest geodesic, then $d\sigma^2 = ds^2 + |\xi|^2 d\theta^2$, see [BY 2]. This construction can be applied to arbitrary vector bundles over a manifold with a (not necessarily metric) connection. An analog of a Sasaki metric on a frame bundle over a Riemannian manifold is given in [Mo]. More general constructions of metrics on fiber bundles are studied in [Fa].

5.8 Theorem. [BY 2] *The index of nullity* $\nu(TM)$ *of the curvature tensor of a Sasaki metric is even. A manifold* M^n, *equipped with a Sasaki metric, is the metric product* $M_1^{n-\nu(TM)/2} \times \mathbb{R}^{\nu(TM)/2}$, *and* TM^n *is the metric product* $TM_1^{n-\nu(TM)/2} \times \mathbb{R}^{\nu(TM)}$.

The proof is based on the existence of $\nu(TM)/2$ parallel linearly independent vector fields on M. The converse of this theorem is not true, i.e., that M

is strongly parabolic does not necessarily imply that a Sasaki metric on TM is also strongly parabolic. The following hypothesis is investigated in [Yam 2]: for a Sasaki metric on the tangent bundle TM, the index of (spherical) k-nullity (where $k > 0$), $\nu_k(M)$, is zero when $\dim M \geq 3$.

For a submanifold $M \subset \bar{M}$ the following two (usually different) metrics on TM can be considered: a Sasaki metric and the metric induced by a Sasaki metric on $T\bar{M}$. For example, for a cylinder $M^2 \subset \mathbb{R}^3$, a Sasaki metric on TM^2 is flat, but the induced metric on TM^2 from $T\mathbb{R}^3 = \mathbb{R}^6$ has nonzero curvature.

5.9 Lemma. [BY 2] *A Sasaki metric on $T\bar{M}$ induces a Sasaki metric on TM iff one of the following two equivalent properties holds:*

1) *TM is a totally geodesic submanifold in $T\bar{M}$,*
2) *M is a totally geodesic submanifold in $\bar{M}$.*

The equivalence of properties 1) and 2) follows from the fact that the projection $p : TM \to M$ is a Riemannian submersion.

For $M \subset \mathbb{R}^m$ the tangent bundle TM can be considered as a submanifold in the Euclidean space $T\mathbb{R}^m = \mathbb{R}^{2m}$, which has index of relative nullity $\mu(TM)$.

5.10 Theorem. [BY 2] *For $M \subset \mathbb{R}^m$ we have $2\mu(M) \geq \mu(TM) \geq \mu(M)$. Moreover, M is a cylindrical submanifold with a $(\mu(TM)-\mu(M))$-dimensional generator, and TM is cylindrical with a $2(\mu(TM) - \mu(M))$-dimensional generator.*

The proof is based on an analysis of second quadratic forms of TM in $T\bar{M}$.

A vector field ξ on a manifold M^n can be identified with an n-dimensional submanifold $\xi(M)$ in TM. A submanifold $\xi(M)$ is totally geodesic in TM iff ξ is parallel; moreover, for compact M, a submanifold $\xi(M)$ is minimal in TM iff ξ is parallel, see [BY 2].

The *volume of a unit vector field* ξ is an n-volume of the submanifold $\xi(M)$ calculated for a Sasaki metric on S_1M:

$$Vol\,(\xi) = \int_M \sqrt{\det(Id + (\nabla\xi)^\top(\nabla\xi)}\, d\,vol,$$

where the covariant derivative $\nabla\xi$ is interpreted as a linear operator on TM. For example, the unit vector fields of *minimal volume* on a round 3-sphere are tangent to leaves of Hopf fibrations [GZ]. For more studies about $Vol(\xi)$ see [Rez 3].

The problem of classifying the totally geodesic (umbilic) submanifolds of TM and S_1M, when M is a space form or CROSS is very important. Another important problem is to estimate the codimension of an isometric embedding of TM with a Sasaki metric into a Euclidean space [BY 2].

A Sasaki metric on the normal bundle $TM^{\perp}$ of a submanifold $M \subset \bar{M}$ was introduced by A. Borisenko, see [BY 1] and [Rec 1,2], and is used for the study of extrinsic geometry of submanifolds in a Riemannian space. For induced local coordinates (x^i, ξ^i) on a normal bundle $TM^{\perp}$ of a submanifold $M \subset \bar{M}$, a *Sasaki metric* is the following:

$$d\sigma^2 = g_{ij}dx^i dx^j + g_{ij}D^{\perp}\xi^i D^{\perp}\xi^j,$$

where $D^{\perp}\xi^i$ is the normal component of the covariant derivative in $\bar{M}$ of a normal vector field (ξ^i) in a tangent direction [BY 2].

5.11 Theorem. [BY 2] *A Sasaki metric on $TM^{\perp}$ is locally symmetric (flat) iff M is a symmetric (flat) Riemannian manifold embedded with a flat normal connection.*

Let $\nu_H(TM^{\perp})$ and $\nu_V(TM^{\perp})$ denote the *horizontal and vertical* (intrinsic) *indexes of nullity* of the curvature tensor of a Sasaki metric.

5.12 Theorem. [BY 1] *a) There exist $\nu_V(TM^{\perp})$ linearly independent normal vector fields on a submanifold M, which are parallel in a normal connection.*

b) A submanifold M in $\mathbb{R}^m$ is foliated by $\nu_H(TM^{\perp})$-dimensional intrinsically flat submanifolds, totally geodesic in M with a flat normal connection.

A normal vector field ξ on a submanifold $M^n \subset \bar{M}$ can be considered an n-dimensional submanifold $\xi(M)$ in $TM^{\perp}$ with Sasaki metric. If $\xi \subset TM^{\perp}$ is parallel in a normal connection, then the submanifold $\xi(M)$ is totally geodesic in $TM^{\perp}$ iff M is a totally geodesic submanifold [Yam 1].

From a geometric point of view it is also interesting to study the *spherical tangent bundle $S_\rho M$* and the *spherical normal bundle $S_\rho M^{\perp}$* which contain the vectors of fixed length ρ and are hypersurfaces in some vector bundles. For example, $S_1 M$ has constant mean curvature and its volume is $Vol(S_1 M^n) = Vol(M^n) \cdot Vol(S^{n-1})$.

The formulas for the Riemannian, Ricci or scalar curvature of Sasaki metrics on TM and $TM^{\perp}$ (resp., $S_\rho M$ and $S_\rho M^{\perp}$) are similar [BY 2]. For example, if TM or $TM^{\perp}$ with a Sasaki metric is Einstein, then M is flat. If the sectional curvature of M is positive, then the sectional curvature of $S_\rho M^{\perp}$ is positive for sufficiently small ρ.

The (strong) parabolicity of a given submanifold $M \subset \bar{M}$ determines the ruled structure of a Sasaki metric on $TM^{\perp}$.

Let ξ_0 be a normal vector at a point q_0 with maximal rank $r(\xi_0) = r(M)$. Then the rank is constant for normal vectors ξ close to ξ_0. The kernel of ξ satisfies $\sum_{a=1}^{p} \xi^a A^a_{ij} x^i = 0$, where A^a_{ij} are the coefficients of the second quadratic forms with respect to the orthogonal basis of normals. Since the rank of the above system is constant, the solution space $L(q, \xi)$ depends regularly on the point and the normal vectors. Thus a horizontal lifting $\tilde{L}(\tilde{q})$ of planes $L(q, \xi)$ to the points $\tilde{q} = (q, \xi)$ of the normal bundle forms a differentiable distribution on the neighborhood of the point $\tilde{q}_0 = (q_0, \xi_0)$.

5.13 Theorem. [Bor 5] *The differentiable horizontal distribution $\tilde{L}(\tilde{q})$ on the normal bundle of a parabolic submanifold M in a Riemannian space $\bar{M}$ is integrable if, at the points of M, the curvature operator $\bar{R}$ of $\bar{M}$ satisfies the condition*

$$\bar{R}(x, y)\xi \begin{cases} = 0, & (x,\ y \in TM,\ \xi \in TM^{\perp}), \\ \parallel J\xi, & (x,\ y \in TM,\ \xi \in TM^{\perp}),\ \textit{for the Kählerian case}. \end{cases} \tag{5.13}$$

The fibers are totally geodesic submanifolds of the normal bundle with a Sasaki metric; and their projections on M are totally geodesic submanifolds of $\bar{M}$.

Koike [Ko 5] investigated the integrability of the lift of a distribution on a submanifold to a tube with a not necessarily constant radius over a Riemannian submanifold.

We also have the converse problem: how is the (strong) parabolicity of a Sasaki metric on $TM^{\perp}$ related to the properties of a given submanifold $M \subset \bar{M}$ [BY 2] ?

5.14 Corollary. [Bor 5] *Let $M \subset \bar{M}$ be a (complete) parabolic submanifold (i.e., $0 < r(M) <$ dim M) satisfying condition* (5.13). *Then for any normal vector $\xi \in T_pM$ with maximal rank $r(\xi) = r(M)$ there exists a (complete) generator $L(\xi) \subset M$, ($T_pL(\xi) =$ ker A_{ξ}), along which the normal vector ξ is stationary, i.e., ξ stays normal under $\bar{\nabla}$-parallel displacement along any geodesic in $L(\xi)$.*

A parabolic submanifold of small codimension is strongly parabolic [Bor 1]:

$$\mu(M) \geq \dim M - \frac{r(M)}{2}(\operatorname{codim} M + 1).$$

The following result completes Corollary 5.7, see also Corollary 7.8'.

5.15 Corollary. ([Bor 5] *for* $q = 1$) *Let* M *be a complete submanifold in a Riemannian space* $\bar{M}$ *satisfying condition* (5.13). *Then* M *is a totally geodesic submanifold, if one of the following holds:*

a) $Ric^q(M)$ *is positive,* M *is compact and* $r(M) \leq \frac{1}{2}(\dim M - q + 1)$,

b) $Ric^q(\bar{M})_{|M}$ *is positive,* $\bar{M}$ *is compact and* $r(M) \leq \frac{1}{2}(\dim \bar{M} - q + 1)$.

Proof. Let $m_0 \in M$ be a point and $\xi_0 \in T_{m_0}M^\perp$ a unit normal vector of maximal rank $r(\xi_0) = r(M)$. By Corollary 5.14 there exists a complete $(\dim M - r(M))$-dimensional ruling $L(\xi_0)$ with $T_{m_0}L(\xi_0) = \ker A_{\xi_0}$, and ξ_0 is stationary along it. Consider a submanifold $\bar{L}$ transversal to $L(\xi_0)$ through m_0. Continue ξ_0 locally as a unit vector field of normals over $\bar{L}$ with the initial property $\nabla^\perp \xi_{|m_0} = 0$. Such normal vectors have maximal rank $r(M)$, and by Corollary 5.14, they define a family of $(\dim M - r(M))$-dimensional rulings $L(\xi)$, which by Corollary 3.31 intersect $L(\xi_0)$. Then we repeat the main part of the proof of [Theorem 2, Bor 5] to complete the proof of Corollary 5.15. □

5.2. Submanifolds with generators in space forms

5.2.1. Submanifolds with generators in Euclidean space

It is important to investigate the relationships between classes of submanifolds with degenerate second fundamental form (strongly parabolic, cylinders) and metrics with degenerate curvature tensor (for example, with positive index of nullity, products of manifolds).

There is a natural commutative diagram where the arrows indicate (exact) inclusion maps:

$$\begin{array}{ccc} \textit{strongly parabolic submanifolds} & \xrightarrow{(1)} & \textit{strongly parabolic metrics} \\ {\scriptstyle (2)}\uparrow & & {\scriptstyle (3)}\uparrow \\ \textit{cylindrical submanifolds} & \xrightarrow{(4)} & \textit{cylindrical metrics} \end{array}$$

1. For a submanifold $M^n \subset \mathbb{R}^N$, the inequality $\nu(M) \geq \mu(M)$ holds with equality for hypersurfaces. The problem of when a Riemannian manifold M^n admits an isometric immersion into $\mathbb{R}^N$ having equality $\nu(M) = \mu(M)$ is studied in [BU 2]; for the basic case of 3-dimensional manifolds with nullity 1, we have the following results:

a) There exist 3-dimensional analytical metrics with $\nu(M) = 1$ which do not admit an isometric embedding with $\mu(M) = 1$ into $\mathbb{R}^N$, for instance,

$$g_{11} = \exp(2x_2) + x_2^2 + x_3^2,\ g_{12} = x_3,\ g_{13} = -x_2,\ g_{22} = g_{33} = 1,\ g_{23} = 0. \tag{5.14}$$

For the metric above, the sectional curvature on planes that are orthogonal to a 1-dimensional nullity foliation is -1.

b) The class of 3-dimensional analytic metrics with $\nu(x) \equiv 1$, which admit isometric embeddings with $\mu(M) = 1$ into $\mathbb{R}^N$ for $N = 4$ and for $N > 4$, depends on two functions with two variables. (The class of 3-dimensional analytic metrics with $\nu(x) \equiv 1$ depends on three functions with two variables).

2. Understanding the congruence classes of isometric immersions f of a connected Riemannian manifold M^n into Euclidean space $\mathbb{R}^{n+1}$ for $n \geq 3$ is the classical rigidity problem for hypersurfaces. In this case, the *type number of M at x*, denoted by $t(x)$, coincides with the *rank of the Gaussian map* of f and with the *rank of a second quadratic form* $r(x)$. It is well-known by the Beez–Killing Theorem that f is rigid if $r(M) \geq 3$ and highly deformable in the flat case $r(M) \leq 1$. The situation for constant $r(M) = 2$ (in this case M has a developable ruled structure with $(n-2)$-dimensional generator) is quite complex. U. Sbrana [Sbr] and E. Cartan [Car 1] gave a detailed local analysis. The deformations are a discrete one-parameter family, and infinite dimensional only in the ruled case $n \geq 4$. The compact case is studied in [Sac 2]. In [DG 3] Sacksteder's theorem is extended as follows.

5.16 Theorem. [DG 3] *Let $f : M^n \to \mathbb{R}^{n+1}$, $(n \geq 3)$, be an isometric immersion of a complete Riemannian manifold that does not contain an open set $M_1^3 \times \mathbb{R}^{n-3}$ with M_1^3 unbounded. Then f admits (nondiscrete) isometric deformations only along ruled strips (i.e., ruled domains in M^n with $(n-1)$-dimensional generators). Furthermore, if f is nowhere completely ruled and the set of totally geodesic points does not disconnect M, then f is rigid.*

If the ambient space is a round sphere S^{n+1}, the local rigidity problem becomes a special aspect of the Euclidean case by considering cones; see [Sbr].

Complete submanifolds M^n are always rigid for $n \geq 4$ but not for $n = 3$; see [DG 2,3]. For analogous problems in terms of the (degenerate) Grassmannian image of a submanifold in Euclidean space, see [BN].

5.17 Example. [Sac 1] A hypersurface $M^3 \subset \mathbb{R}^4$ given by the equation

$$x_4 = x_3 \cos(x_1) + x_2 \sin(x_1)$$

has index of relative nullity $\mu(x) \equiv 1$, but it is not a cylinder.

5.18 Theorem. [AB] *Let M^{n+1} be a hypersurface in the Euclidean space $\mathbb{R}^{n+2}$ satisfying the condition*

$$\forall\, m \in M\ \exists \text{ an } n\text{-dimensional generator contained in } M. \tag{5.15}$$

a) Then M is foliated by $(n-1)$-dimensional generators.
b) Let m be a point in M through which there are at least two n-dimensional generators. Then there exists an open neighborhood U of m in M which is the Riemannian product of a ruled surface N^2 and $\mathbb{R}^{n-2}$. Moreover, if through every point of M there exist two n-dimensional generators, then M is the product of $\mathbb{R}^{n-2}$ and a double ruled surface N^2 in $\mathbb{R}^3$.

It is not difficult to see that $\mu(M) \geq n-2$ follows from condition (5.15). Note that a minimal hypersurface M^n in $\mathbb{R}^{n+1}$, which admits a foliation by Euclidean $(n-1)$-planes, is either totally geodesic or the product $M_1^2 \times \mathbb{R}^{n-2}$, where a surface M_1^2 is the standard helicoid in $\mathbb{R}^3$ [BV].

5.19 Example. [AB] The hypersurface M^3 of $\mathbb{R}^4$ given by

$$\bar{r}(u, v, w) = (u \cos w,\ u \sin w + \cos w,\ w + v \sin w,\ w) \tag{5.16}$$

is foliated by planes ($\bar{r}$ is linear in u, v) but is not a product of a surface and $\mathbb{R}^1$, the leaves tangent to $\ker h$, because the 2-dimensional distribution $(\ker h)^\perp$ is not integrable.

3. The so-called *Cylinder Theorems* for strongly parabolic embeddings into Euclidean space with additional conditions on the sign of a curvature were obtained first by [O'N 1] for $K_M = 0$, by [Har] for $K_M \geq 0$ and then by [Bor 4] for $Ric_M \geq 0$. For more general results see Section 6.1.

5.20 Theorem. [Bor 4] *Let $M \subset \mathbb{R}^N$ be a complete parabolic submanifold (i.e., $0 < r(M) < \dim M$) satisfying the condition $Ric_M \geq 0$. Then M is cylindrical with a* $(\dim M - r(M))$*-dimensional generator.*

Borisenko [Bor 4] also obtained cylinder theorems for parabolic submanifolds with "large" rank $r(M) \geq \dim M - 3$ and additional conditions for an asymptotical subspace, but without any assumption on the sign of the curvature.

5.21 Definition. [Bor 4] The subspace $A_m \subset T_m M$ of maximal dimension, which consists of asymptotical vectors, is called *asymptotical*. The dimension of the asymptotical subspace is called the *planar order* at the point m.

5.22 Theorem. [Bor 4] *Let $M^n \subset \mathbb{R}^N$ be a complete submanifold with constant index of relative nullity, rank $r(M)$ equal to a) 2, b) 3; and planar order at each point equal to a) $n-2$, b) $n-3$, respectively. Then M^n is cylindrical with an a)* $(n-2)$*-dimensional generator, b)* $(n-3)$*-dimensional generator, respectively.*

4. The cylindricity of embeddings of metric products $M_1 \times M_2$ into $\mathbb{R}^N$ with small codimension are considered in [Moo 1], [AMa] and [BDT].

Let $M^n = M_1^{n_1} \times M_2^{n_2}$, $(n_1, n_2 \geq 2)$, be a product of two connected complete Riemannian manifolds. If neither M_i is flat everywhere or contains a *Euclidean strip*, that is, an open submanifold which is isometric to the product $I \times \mathbb{R}^{n_i - 1}$ where $I \subset \mathbb{R}$ denotes an open interval, then any isometric immersion $f : M^n \to \mathbb{R}^{n+2}$ is a product of hypersurface immersions [AMa], [Moo 1]. This means that there exist an orthogonal factorization $\mathbb{R}^{n+2} = \mathbb{R}^{n_1+1} \times \mathbb{R}^{n_2+1}$ and isometric immersions $f_i : M_i^{n_i} \to \mathbb{R}^{n_i+1}$, $(i = 1, 2)$, such that $f(x_1, x_2) = (f_1(x_1), f_2(x_2))$. This outstanding global theorem is proved for any number of factors whenever the codimension equals that number. In [BDT] the problem when $f : M^n = M_1^{n_1} \times M_2^{n_2} \to \mathbb{R}^{n+2}$ may fail to be a product of hypersurface immersions is studied.

5.23 Theorem. [BDT] *Let $f : M^n = M_1^{n_1} \times M_2^{n_2} \to \mathbb{R}^{n+2}$, $(n_i \geq 2)$, be an isometric immersion of a complete connected Riemannian manifold where no factor is everywhere flat. Then there is a dense open subset each of whose points lies in a product neighborhood $U = U_1 \times U_2$, with $U_i \subset M_i$ open, such that $f_U : U_1 \times U_2 \to \mathbb{R}^{n+2}$ is one of the following types:*

1) *f_U is a product of immersions.*

2) *Each U_i is isometric to $I_i \times \mathbb{R}^{n_i - 1}$, $(i = 1, 2)$ and $f_U = g \times Id$, where $g : I_1 \times I_2 \to \mathbb{R}^4$ is an isometric immersion and $Id : \mathbb{R}^{n-2} \to \mathbb{R}^{n-2}$ is the identity map.*

3) *Only one* U_j *is isometric to* $I_j \times \mathbb{R}^{n_j-1}$ *and* $f_U = \tilde{f} \times Id : (U_i \times I_j) \times \mathbb{R}^{n_j-1} \to \mathbb{R}^{n+2}$, $(i \neq j)$, *where* $Id : \mathbb{R}^{n_j-1} \to \mathbb{R}^{n_j-1}$ *is the identity map and* $\tilde{f} : U_i \times I_j \to \mathbb{R}^{n_i+3}$ *is a composition* $\tilde{f} = h \circ g$ *of the isometric immersions* $g : U_i \times I_j \to V$, *and* $h : V \to \mathbb{R}^{n_i+3}$ *where* $V \subset \mathbb{R}^{n_i+2}$ *is open.*

For type 1) we either have a product of hypersurfaces or one of the factors is totally geodesic. Types 2) and 3) are not disjoint since the immersion g in 2) may in fact be a composition. A complete local classification of flat surfaces in $\mathbb{R}^4$ which are nowhere compositions of isometric immersions is provided in [CDa 3]. The example in [AMa] relates immersions of types 1) and 2). The case when one of the factors of M is everywhere flat is also studied.

5.24 Definition. [BDT] An isometric immersion $f : N^{n+m} \to \mathbb{R}^N$ is called an *m-cylinder* whenever there exists a Riemannian manifold M^n such that N^{n+m}, $\mathbb{R}^N$ and f have the orthogonal factorizations $N^{n+m} = M^n \times \mathbb{R}^m$, $\mathbb{R}^N = \mathbb{R}^{N-m} \times \mathbb{R}^m$, and $f = \tilde{f} \times \mathrm{Id}$, where $\tilde{f} : M^n \to \mathbb{R}^{N-m}$ is an isometric immersion and $\mathrm{Id} : \mathbb{R}^m \to \mathbb{R}^m$ is the identity map.

We recall a characterization of complete cylinders due to P. Hartman [Har]:

5.25 Lemma. *Let* $f : M^n \to \mathbb{R}^N$ *be an isometric immersion of a connected complete Riemannian manifold with nonnegative Ricci curvature such that* $f(M)$ *contains* m *linearly independent lines through one point. Then* f *is an* m*-cylinder.*

5.26 Theorem. [BDT] *Let* M^n *be a complete connected Riemannian manifold of nonnegative Ricci curvature without flat points and let* $f : M^n \times \mathbb{R}^m \to \mathbb{R}^{n+m+2}$ *be a* C^1*-regular isometric immersion. Then* f *is either an* m*-cylinder or it is an* $(m-1)$*-cylinder,* $f = \tilde{f} \times Id : (M^n \times \mathbb{R}) \times \mathbb{R}^{m-1} \to \mathbb{R}^{n+m+2}$, *and there exists a flat Riemannian manifold* N^{n+2} *and isometric* $g : M^n \times \mathbb{R} \to N^{n+2}$ *and* $h : N^{n+2} \to \mathbb{R}^{n+3}$ *such that* $\tilde{f} = h \circ g$ *is a composition.*

Furthermore, when M^n *is simply connected, we can take* N^{n+2} *in the latter case to be an open subset of* $\mathbb{R}^{n+2}$ *and then* $g = \tilde{g} \times Id$, *where* $Id : \mathbb{R} \to \mathbb{R}$ *is the identity map and* $\tilde{g} : M^n \to \mathbb{R}^{n+1}$ *is an embedding whose image is a convex hypersurface.*

The above result is false without the assumption of 1-regularity, see [BDT] and [Hen 2]. For $m = 1$, a weaker result without the 1-regularity assumption is given in [Nor 2].

A smooth ruled submanifold $M^{\nu+n} \subset \mathbb{R}^N$ is given (locally) in parameter form by

$$\bar{r}(u_1, \ldots, u_n; v_1, \ldots, v_\nu) = \bar{\rho}(u_1, \ldots, u_n) + \sum_{i=1}^{\nu} v_i \bar{a}_i(u_1, \ldots, u_n), \quad (5.17)$$

where $\{\bar{a}_i(u_1, \ldots, u_n)\}$ are bases of generators and $\bar{\rho}(u_1, \ldots, u_n)$ is a base submanifold. In particular, for a cylindrical submanifold, the vectors $\{\bar{a}_i\}$ are constant.

A *one-parameter family of generators* is a submanifold $M^{\nu+1} \subset \bar{M}$ of the form

$$\tilde{r}(u_1, \ldots, u_n; t) = \bar{r}(u_1(t), \ldots, u_n(t); v_1, \ldots, v_\nu), \quad (5.18)$$

where $\{u_i(t)\}$ is a smooth curve in $\mathbb{R}^n$ [BU 1]. From the Gauss equation (5.4) for a ruled submanifold $M^{\nu+n} \subset \mathbb{R}^N$, it follows that $K_{mix} \leq 0$. Thus, if $M^{\nu+1}$ is a ruled submanifold in the Euclidean space $\mathbb{R}^N$ with zero scalar curvature, then M is developable.

5.27 Theorem. [BU 1] *Let $M^{\nu+n}$ be a ruled submanifold in a Euclidean space $\mathbb{R}^N$. Then M is cylindrical iff all one-parameter families of generators have zero scalar curvature.*

Note that the developable ruled submanifold $M^{\nu+n} \subset \mathbb{R}^N$ with integrable distribution $TL^{\perp}$ is cylindrical.

5.2.2. Submanifolds with generators in CROSS

The ruled submanifolds in S^m and $\mathbb{R}^m$ are related by a geodesic map of the semisphere $S^m_+ = \{x_0 > 0\}$,

$$q = (x_0, \ldots, x_n) \in S^m_+ \to \tilde{q} = (\frac{x_1}{x_0}, \ldots, \frac{x_n}{x_0}) \in \mathbb{R}^m. \quad (5.19)$$

5.28 Lemma. [Bor 9] *Let M be a submanifold in a semisphere $S^m_+ = \{x_0 > 0\}$ and $\tilde{M}$ its image in $\mathbb{R}^m$ under a geodesic map. Suppose that ξ is normal to M at a point q and $\tilde{\xi}$ is normal to $\tilde{M}$ at the point $\tilde{q}$. Then*

1) *the orthogonal projection of ξ onto the hyperplane $x_0 = 1$ is the normal $\tilde{\xi}$,*

$$\xi = (\xi_0, \ldots, \xi_n) \to \tilde{\xi} = (\xi_1, \ldots, \xi_n)/\sqrt{1 - \xi_0^2}, \quad (5.20)$$

(note that $\xi_0 \neq 1$ in S^m_+),

2) $A_{ij}(\tilde{\xi}, \tilde{q}) = cA_{ij}(\xi, q)$, *where* $A_{ij}(\tilde{\xi}, \tilde{q})$ *and* $A_{ij}(\xi, q)$ *are coefficients of second quadratic forms of* M *and* $\tilde{M}$ *for the normal vectors* ξ *and* $\tilde{\xi}$.

Thus locally the (strong) parabolic submanifolds in S^m and $\mathbb{R}^m$ are in one-to-one correspondence, see also [AkG]. The "inner" projection in the sphere S^m is also important.

5.29 Lemma. [Bor 9] *Assume that a submanifold* $M \subset S^m$ *has a regular projection* $\tilde{M}$ *onto some* $(\dim M + t)$*-dimensional great sphere. Then* $r(\tilde{M}) = r(M)$.

As in Lemma 5.28, the second quadratic forms of M and $\tilde{M}$ in Lemma 5.29 are proportional.

In contrast to a sphere S^m (or RP^m), where totally geodesic submanifolds are of one type (e.g., great spheres with all possible dimensions) other CROSS have different types of totally geodesic submanifolds:

$$\begin{aligned} CP^m &: \{RP^n,\ CP^n\},\ n \le m; \\ HP^m &: \{RP^n,\ CP^n,\ HP^n,\ S^3\},\ n \le m; \\ CaP^2 &: \{RP^n,\ CP^n,\ HP^n,\ S^i\},\ n \le 2,\ i = 3,\ 5,\ 6,\ 7,\ 8. \end{aligned}$$

Such submanifolds in KP^m, $(K = \mathbb{C},\ \mathbb{H})$, are related with the help of the map *exp* to some subspaces $\bar{K}^{\bar{n}+1}$ which are contained in K^{n+1} [Bes 1]:

1) it is possible to have the restriction for type of numbers $\bar{K} \subset K$,

2) the dimension can be decreased $\bar{n} \le n$.

The standard totally geodesic foliations on CROSS are the *generalized Hopf fibrations* on Euclidean spheres, see (1.2):

$$S^1 \subset S^{2n+1} \xrightarrow{\pi} CP^n, \quad S^3 \subset S^{4n+3} \xrightarrow{\pi} HP^n, \quad S^7 \subset S^{15} \xrightarrow{\pi} S^8 \tag{5.21}$$

and a foliation on complex projective space [Es 3]

$$CP^1 \subset CP^{2n+1} \xrightarrow{\pi} HP^n. \tag{5.21'}$$

For a submanifold $\tilde{M}$ in the base of the Hopf fibration, the inverse image $\pi^{-1}(\tilde{M})$ is a ruled submanifold in a sphere or in CP^{2n+1} with equidistant 1-, 3-, 7- or 2-dimensional generators (fibers).

5.30 Definition. Consider the following classes of ruled submanifolds in S^m and CP^m with complex structure J:

1. $M^{\nu+n} \subset S^m$ with generator $L = S^\nu$,
2. $M^{\nu+n} \subset CP^m$ with generator $L = RP^\nu$ and the property $J(TL) \perp TM$,
3. $M^{\nu+n} \subset CP^m$ with generator $L = CP^{\nu/2}$ and the property $J(TM) = TM$,

Also certain classes of ruled submanifolds in HP^m and CaP^2 may be considered. The curvature tensor of an ambient Riemannian manifold along ruled submanifolds (see Definition 5.30) has the same simple form

$$\bar{R}(x, y)x = -ky(x, x), \quad (x \in TL, \; y \in TL^\perp). \tag{5.22}$$

5.31 Definition. [Bor 2] A submanifold M^l in CROSS is called *t-uniquely projectible*, if there exist totally geodesic submanifolds

$$1)\; S_0^{l+t} \subset S^n, \quad 2)\; CP_0^{(l+t)/2} \subset CP^n, \quad 3)\; HP_0^{(l+t)/4} \subset HP^n,$$

such that the tangent spaces to the totally geodesic submanifolds

$$1)\; S^{n-(l+t)}(q), \quad 2)\; CP^{n-(l+t)/2}(q), \quad 3)\; HP^{n-(l+t)/4}(q),$$

which contain the point $q \in M$ and are orthogonal to S_0^{l+t}, $CP_0^{(l+t)/2}$, and $HP_0^{(l+t)/4}$, resp., have zero-intersections with the tangent space T_qM. For $t = 0$, a submanifold M^l is called *uniquely projectible.*

The image $\tilde{M}$ of M under the above map $q \in M \to \tilde{q} \in S_0^{l+t}$, $CP_0^{(l+t)/2}$, $HP_0^{(l+t)/4}$, resp. is a regular submanifold with the same rank; see Lemma 5.29.

With the help of a result by T. Frankel [Fr 1] (see Section 3.4.1) the estimates of $\dim L$ for a ruled submanifold and the tests of $h = 0$ for a parabolic submanifold in CROSS may be obtained in terms of the dimension of $\bar{M}$ or the codimension of M. But the key role is played by the number t from the definition of unique projectiveness.

5.32 Theorem. [Bor 2] *Let M^l be a compact t-uniquely projectable submanifold in CROSS satisfying the condition*

$$r(M) \le \begin{cases} (l-t)/2, & if\ \bar{M} = S^n, \\ (l-t)/2 - 3/2, & if\ \bar{M} = CP^n, \\ (l-t)/2 - 9/2, & if\ \bar{M} = HP^n. \end{cases}$$

Then $r(M) = 0$ and M^l is a totally geodesic submanifold of the same type.

From Theorem 5.32, with t replaced by $\operatorname{codim} M$, follows a test (without the condition of unique projectiveness) of totally geodesic submanifolds in

CROSS, see [Bor 2] and also Chapter VII. The inequalities 2, 3 in Theorem 5.32 are proved with the help of the following lemma.

5.33 Lemma. [Bor 2] *For submanifolds $\tilde{M}$ and $M = \pi^{-1}(\tilde{M})$, where π are the Hopf fibrations* (5.21) *with s-dimensional fiber, the inequality $r(M) \leq r(\tilde{M}) + 2s$ holds.*

A compact uniquely projectable Kählerian submanifold in CP^n is totally geodesic [Bor 2].

In [Rov 2,8] the following classes of ruled submanifolds in CROSS are introduced:

RS_1 – *uniquely projectible* (in the sense of a neighborhood of one generator),

RS_2 – *with positive mixed sectional curvature,*

RS_0 – *ruled developable submanifolds* (or strongly parabolic submanifolds)

RS_1 and RS_2 generalize the well-known class RS_0 and are related to the classes of parabolic submanifolds:

PS_1 - *uniquely projectible,*

PS_2 – *with positive sectional curvature.*

See the commutative diagram below, where the arrows represent (exact) inclusion maps:

$$\begin{array}{ccc} RS_1 & \xrightarrow{(1)} & RS_2 \\ {\scriptstyle (3)}\uparrow & & {\scriptstyle (4)}\uparrow \\ PS_1 & \xrightarrow{(2)} & PS_2 \end{array}$$

5.34 Theorem. [Rov 8] *Assume that a complete submanifold M in S^N or CP^N belongs to RS_2 (resp., PS_2). Then any complete generator $L \subset M$ has a neighborhood which belongs to RS_1 (resp., PS_1). The dimension of a generator of such a submanifold is estimated by*

$$\dim L \leq \begin{cases} \rho(\operatorname{codim} L) - 1, \\ 2, \quad for\ the\ K\ddot{a}hlerian\ case. \end{cases}$$

For a submanifold M of RS_1 or PS_1, the above inequalities can be proved by projecting from M onto a totally geodesic subspace of the same dimension, and then using Theorem 1.32 ([Ab 3] for the Kählerian case). Moreover, for a ruled submanifold in Theorem 5.34 the structural tensor of a foliation $\{L\}$ has no eigenvectors [Rov 8] (compare with Lemma 3.40 and Lemma 3.41). This idea is developed in Chapters VI and VII.

5.35 Corollary. [Rov 4] *Let M be a complete parabolic submanifold in $S^m(1)$, and $CP^m(1)$ with $J(TM) = TM$ and suppose that the Ricci curvature Ric_M is greater than* $\dim M - 2$, *and* $\frac{1}{4}\dim M$, *respectively. If*

$$r(M) \leq \begin{cases} \dim M - \nu(\dim M), & for S^m(1)), \\ \dim M - 1, & for CP^m(1) \end{cases}$$

then M is a totally geodesic submanifold.

In the above statement $\nu(n)$ is defined as in [Fe 2]:

$$\nu(n) = \max\{t : \rho(n-t) \geq t+1\}.$$

Some values of $\nu(n)$ are given by

$$\nu(n) = n - (\text{highest power of } 2 \leq n), \quad \text{for } n \leq 24,$$

in particular, $\nu(n) \leq 8d - 1$ for $n < 16^d$ and $\nu(2^d) = 0$. For example, $\nu(1) = 0,\ \nu(2) = 0,\ \nu(3) = 1,\ \nu(4) = 0,\ \nu(5) = 1,\ \nu(6) = 2,\ \nu(7) = 3,\ \nu(8) = 0$, etc.

For a stronger inequality for the Ricci curvature, but only at one point $m \in M$ do we have the following theorem (compare with Corollary 6.9 in Chapter VI and [Bor 4] where $Ric_M \geq \dim M - 1$).

5.36 Theorem. *Let $M \subset S^m(1)$ be a complete submanifold and suppose that for some point $m \in M$ with rank $r(m) = r(M) <$* $\dim M$, *the Ricci curvature Ric_M is not smaller than* $\dim M - 1$. *Then M is a great sphere.*

For a submanifold $M(k) \subset \bar{M}(k)$ we have $\mu(M) \geq \dim M - \operatorname{codim} M$ [CK]. In view of Theorem 5.36, an isometric immersion between complete space forms $M(k) \subset \bar{M}(k)$ with $k > 0$ and $\dim \bar{M} < 2\dim M$ is totally geodesic [Fe 5], [Bor 10]. For the case of $Ric_M \geq \overline{Ric}_{|M}$, Theorem 5.36 (and an analogous theorem for a totally real submanifold in CP^N) is proved in [Ab 4].

The proof of Theorem 5.36 (given below) is based on the following lemma.

5.37 Lemma. *Let M be a complete parabolic submanifold in a sphere S^m. Then for any unit normal vector ξ with maximal rank $r(\xi)$ equal to $r(M) <$* $\dim M$ *the second quadratic form A_ξ has zero signature (the same number of positive and negative eigenvalues).*

An analogous fact for strongly parabolic submanifolds with positive index of relative nullity $\mu(M)$ is proved in [DG 1].

Proof. Let $\xi_0 \in T_m M^\perp$ be any unit normal vector with rank $r(\xi) = r(M)$ and let $L_0 \subset M$ be a $(\dim M - r(M))$-dimensional great sphere (a ruling), along which the normal vector ξ_0 is stationary and consequently the matrix A_ξ has constant rank and signature (see Corollary 5.14). We extend the normal vector ξ_0 to a smooth field of unit normal vectors $\{\xi\}$ on $U \subset M$, the transversal to L_0 $r(M)$-dimensional neighborhood of a point m, $T_m U \bot T_m L_0$. We may assume that $r(\xi) = r(M)$. Let $\{L\}$ be a family of $(\dim M - r(M))$-dimensional rulings on M, which are obtained from $\{\xi\}$ in a way analogous to that of the ruling L_0. Then a neighborhood $W \subset M$ of a point m has a smooth foliation by parts of $\{L\}$. Let $-W$ be the antipodal set for W on S^N. It is obvious that $-W$ is an open $(\dim M)$-dimensional submanifold, a neighborhood of the point $-m$ in M. Thus

$$A_{\xi_m} = A_{-\xi_m} = -A_{\xi_{-m}}. \tag{5.23}$$

As was noted above, the quadratic forms A_{ξ_m} and $A_{\xi_{-m}}$ have equal signatures. Thus from (5.23) it follows that the quadratic form A_{ξ_m} has zero signature. □

Proof of Theorem 5.36. Assume on the contrary that the rank $r(M)$ of a submanifold M is positive. In view of the given condition for Ricci curvature, the second fundamental form of a mean curvature vector at the point m is positive definite on the subspace orthogonal to the vectors with strong parabolicity (see the proof of Lemma 6.1). Consequently, the parabolicity at point m is equal to strong parabolicity: $\mu(m) = \dim M - r(m)$, and for all normal vectors $\xi \in T_m M^\perp$ with rank $r(\xi) = r(M)$ which are close to a mean curvature vector, the inequality $A_\xi \geq 0$ holds. This contradicts Lemma 5.37. Thus $r(M) = 0$ and M is a totally geodesic submanifold. □

The simplest ruled submanifold in Euclidean space is a cylinder $f(M_1 \times \mathbb{R}^\nu) \subset \mathbb{R}^N$. There are no cylinders in CROSS.

5.38 Definition. The *Segre embedding* $M^{\nu+n} = f(S^\nu \times S^n) \subset S^{\nu+n+\nu n}$ is given by $f : (u_1, \dots, u_\nu, v_1, \dots, v_n) \to (u_1 v_1, \dots, u_i v_j, \dots, u_\nu v_n)$, where $\{u_i\}$ and $\{v_j\}$ are the coordinates of points on the unit spheres S^ν and S^n. The *complex Segre embedding* is given by an analogous rule.

In such spaces (CROSS) the Segre embedding serves as a standard example of a ruled submanifold that is a metric product. The rulings of the Segre embedding may be chosen in two canonical ways.

5.39 Example. *The simplest Segre embedding, i.e., Clifford torus* $M^2 = f(S^1 \times S^1) \subset S^3$, is a surface with flat metric.

In Chapters VI and VII we will generalize the extrinsic method of Theorem 5.34 and give the tests for Segre embedding.

Note that any smooth *spherical fibration* (in the topological sense) $S^k \subset E \xrightarrow{\pi} B$ with a compact base B admits a smooth ruled embedding into a sphere S^N with large N, i.e., an embedding whose fibers are great k-dimensional spheres (rulings) [Pet]; such embeddings are called *fibrations by great spheres.* There are infinitely many topologically inequivalent smooth 3-sphere fibrations of the 7-sphere (whose bases are not diffeomorphic to a standard 4-sphere) [Mln]. Such fibrations may be realized as ruled submanifolds $M^{3+4} \subset S^{N+7}$ with large N. But for $N = 0$ it was shown in [GWY] that every smooth fibration of S^7 by great 3-spheres is topologically equivalent to the Hopf fibration, that is, a topological structure of a compact ruled submanifold depends on the codimension of an immersion into a sphere.

If $M \subset S^N$ is foliated by great spheres, then there is a hierarchy of three questions, stated here in increasing order of difficulty [Pet]: 1) *Given two such foliations, are they topologically equivalent?* 2) *If they are topologically equivalent, is it possible to deform one into another through a one-parameter family of such foliations?* 3) *What is the homotopy type of the space of all such foliations*?

These questions were completely answered for great circle foliations of the 3-sphere [GW] (see Section A.1). The more complicated case where $M = S^p \times S^q$ is embedded into S^{p+q+1} in the standard way is considered in [Pet] and some special cases are studied completely. For example, if p or q is even, then $M = S^p \times S^q$ admits no great circle foliations.

5.40 Theorem. [Pet] *The space of all oriented great 3-sphere foliations of* $M = S^3 \times S^3$ *embedded in* S^7 *is a deformation retract of the "subspace of Hopf fibrations" and has the homotopy type of the disjoint union of four copies of* RP^3.

In the course of proving Theorem 5.40 it was also shown that every such foliation *has an orthogonal pair of leaves and can be extended to a smooth great 3-sphere foliation of* S^7.

5.2.3. Submanifolds with generators in a Lobachevsky space

Foliations (fibrations) on a Lobachevsky space $\mathbb{H}^n$ are investigated in [Bro], [Eps], [KW], [EG], [Gor], [Pos], [Ze 1] and [BW]. Ruled submanifolds in $\mathbb{H}^n$ are investigated in [AH], [Nom 2], [Fe 4], [Th 4], [Bor 7], [Lis], [BDJ], [BD], [AP], [CaM], [Bro] and [Bas].

There are many strongly parabolic submanifolds in $\mathbb{H}^n$; recall that they have a ruled structure. Any strong parabolic submanifold in a Euclidean space $\mathbb{R}^n$ with singular points outside of the unit ball corresponds (by using Cayley-Klein interpretation) to a complete strongly parabolic submanifold in $\mathbb{H}^n$. In [Th 3] the nondeveloped ruled submanifolds $M^{\nu+1} \subset \mathbb{H}^n$ with ν-dimensional ruling are investigated.

K. Abe and A. Haas [AH] are interested in classifying all isometric immersions of a hyperbolic space form $\mathbb{H}^{n-1}$ into another hyperbolic space form $\mathbb{H}^n$. Since the rank of the shape operator A is at most one, the subset $U = \{x \in \mathbb{H}^n : \operatorname{rank}(A) = 1\}$ consists of at most countably many open connected components in $\mathbb{H}^n$, each of which is C^∞-foliated by complete totally geodesic hypersurfaces (the relative nullity foliation) $\{L_\alpha\}_{\alpha\in A}$.

5.41 Definition. [AH] Let U be an open subset of $\mathbb{H}^n$ whose connected components are U_i, $(i = 1, 2, \dots)$, each of which is endowed with a C^∞-foliation $\{L_\alpha\}_{\alpha\in A_i}$ by complete totally geodesic hypersurfaces. Let $\{L_\alpha\}_{\alpha\in A}$ be the disjoint union of $\{L_\alpha\}_{\alpha\in A_i}$, where $A = \cup A_i$. The triple $(U, \{L_\alpha\}_{\alpha\in A}\mathbb{H}^n - U)$ is called a *C^∞-lamination* on $\mathbb{H}^n$.

5.42 Theorem. [AH] *Given a differentiable lamination on a hyperbolic plane $\mathbb{H}^2$, there is a family of isometric immersions of $\mathbb{H}^2$ into a hyperbolic space $\mathbb{H}^3$ such that the induced relative nullity foliations are determined by the lamination.*

For any smooth curve orthogonal to the generators of a strongly parabolic submanifold $M \subset \mathbb{H}^n(-1)$ with $\mu(M) = \dim M - 1$, the normal curvature is ≤ 1 [Fe 2].

There exist strongly parabolic submanifolds in $\mathbb{H}^N$, which are analogous to cylinders in $\mathbb{R}^N$.

5.43 Definition. [Bor 7] Let $\mathbb{H}^{l+p-\nu}$ be a horosphere, either an equidistant submanifold or a totally geodesic submanifold in $\mathbb{H}^{l+p-\nu+1}$, and a complete submanifold $F^{l-\nu}$ in $\mathbb{H}^{l+p-\nu}$. The union of the complete geodesics orthogonal to $\mathbb{H}^{l+p-\nu}$ in $\mathbb{H}^{l+p-\nu+1}$ form a submanifold $F^{l-\nu+1}$. The union of the subspaces

$\mathbb{H}^{\nu-1}$ orthogonal to $\mathbb{H}^{l+p-\nu+1}$ through all points of $F^{l-\nu+1}$ form a complete ruled submanifold $M^l \subset \mathbb{H}^{l+p}$ with ν-dimensional ruling, called a *cylindrical submanifold.*

The cylindrical isometric immersions of $\mathbb{H}^2$ into a hyperbolic space $\mathbb{H}^3$ are investigated in [Nom 2]; see also [AP].

A cylindrical submanifold M^l in $\mathbb{H}^{l+p}$ is strongly parabolic with index of relative nullity $\mu(M) \geq \nu$; the orthogonal distribution (to rulings) is integrable, and each base $F^{l-\nu}$ is totally umbilic. If a base $F^{l-\nu}$ belongs to an equidistant or a totally geodesic submanifold (first case), then M^l is represented by a cylinder with a $(\nu - 1)$-dimensional generator over a cone submanifold $F^{l-\nu+1}$ with vertex outside of a cylinder when the vertex is ∞. If a base belongs to a horosphere (second case), then M^l is represented by a cylinder with a $(\nu-1)$-dimensional generator over a cone submanifold $F^{l-\nu+1}$ with the vertex on the absolute.

The normal curvature k_n of any geodesic on a base is constant and $0 \leq k_n < 1$ for the first case, or $k_n = 1$ for the second case, when $F^{l-\nu}$ lies on a horosphere. The metric of a cylindrical submanifold M^l in $\mathbb{H}^{l+p}$ is semi-reducible and may be written in the form

$$ds^2 = ds_1^2 + \varphi^2(x_1, \ldots, x_\nu) ds_2^2, \tag{5.24}$$

where ds_1^2 is a metric in $\mathbb{H}^\nu$ which depends on the variables $x_1, \ldots, x_\nu$; ds_2^2 is a metric on some base $F^{l-\nu}$ which depends on the variables $x_{\nu+1}, \ldots, x_l$; and the function φ is equal to $\exp(-x_1)$ or $\cosh(x_1)$ in the special coordinate system [Bor 7].

Next we consider the relationship between cylindrical submanifolds and metrics.

5.44 Definition. [Bor 7], [Lis] A complete metric on a Riemannian manifold M with constant index of (intrinsic) k-nullity $\nu_k(M) > 0$ (see 2.12 for examples) is called *k-cylindrical,* if there exists a $(\dim M - \nu_k(M))$-dimensional submanifold orthogonal to a distribution of k-nullity, which for $\nu_k(M) < \dim M - 1$ is totally umbilic and for $\nu_k(M) = \dim M - 1$ is a curve with constant curvature.

5.45 Theorem. [Bor 7] *A k-cylindrical metric with $k < 0$ is foliated by $\nu_k(M)$-dimensional complete totally geodesic submanifolds (k-nullity foliation) and may be written in the form* (5.24) *with the function φ equal to* $\exp(-x_1)$ *or* $\cosh(x_1)$.

5.46 Theorem. [Bor 7] *Let $M^l \subset \mathbb{H}^n(-1)$ be a strongly parabolic submanifold with $l > \mu(M) > 0$ and suppose there exists a submanifold orthogonal to the relative nullity distribution, which, for $\mu(M) < l-1$, is totally umbilic and for $\mu(M) = l-1$ is a curve with constant curvature. Then M is a cylindrical submanifold that is represented (in a Cayley-Klein interpretation) by a cylinder with a $\mu(M)$-dimensional generator over a cone submanifold with the vertex on the absolute or outside of it.*

In [Lis], the subclass of strongly parabolic metrics, which corresponds in a one-to-one fashion to some subclass of strongly parabolic submanifolds in a Lobachevsky space, is studied.

5.47 Theorem. [Lis] *Let M^l have a k-cylindrical metric with $k < 0$ and suppose that one of the following conditions holds:*

1) $\varphi = \exp(-x_1)$ *and a base ds_2^2 admits an isometric embedding into $\mathbb{R}^{l-\nu+p}$ with zero index of relative nullity,*

2) $\varphi = \cosh(x_1)$ *and a base ds_2^2 admits an isometric embedding into $\mathbb{H}^{l-\nu+p}$ with zero index of relative nullity.*

Then M admits an isometric embedding into $\mathbb{H}^{l+p}$ as a cylindrical submanifold (with a ν-dimensional generator and the same codimension as for an embedding of a base).

5.3. Submanifolds with nonpositive extrinsic q-Ricci curvature

5.3.1. Radius of immersed manifold and nonexistence of immersions

For a symmetric bilinear map $h : \mathbb{R}^n \times \mathbb{R}^n \to \mathbb{R}^p$ and an orthonormal system of $q+1$ vectors $\{x_i\}_{0 \le i \le q} \subset \mathbb{R}^n$, we define the *extrinsic q-Ricci curvature* by

$$Ric_h^q(x_0, x_1, \dots, x_q) = \sum_{i=1}^{q} [(h(x_0, x_0), h(x_i, x_i)) - h^2(x_0, x_i)],$$

where $(\ ,\ )$ is a scalar product in $\mathbb{R}^p$. For $q = 1$ it is called the *extrinsic sectional curvature*, see [Bor 1,2], [BS] and [CK]. In view of the equality

$$Ric_h^q(x_0, x_1, \dots, x_q) = \sum_{i=1}^{q} Ric_h^1(x_0, x_i)$$

the inductive formula analogous to (3.34')

$$Ric_h^{q+1}(x_0, x_1, \dots, x_{q+1}) = \frac{1}{q}\sum_{i=1}^{q+1} Ric_h^q(x_0, x_1, \dots, \widehat{x_i}, \dots, x_{q+1})$$

holds, where the symbol $\widehat{\ }$ means the absence of some vector. Thus for every $1 \le q < n-1$ it follows from $Ric_h^q \le 0$, that $Ric_h^{q+1} \le 0$, and from $Ric_h^q \equiv 0$, it follows that $Ric_h^1 \equiv 0$.

5.48 Lemma. *([Ots] for $q = 1$) Let $h : \mathbb{R}^n \times \mathbb{R}^n \to \mathbb{R}^p$ be a symmetric bilinear map with $Ric_h^q \le 0$. Then any subspace $V \subset \mathbb{R}^n$ with $\dim V > p + q - 1$ contains an asymptotical vector, i.e., there exists a nonzero $x \in V$ such that $h(x, x) = 0$.*

Proof. Assume on the contrary that there exists a subspace $V \subset \mathbb{R}^n$ with $\dim V = n_1 > p+q-1$ and without asymptotical vectors. If $h_1 : V \times V \to \mathbb{R}^p$ is the restriction of h, then the function $f(x) = h_1^2(x, x)$ is positive on the unit sphere S^{n_1-1}. The minimum of f is reached at some unit vector x_0. Let $F(x) = f(x) - \lambda(x, x)$. From the necessary condition for extremum applied to the vector x_0, we obtain

$$\begin{aligned} \text{a) } \frac{1}{2} dF(x_0)x &= 2(h_1(x_0, x_0), h_1(x_0, x)) - \lambda(x_0, x) = 0, \\ \text{b) } \frac{1}{2} d^2F(x_0)(x, x) &= 2(h_1(x_0, x_0), h_1(x, x)) + 4h_1^2(x_0, x) - \lambda(x, x) \ge 0, \end{aligned} \tag{5.25}$$

where $x \in V$ is an arbitrary vector. In view of (5.25 a) the subspace $V_1 = \{x \in V : h(x_0, x) = 0\}$ is orthogonal to x_0 and $\dim V \ge n_1 - p$. Since $n_1 - p \ge q$, we can take the orthonormal system of q vectors $\{x_i\}_{1 \le i \le q} \subset V_1$. From $h(x_0, x_i) = 0$ and (5.25 b) it follows that $(h(x_0, x_0), h(x_i, x_i)) \ge \frac{1}{2}\lambda > 0$, which contradicts the given inequality $Ric_h^q \le 0$:

$$Ric_h^q(x_0, x_1, \dots, x_q) = \sum_{i=1}^{q}(h(x_0, x_0), h(x_i, x_i)) \ge \frac{q}{2}\lambda > 0. \qquad \square$$

Lemma 5.48 can be extended for Hilbert's case $n = \infty$, see [Okr 1] for $q = 1$.

Note that for a submanifold $M \subset \bar{M}$ with the second fundamental form h

and an orthonormal system of $q+1$ vectors $\{x_i\}_{0\le i\le q} \subset T_m M$, we have

$$Ric^q(x_0,\ldots,x_q) - \overline{Ric}^q(x_0,\ldots,x_q) = \sum_{i=1}^{q}[(h(x_0,x_0),h(x_i,x_i)) - h^2(x_0,x_i)]. \tag{5.26}$$

In particular, for a submanifold M in a Euclidean space, we have the following equality

$$Ric^q(x_0,x_1,\ldots,x_q) = Ric^q_h(x_0,x_1,\ldots,x_q). \tag{5.26$'$}$$

Consider the application of Lemma 5.48 to submanifolds.

J. Moore [Moo 3] proved that when $\bar{M}$ is a complete simply connected Riemannian space with sectional curvature $a \le \bar{K} \le b \le 0$ and M is a compact Riemannian manifold with $K \le a-b$, then M possesses no isometric immersion in $\bar{M}$ unless $\dim\bar{M} \ge 2\dim M$ (for $b=0$ see [O'N 6]). On the other hand, H. Jacobowitz [Jac] showed that an isometric immersion of an n-dimensional compact Riemannian manifold, with sectional curvature smaller than $\frac{1}{d^2}$, into $\mathbb{R}^{2n-1}$ can never be contained in a ball of radius d. T. Ishihara [Is] proved a nonembedding theorem, using the Otsuki Lemma which generalized both these results. Below we generalize the result by [Is] (and hence by [Moo 3] and [Jac]) for the case of partial Ricci curvature.

In [Is], the positive continuous function $C(b,d)$, $(b \le 0,\ d > 0)$, is defined as follows :

$$C(b,d) = \begin{cases} \frac{1}{d} & if\ b=0, \\ \sqrt{-b}\coth(d\sqrt{-b}) & if\ b<0. \end{cases}$$

This function is monotonically decreasing with respect to b as well as d; obviously, $C^2(b,d) > -b$. Let M be a compact submanifold in $\bar{M}$ with distance function $\bar{d}$. For any point $x \in \bar{M}$ and any $r>0$ set $B(x,r) = \{y \in \bar{M} : \bar{d}(x,y) \le r\}$. Then we let

$$r(M) = \inf\{r : M \subset B(x,r)\} = \inf\{\max\{\bar{d}(x,y) : y \in M\},\ x \in \bar{M}\}.$$

There is a point $x_0 \in \bar{M}$ such that $M \subset B(x_0, r(M))$. Moreover, there is a point $y_0 \in M$ such that $\bar{d}(x_0,y_0) = r(M)$. We will call $r(M)$ the *radius of M* in $\bar{M}$ and $B(x_0, r(M))$ a *minimum ball containing M* [Is]. Generally, there are several minimal balls containing M. For example, there are two balls (semispheres) when M is a great circle in a 2-sphere $\bar{M}$. But there is only one minimal ball for a compact manifold immersed in a Euclidean space [Is].

5.49 Theorem. ([Is] *for* $q = 1$) *Let* $\bar{M}^{n+p-q}$ *be a complete simply connected Riemannian space with sectional curvature* $a \le \bar{K} \le b \le 0$ *and* M^n *a compact Riemannian manifold with diameter* $< d$. *Assume that at every point of* M^n *there is a p-dimensional subspace in the tangent space, along which the partial Ricci curvature*

$$\frac{1}{q} Ric^q_M \le a + C^2(b, d). \tag{5.27}$$

Then M^n *cannot be isometrically immersed in* $\bar{M}^{n+p-q}$.

This theorem is an immediate consequence (a compact submanifold $M \subset \bar{M}$ with diameter d should be contained in some ball of radius d) of the following theorem.

5.50 Theorem. ([Is] *for* $q = 1$) *Let* $\bar{M}^{n+p-q}$ *satisfy the same conditions as in Theorem* 5.49 *and let d be a positive constant. Let* M^n *be a compact Riemannian manifold such that at every point of* M^n *there is a p-dimensional subspace in the tangent space along which inequality* (5.27) *holds. Then no isometric immersion of* M^n *into* $\bar{M}^{n+p-q}$ *is contained in a ball of radius* $< d$.

Proof of Theorem 5.50. Suppose that M is isometrically immersed in $\bar{M}$ and contained in a ball of radius $< d$. Then we have $r(M) < d$. Take $p_0 \in \bar{M}$ and $p_1 \in M$ satisfying $r(M) = \bar{d}(p_0, p_1)$. Let $\gamma : [0, 1] \to \bar{M}$ be a minimal geodesic with $\gamma(0) = p_0$, $\gamma(1) = p_1$, see Figure 20. For each unit tangent vector $y \in T_{p_1}M$, there is a unique Jacobi vector field Y along γ such that $Y(0) = 0$ and $Y(1) = y$. Corresponding to Y, we have a one-parameter family of geodesics from p_0 to M, specifically, $\gamma_s(t) = \gamma(s, t)$, $(|s| < \varepsilon,\ t \in [0, 1])$, satisfying $\gamma_0(t) = \gamma(t)$, $\frac{\partial}{\partial s}\gamma(0, t) = Y(t)$. We denote by $\mathcal{E}(\gamma_s) = \frac{1}{2}\int_0^1 (\gamma_s', \gamma_s')\, dt$, the energy of γ_s. Then from the definition of γ it follows that $\mathcal{E}(\gamma_s) \le \mathcal{E}(\gamma)$. Hence

$$0 \ge \left(\frac{d^2}{ds^2}\mathcal{E}(\gamma_s)\right)_{s=0} = I(Y, Y) + (h(y, y), \gamma'(1)), \tag{5.28}$$

where $I(\ ,\ )$ is the index form, see Section 3.1. Taking a proper Jacobi field on a space of constant curvature b, Moore [Moo 3] proved that

$$I(Y, Y) \ge r(M)C(b, r(M)).$$

Hence from (5.28) we get $(h(y, y), \gamma'(1)) < -r(M)C(b, r(M))$. Since the function $C(b, d)$ is monotonically decreasing, it follows that $C(b, r(M)) >$

$C(b, d)$. In view of $|\gamma'(1)| = r(M)$, we obtain for all unit vectors $y \in T_{p_1}M$ the inequality

$$|h(y, y)| \geq C(b, r(M)) > C(b, d).$$

Since the second fundamental form h at p_1 has no asymptotical vectors on V, it follows that, as in the proof of Lemma 5.48, we get $q + 1$ orthonormal vectors $x_0, \dots, x_q$ in $V \subset T_{p_1}M$ satisfying the properties

$$h(x_0, x_i) = 0, \; h(x_0, x_0) \cdot h(x_i, x_i) \geq \frac{1}{2}\lambda > C^2(b, d).$$

From this it follows that

$$Ric_h^q(x_0, \dots, x_q) > qC^2(b, d).$$

On the other hand, for any $q + 1$ orthonormal vectors $x_0, \dots, x_q$ in a given p-dimensional subspace $V \subset T_{p_1}M$, by the conditions of the theorem, we have

$$Ric_h^q(x_0, \dots, x_q) = Ric_M^q(x_0, \dots, x_q) - Ric_{\bar{M}}^q(x_0, \dots, x_q) \leq qC^2(b, d),$$

which is a contradiction. □

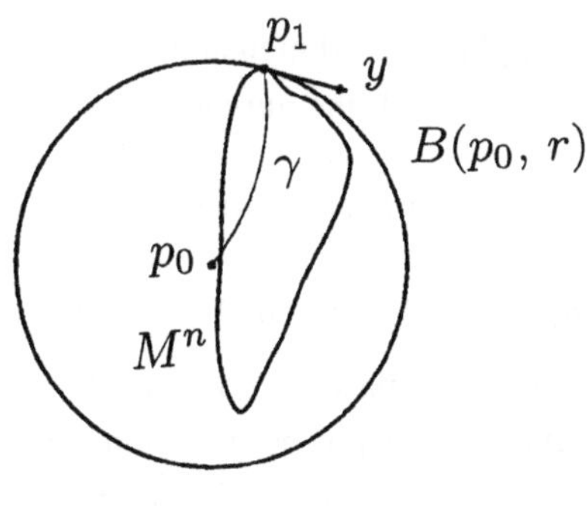

Figure 20.

5.51 Corollary. *a) Let M^n be a compact Riemannian manifold with nonpositive q-Ricci curvature. Then M^n cannot be isometrically immersed into $\mathbb{R}^{2n-q}$.*

b) Let M^n be a Riemannian manifold with negative q-Ricci curvature. Then M^n cannot be isometrically immersed into $\mathbb{R}^{2n-q-1}$.

For $q = 1$ Corollary 5.51 is proved in [CK]; for some interpretation see [BS].

Proof. a) Assume on the contrary that $M^n \subset \mathbb{R}^{n+p}$, where $p \leq n-q$. By Lemma 5.48 for each point $m \in M$ there exists an asymptotical vector $x \in T_m M$. But for a compact submanifold $M^n \subset \mathbb{R}^{n+p}$, there exists a point $m_0 \in M$ and a unit normal vector $\xi_0 \in T_{m_0}M^{\perp}$ with positive second quadratic form A_{ξ_0} [KN]. Hence, there are no asymptotical vectors at the point m_0, which is a contradiction.

b) Assume on the contrary that $M^n \subset \mathbb{R}^{n+p}$, where $p < n-q$. By Lemma 5.48, for any point $m \in M$ there exists a unit asymptotical vector $x_0 \in T_m M$. Note that the subspace $V = \{x \in T_m M : h(x_0, x) = (x_0, x) = 0\}$ has dimension $\dim V \geq q$. Then for any orthonormal system of q vectors $\{x_i\}_{1 \leq i \leq q} \subset V$ we have $Ric^q(x_0, x_1, \dots, x_q) = Ric^q_h(x_0, x_1, \dots, x_q) = 0$, which is a contradiction. □

Note that the product $M^{2n} = M^n(-1) \times M^n(-1)$ of hyperbolic space forms has sectional curvature $K_M \in [-1, 0]$ and constant Ricci curvature $Ric_M = -(n-1)$; hence $Ric^{n+1}_M \leq -1$ and $Ric^n_M \leq 0$. By Corollary 5.51, such M^{2n} cannot be isometrically locally immersed into $\mathbb{R}^{3n-2}$.

Another generalization $\bar{\gamma}_s$ of the notion of extrinsic sectional curvature (as the intrinsic analog of the s-dimensional sectional curvature γ_s, in particular, as the Lipschitz-Killing curvature of a Riemannian manifold) is given in [Bor 1].

5.52 Definition. [Bor 1] For an s-dimensional subspace $V \subset T_p M$ containing an orthonormal basis $\{e_1, \dots e_s\}$, the number $\bar{\gamma}_s(V)$ is defined by the formula

$$\bar{\gamma}_s(V) = \frac{1}{2^{s/2}k!}\Big(\sum_{i,j \in S_s} \varepsilon_i \varepsilon_j [(h(u_{i_1}, u_{j_1}), h(u_{i_2}, u_{j_2})) - (h(u_{i_1}, u_{j_2}), h(u_{i_2}, u_{j_1}))]$$
$$\dots \times [(h(u_{i_{s-1}}, u_{j_{s-1}}), h(u_{i_s}, u_{j_s})) - (h(u_{i_{s-1}}, u_{j_s}), h(u_{i_s}, u_{j_{s-1}}))]\Big).$$

where $\{u_i\}$ is an orthonormal base of normals, S_s is the set of permutations of order s and ε_i is the sign of the permutation $i = (i_1, \dots, i_s)$. The number $\bar{\gamma}_s(M) = \sup\{\bar{\gamma}_s(V) : V \subset TM, \dim V = s\}$ is called the *extrinsic s-dimensional curvature* of M.

If $s = 2$, then for a submanifold with nonpositive extrinsic s-dimensional curvature, the inner sectional curvature is not greater than the sectional curvature of the ambient space. If the ambient space is Euclidean, then the extrinsic s-dimensional curvature coincides with the intrinsic one, see [CK].

The next result generalizes the Otsuki Lemma in a different way from that of Lemma 5.48.

5.53 Lemma. [Bor 1] *Let $h : \mathbb{R}^n \times \mathbb{R}^n \to \mathbb{R}^p$ be a symmetric bilinear map satisfying the property $\bar{\gamma}_s(V) \le 0$ for all $V \subset \mathbb{R}^n$ with* $\dim V = s$, *where* (,) *is a scalar product in $\mathbb{R}^p$. If h is without asymptotical vectors, then $p \ge \frac{n-1}{s-1}$.*

The proof of Lemma 5.53 is similar to the proof of Lemma 5.48. As a corollary, Borisenko [Bor 1] obtained that a) a compact Riemannian manifold M^n with $\gamma_s \le 0$ cannot be isometrically immersed into $\mathbb{R}^{n+p}$, where $p \le \frac{n-1}{s-1}$, b) a Riemannian manifold M^n with $\gamma_s < 0$ cannot be locally isometrically immersed into $\mathbb{R}^{n+p}$ where $p \le \frac{n-1}{s-1} - 1$.

A submanifold $M^n \subset \bar{M}^{n+p}$ with $\bar{\gamma}_s(M) \le 0$ and small codimension is (strongly) parabolic [Bor 1]:

$$\mu(M) \ge n - (s-1)p(p+1), \quad r(M) \le 2(s-1)p.$$

5.3.2. The index of relative nullity

5.54 Lemma. ([Flo 2] *for $q = 1$) Let $h : \mathbb{R}^n \times \mathbb{R}^n \to \mathbb{R}^p$ be a symmetric bilinear map with $Ric_h^q \le 0$. Then $\mu(h) \ge n - 2p - q + \delta_{1q}$.*

Note that from $Ric_h^q \equiv 0$, where $q < n-1$, it follows that $Ric_h^1 \equiv 0$ and then $\mu(h) \ge n - p$ [CK]. Lemma 5.54 is a bridge to submanifolds with generators because from it we obtain the estimate for the index of relative nullity $\mu(M)$ of a submanifold $M^n \subset \bar{M}^{n+p}$ with nonpositive extrinsic q-Ricci curvature and "small" codimension (for $q = 1$, see [Flo 1,2] and [Bor 1]):

$$\mu(M) \ge n - 2p - q + \delta_{1q}. \tag{5.29}$$

The proof of Lemma 5.54 follows immediately from the lemmas given below.

5.55 Definition. [Moo 2] We say that $y \in \mathbb{R}^n$ is a *regular element* of a bilinear map $\beta : \mathbb{R}^n \times \mathbb{R}^m \to \mathbb{R}^p$, if $\dim Im\,\beta(y) = \max\{\dim Im\,\beta(z) : z \in \mathbb{R}^n\}$. We denote the set of regular elements of β by $RE\ (\beta)$.

Note that the set $RE\ (\beta)$ is open and dense in $\mathbb{R}^p$.

5.56 Lemma. [Moo 2] *Let $\beta : \mathbb{R}^n \times \mathbb{R}^m \to \mathbb{R}^p$ be a bilinear map and $y_0 \in RE\ (\beta)$. Then $\beta(y, \ker(\beta(y_0)) \subset Im\,\beta(y_0)$ for all $y \in \mathbb{R}^n$.*

Proof. Let $z_1, \dots, z_r$ be vectors in $\mathbb{R}^m$ with $r = \dim Im\,\beta(y_0)$ and $Im\,\beta(y_0) = span\{\beta(y_0, z_j),\ 1 \le j \le r\}$. It is easy to see that the vectors $\{\beta(y_0 + ty, z_j)\}_{1 \le j \le r}$ are linearly independent for all but a finite number of values of t. Hence, they generate a family of r-dimensional subspaces that varies continuously with t if $|t| < \varepsilon$ for some $\varepsilon > 0$. But if $z \in \ker \beta(y_0)$, then $\beta(y_0 + ty, z) = t\,\beta(y, z)$. Therefore, by continuity, $\beta(y, z) \in Im\,\beta(y_0)$. □

We denote the set of asymptotic vectors of h by $As(h)$. Recall, that $T \subset \mathbb{R}^n$ is an *asymptotic subspace* of h if $h(x, y) = 0$ for all $x, y \in T$.

The following result generalizes Lemma 5.48.

5.57 Lemma. ([Flo 2] *for* $q = 1$) *Let* $h : \mathbb{R}^n \times \mathbb{R}^n \to \mathbb{R}^p$ *be a symmetric bilinear map with* $Ric_h^q \leq 0$. *Then there exists an asymptotic subspace* $T \subset \mathbb{R}^n$ *of* h *such that* $\dim T \geq n - p - q + \delta_{1q}$.

Proof. If $n - p - q + \delta_{1q} = 1$, then by Lemma 5.48 there exists an asymptotic vector, and this concludes the proof. Thus assume $n - p - q + \delta_{1q} \geq 2$. For each vector $x_0 \in As(h)$, a regular element for h, we define a linear transformation $h(x_0) : \mathbb{R}^n \to \mathbb{R}^p$ by $h(x_0)y = h(x_0, y)$, set $V_1 = V_1(x_0) = \ker h(x_0)$ and $W_1 = W_1(x_0) = \{Im\,h(x_0)\}^{\perp}$, and define $h_1 = h_{|V_1 \times V_1}$. With the above notation we claim that $Im\,h_1 \subset W_1$.

From the above assumption, it follows that $\dim V_1 \geq n - p \geq q + 2 - \delta_{1q}$. To prove the claim, take the orthonormal system of vectors $\{z_i\}_{\delta_{1q} \leq i \leq q} \subset V_1$, which are orthogonal to x_0. Then, since $x_0 \in As(h)$, we have for any unit $y \in \mathbb{R}^n$ and for all t,

$$(h(x_0 + ty, x_0 + ty), h(z_i, z_i)) - (h(x_0 + ty, z_i), h(x_0 + ty, z_i))$$
$$-2t(h(x_0, y), h(z_i, z_i)) + t^2[(h(y, y), h(z_i, z_i)) - (h(y, z_i), h(y, z_i))].$$

We can assume that $y \perp V_1$ (see the coefficient for t). Thus, there is a unit vector $x_t = \frac{1}{\sqrt{1+t^2}}(x_0 + ty)$ that is orthogonal to each z_i. Hence for all t,

$$(1 + t^2)Ric_h^q(x_t, z_1, \ldots, z_q)$$
$$= 2t(h(x_0, y), \sum_{i=1}^{q} h(z_i, z_i)) + t^2 Ric_h^q(y, z_1, \ldots, z_q).$$

In view of $Ric_h^q \leq 0$, we have $(h(x_0, y), \sum_{i=1}^{q} h(z_i, z_i)) = 0$, $(y \in \mathbb{R}^n)$, that is, $\sum_{i=1}^{q} h(z_i, z_i) \subset W_1$. Note that for $q = 1$ we have $h(z, z) \subset W_1$ for all $z \in V_1$, and in view of the symmetry of h, the claim is proved.

So assume $q > 1$. Since the analogous property $\sum_{i=0}^{q-1} h(z_i, z_i) \subset W_1$ holds, we have $h(z_0, z_0) - h(z_q, z_q) \subset W_1$. In the same way we obtain $h(z_0, z_0) - h(z_i, z_i) \subset W_1$ for each i and hence $h(z_0, z_0) = \frac{1}{q}\sum_{i=1}^{q}[h(z_0, z_0) -$

$h(z_i, z_i)] + \frac{1}{q}\sum_{i=1}^{q} h(z_i, z_i) \subset W_1$. Since z_0 is an arbitrary unit vector in V_1, it follows that in view of the symmetry of h that $Im\, h_1 \subset W_1$.

The above claim allows us to proceed inductively as follows. Set $V_0 = \mathbb{R}^n$ and $W_0 = \mathbb{R}^p$. Given $k \geq 0$ for the symmetric bilinear map $h_k = h_{|V_k \times V_k} : V_k \times V_k \to W_k$ with nonpositive $Ric^q_{h_k}$, define

$$r_k = \max\{\dim Im\, h_k(x) : x \in As\,(h_k)\}$$

and suppose that if $k \geq 1$,

$$n_k = \dim V_k = n - \sum_{i=0}^{k-1} r_i, \quad p_k = \dim W_k = p - \sum_{i=0}^{k-1} r_i.$$

Picking $x_k \in As\,(h_k)$ such that $\dim Im\, h_k(x_k) = r_k$, set

$$V_{k+1} = V_{k+1}(x_0, \ldots, x_k) = \ker h_k(x_k)$$

and then $n_{k+1} = \dim V_{k+1} = n - \sum_{i=0}^{k} r_i$. The above claim implies that $Im\, h_{k+1} \subset W_{k+1}$, where $W_{k+1} = W_{k+1}(x_0, \ldots, x_k) = \{Im\, h_k(x_k)\}^{\perp} \subset W_k$ and $h_{k+1} = h_{|V_{k+1} \times V_{k+1}}$. Since

$$0 \leq p_{k+1} = \dim W_{k+1} = p - \sum_{i=0}^{k} r_i,$$

there exists a positive integer m such that $r_m = 0$. This tells us that $As(h_m) = \ker h_m$. Set $T = \ker h_m$. By Lemma 5.48 for each subspace $S \subset V_m$ such that $\dim S \geq p_m + q - \delta_{1q}$, we have that $S \cap T = S \cap As(h_m) \neq 0$. Hence, $\dim T \geq n_m - p_m + q - \delta_{1q} = n - p + q - \delta_{1q}$. Moreover, since $h_m = h_{|V_m \times V_m}$, T is an asymptotic subspace of h. This concludes the proof of Lemma 5.57. □

Proof of Lemma 5.54. Let $T \subset \mathbb{R}^n$ be an asymptotic subspace of h such that $\dim T = n - p - q + \delta_{1q}$, see Lemma 5.57. Then the orthogonal complement T' to T in $\mathbb{R}^n$ has $\dim T' = p + q - \delta_{1q} \geq q + 1 - \delta_{1q}$. Define a bilinear map $\beta : T' \times T \to \mathbb{R}^p$ by $\beta = h_{|T' \times T}$. Take $y_0 \in RE\,(\beta)$, $z \in T$, $z' \in \ker \beta(y_0) \subset T$ and an orthonormal system of vectors $\{y_i\}_{1 \leq i \leq q} \in T'$, which in the case of $q > 1$, are also orthogonal to y_0. Using only the assumption on T, we have for all $s, t \in \mathbb{R}$ that

$$\begin{aligned}
&(h(y_0 + tz, y_0 + tz), h(y_i + sz', y_i + sz')) - h^2(y_0 + tz, y_i + sz') \\
&= (h(y_0, y_0) + 2t\, h(y_0, z), h(y_i, y_i) + 2s\, h(y_i, z')) - h^2(y_0 + tz, y_i + sz').
\end{aligned}$$

Since $h(y_0, z') = 0$, we get that

$$\begin{aligned}(h(y_0+tz, y_0+tz), h(y_i+sz', y_i+sz')) - h^2(y_0+tz, y_i+sz')\\ = \{(h(y_0,y_0), h(y_i,y_i)) - h^2(y_0,y_i)\} - t^2h^2(z,y_i)\\ +2t\{(h(y_0,z), h(y_i,y_i)) - (h(y_0,y_i), h(z,y_i))\}\\ +2s\{(h(y_0,y_0), h(y_i,z')) + 2t(h(y_0,z), h(y_i,z'))\},\end{aligned}$$

which is linear in s. For unit vectors $y(t) = \frac{1}{\sqrt{1+t^2}}(y_0 + tz)$ and $y_i(s) = \frac{1}{\sqrt{1+s^2}}(y_i + sz')$ we then get

$$\begin{aligned}(1+t^2)(1+s^2)Ric_h^q(y(t), y_1(s), \ldots, y_q(s)) = Ric_h^q(y_0, y_1, \ldots, y_q)\\ -t^2\sum_{i=1}^q h^2(z,y_i) + 2t\{(h(y_0,z), \sum_{i=1}^q h(y_i,y_i)) - \sum_{i=1}^q (h(y_0,y_i), h(z,y_i))\}\\ +2s\{(h(y_0,y_0), h(\sum_{i=1}^q y_i, z')) + 2t(h(y_0,z), h(\sum_{i=1}^q y_i, z'))\},\end{aligned}$$

which is linear in s. This implies, in view of $Ric_h^q \leq 0$, that for all $t \in \mathbb{R}$

$$(h(y_0,y_0), h(\sum_{i=1}^q y_i, z')) + 2t(h(y_0,z), h(\sum_{i=1}^q y_i, z')) = 0,$$

which tells us that $(h(y_0,y_0), h(\sum_{i=1}^q y_i, z')) = 0$, and hence $(h(y_0,y_0), h(y,z')) = 0$ for all $y \in T'$. Since z and z' are arbitrary it follows that

$$\beta(y, \ker\beta(y_0)) \perp Im\,\beta(y_0).$$

This, together with Lemma 5.56, tells us that $h(y,x) = 0$ for all $y \in T'$ and $x \in \ker\beta(y_0)$. But since $\ker\beta(y_0) \subset T$, we obtain $\ker\beta(y_0) \subset \ker h$. Then

$$\mu(h) \geq \dim\ker\beta(y_0) = \dim T - \dim Im\,\beta(y_0) \geq n - 2p - q + \delta_{1q},$$

which concludes the proof. □

Dajczer and Rodriquez [DR 2] have shown that *any isometric immersion of a Kähler manifold into a complex space form $\bar{M}^N(c)$, $(c \neq 0)$, with positive index of relative nullity must be holomorphic.* From the proof of that theorem and Lemma 5.54 we get the following corollary.

5.58 Corollary. ([Flo 2] *for* $q = 1$) *Let* M^{2n} *be a Kähler manifold and* $m_0 \in M$ *such that* $Ric^q(m_0) \leq qc$, *where* c *is a positive constant. Then there is no isometric immersion of* M^{2n} *into a real space form* $\bar{M}^{2n+p}(c)$ *for* $p < n-q+\delta_{1q}$.

Proof. Suppose that such an immersion exists and call it f. Composing f with the totally geodesic and totally real inclusion i of the $(2n+p)$-dimensional real space form into a complex space form $C\bar{M}^{2n+p}(c)$, we conclude from Lemma 5.54 that the index of relative nullity $\mu(i \circ f)_{m_0} > 0$. But the proof of the main result in [DR 2] tells us that $T_{m_0}M$ must be invariant with respect to the complex structure J of M. This is a contradiction because the immersion $i \circ f$ is totally real. □

Recall, see Corollary 1.39a), that any isometric immersion of a complete Riemannian manifold M^n into a round sphere S^N with index of relative nullity $\mu(M) \geq \rho(n - \mu(M))$ must be totally geodesic.

From [Ab 3] and [DR 2] it follows that any isometric immersion of a complete Kähler manifold into CP^N with positive index of relative nullity must be totally geodesic. From the above results and Lemma 5.54 we get the following corollary.

5.59 Corollary. ([Fe 2], [Flo 2] for $q = 1$) *a) Let* $f : M^n \to S^{n+p}$ *be an isometric immersion of a complete Riemannian manifold with* $Ric_h^q \leq 0$. *If* $2p < n - \nu(n) - q + \delta_{1q}$ *then* f *is a totally geodesic inclusion.*

b) Let $f : M^{2n} \to CP^{n+p}$ *be an isometric immersion of a complete Kähler manifold with* $Ric_h^q \leq 0$. *If* $2p < n - q + 1$ *then* f *is a totally geodesic inclusion.*

5.3.3. Extremal theorems

The next theorem follows from (5.29), Theorem 5.6 and Corollary 5.7 (see also Theorem 1.38).

5.60 Theorem. *Let* $M^n \subset \bar{M}^{n+p}$ *be a complete curvature invariant submanifold with* $Ric_h^q \leq 0$. *Then* M *is a totally geodesic submanifold if any of the following requirements are satisfied:*

1) M *is compact with* $Ric^s(M) > 0$ *and* $4p \leq n - s - 2q + 2\delta_{1q}$,

2) $\bar{M}$ *is compact with* $Ric^s(\bar{M})_{|M} > 0$ *and* $5p \leq n - s - 2q + 2\delta_{1q}$,

3) $2p < n - \nu(n) - q + \delta_{1q}$ *and for some* $k = const > 0$

$$\bar{R}(x, y)x = -ky(x, x), \quad (x,\ y \in TM). \tag{5.30}$$

See Corollary 1.39 for a submanifold M of the sphere $S^{n+p}(k)$ for which 3) holds with $q = 1$; for a submanifold satisfying a stronger condition (5.13) than (5.8) with $q = s = 1$, see [Bor 5]. An analogous result in the case when 3) is assumed for submanifolds in a Hilbert hypersphere is true without any additional assumption on their finite codimension; for $q = 1$, see [Okr 1].

Proof. The relative nullity (totally geodesic) foliation on the regularity domain has complete leaves and dimension $\mu(M) \geq n - 2p - q + \delta_{1q}$, see Theorem 5.6 and (5.29).

1) From $Ric^s(M) > 0$ it follows that $\mu(M) < n - \mu(M) + s$ (see Section 3.4.1) and hence $4p > n - s - 2q + 2\delta_{1q}$ or M is a totally geodesic submanifold.

2) From $Ric^s(\bar{M})_{|M} > 0$ it follows that $\mu(M) < n + p - \mu(M) + s$ (see Section 3.4.1), and hence $5p > n - s - 2q + 2\delta_{1q}$, or M is a totally geodesic submanifold.

3) Since $K_{mix} = k$, then by Theorem 1.38, $\mu(M) \leq \nu(n)$. Hence $2p \geq n - \nu(n) - q + \delta_{1q}$, or M is a totally geodesic submanifold. □

Theorem 5.61 below follows directly from Theorem 5.60.

5.61 Theorem. *Let $M^n \subset \bar{M}^{n+p}$ be a compact simply connected curvature invariant submanifold. Then M^n is a totally geodesic submanifold isometric to the unit sphere, if one of the following properties holds:*

1) $Ric^s(M) \leq s \leq Ric^s(\bar{M})_{|M}$ *for some* $s < n - 1$, *and* $2p < n - \nu(n) - s + \delta_{1s} - 1$,

2) $K(\bar{M})_{|M} \equiv 1,\ K_M \leq 1,\ inj(M) \geq \pi$ *and* $2p < n - 1$.

The property $inj(M) \geq \pi$ in case 2) follows from the inequality $K_M > 0$ when n is even, and from the inequality $K_M \geq \frac{1}{4}$ when n is odd, see Appendix A.

Proof. 1) Since the extrinsic partial Ricci curvature Ric_h^s is nonpositive, by Lemma 5.54, $\mu(M) > \nu(n)+1 \geq 2$. For any orthonormal vectors $\{x_0, \ldots, x_s\} \subset TM$ with $x_0 \in \ker h$, we have $Ric^s(x_0, x_1, \ldots, x_s) = s = \overline{Ric}^s(x_0, x_1, \ldots, x_s)$ and hence the relative nullity foliation $\{L\}$ on M has $K_L = K_{mix} = 1$. Then from Theorem 5.60, case 3) (see the proof) it follows that $\mu(M) = n$, i.e., M is a totally geodesic submanifold. Since the sectional curvature $K_M = 1$, a simply connected M is isometric to the unit n-sphere.

2) Since the extrinsic sectional curvature is nonpositive, from Lemma 5.54 (with $q = 1$) it follows that $\mu(M) \geq 2$; hence M contains a $\mu(M)$-dimensional

totally geodesic submanifold L (a generator), with $K_L = 1$, which is simply connected by the condition $inj(M) \geq \pi$. Since L contains a closed geodesic γ of M with the length 2π and index $n-1$, M is isometric to the unit sphere, see Theorem A.20 by Toponogov. Since the extrinsic sectional curvature of M is zero, $\mu(M) \geq n-p$ [CK]. From above it follows that $\mu(M) \geq \frac{n}{2}+1$ and by Corollary 5.15 (with $q=1$), $\mu(M)=n$, i.e., M is a totally geodesic submanifold. □

If the curvature of a manifold $\bar{M}$ has stronger restrictions, then we obtain the following extremal theorem.

5.62 Theorem. *Let M^n be a compact curvature invariant submanifold in a complete simply connected Riemannian space $\bar{M}^{n+p}$ satisfying the conditions $\frac{9}{4} \geq K_{\bar{M}} \geq 1$, $Ric^s(M) \leq s$ for some $s < n-1$ and $2p \leq n-s-2+\delta_{1s}$. Then $\bar{M}^{n+p}$ is isometric to the unit sphere.*

Proof. Since the extrinsic curvature Ric_h^s is nonpositive, it follows from Lemma 5.54 that $\mu(M) \geq 2$. For any orthonormal vectors $\{x_0, \dots, x_s\} \subset TM$ with $x_0 \in \ker h$, we have $Ric^s(x_0, x_1, \dots, x_s) = s = \overline{Ric}^s(x_0, x_1, \dots, x_s)$ and hence the relative nullity foliation $\{L\}$ on M has $K_L = K_{mix} = 1$. In particular, $\bar{M}$ contains a $\mu(M)$-dimensional totally geodesic submanifold L (a ruling from M) with sectional curvature one.

5.63 Lemma. [Bor 5] *Let $\bar{M}$ be a compact Riemannian C^4-manifold with curvature $\frac{1}{4} < K_{\bar{M}} \leq 1$. Then a complete totally geodesic L^μ, ($\mu \geq 2$), is simply connected.*

In view of the curvature restrictions on $\bar{M}$ and by Lemma 5.63, the ruling L is simply connected and hence is isometric to the unit sphere $S^{\mu(M)}$. Thus $\bar{M}$ contains a closed geodesic with the length 2π. Let m_1, m_2 and m_3 be the points on this geodesic, which define the triangle with equal sides $\frac{2}{3}\pi$. Since $\bar{M}$ is simply connected and its curvature is restricted as given above, every geodesic with length $\frac{2}{3}\pi$ is the shortest one, see [Top 3].

5.64 Theorem. *([Top 3]) Let $\bar{M}$ be a compact Riemannian C^4-manifold with sectional curvature $K_{\bar{M}} \geq 1$. If there exists a triangle in $\bar{M}$ with perimeter 2π whose sides are the shortest geodesics, then $\bar{M}$ is isometric to the unit sphere.*

Then we apply Theorem 5.64 to this triangle with vertices m_1, m_2, m_3 and obtain that $\bar{M}$ is isometric to the unit sphere. □

The next theorem follows directly from Theorem 5.61 and Theorem 5.62.

5.65 Theorem. *Let M^n be a compact curvature invariant submanifold in a complete simply connected Riemannian space $\bar{M}^{n+p}$ with $\frac{9}{4} \geq K_{\bar{M}} \geq 1$ and suppose that one of the following conditions holds:*

1) $Ric^s(M) \leq s$ *for some* $s < n-1$, *and* $2p < n - \nu(n) - s + \delta_{1s} - 1$,

2) $K_M \leq 1$, $inj(M) \geq \pi$ *and* $2p < n-1$.

Then $\bar{M}^{n+p}$ is isometric to the unit sphere and M^n is a totally geodesic submanifold.

Theorems 5.60, 5.61, 5.62 and 5.65 generalize Theorems 3–6 in [Bor 5], which were first obtained in [Bor 2] for submanifolds in CROSS.

5.4. Ruled submanifolds with conditions on mean curvature

Some authors investigated ruled submanifolds in space forms with additional conditions on its mean curvature vector. The well-known Theorem of E. Catalan states that the only ruled minimal surfaces in $\mathbb{R}^3$ are the plane and the helicoid, see [Law 2] for the case of S^3 and [CDa 4] for the hyperbolic space $\mathbb{H}^3$. These classical results can be generalized in three ways.

The classification of minimal ruled submanifolds $M^{\nu+n}$ with $n = 1$ in space forms was obtained independently by [Lum 3], [Th 2] and [BDJ] (for hypersurfaces in $\mathbb{R}^N$, see [Au 3] and [BV]). They showed that such minimal ruled submanifolds are generalized helicoids. This extends the theorem of Catalan in one direction.

5.66 Definition. [BDJ] Let $c = c(t)$ be a curve in a space form $\bar{M}(k)$ with constant curvature and Frenet frame $e_1, \ldots, e_m$. Set $\nu = [\frac{m}{2}]$ and define the map $f : \mathbb{R}^{\nu+1} \to \bar{M}(k)$ by $f(s, t_1, \ldots, t_\nu) = \exp \sum_{j=1}^{\nu} t_j e_{2j}(s)$, which is called (in any parameterization) a *helicoid* (associated to the curve c).

5.67 Lemma. [BDJ] *The helicoid f associated to the curve $c : \mathbb{R} \to \bar{M}(k)$, wherever it is regular, describes a minimal immersion.*

5.68 Lemma. [BDJ] *The helicoid f admits a one-parameter subgroup $A(s)$ of rigid motions of $\bar{M}(k)$ such that*

$$A(s) f(t, t_1, \ldots, t_\nu) = f(s+t, t_1, \ldots, t_\nu), \quad (s,\ t,\ t_1, \ldots, t_\nu \in \mathbb{R}).$$

5.69 Example. It is not true that any ruled minimal immersion invariant under a one-parameter subgroup $A(s)$ of rigid motions of $\bar{M}(k)$ is a helicoid. A simple example can be obtained as follows. Consider the action of the one-parameter subgroup $O(4; R)$ in the 2-plane of $\mathbb{R}^4$ spanned by $(1,0,0,0)$ and $(0,0,1,0)$, given by

$$f(s,t_1,t_2) = (t_1\cos s,\ t_1\sin s,\ t_2\cos s,\ t_2\sin s).$$

For $(t_1,t_2) \neq (0,0)$ this clearly describes a ruled immersion that one can easily show to be minimal. By fixing t_1 and t_2, we obtain a plane curve which is normal to the leaves of f. We conclude that f cannot be a helicoid, otherwise it would be contained in a 3-dimensional subspace of $\mathbb{R}^4$ and this is not the case.

5.70 Definition. [BDJ] Let $A(s)$ be a one-parameter subgroup of rigid motions of $\bar{M}(k)$ and S a ν-dimensional complete totally geodesic submanifold of $\bar{M}(k)$ orthogonal to the orbits of $A(s)$. Then the map $f : \mathbb{R}\times S \to \bar{M}(k)$ defined by $f(s,p) = A(s)p$ wherever it is regular, describes a ruled immersion. These maps (that are minimal wherever they are regular) will be called *generalized helicoids*.

The next proposition gives an idea of how generalized helicoids look.

5.71 Lemma. [BDJ] *Let $A(s)$ be a one-parameter subgroup of rigid motions of $\bar{M}(k)$ and $S\subset\bar{M}(k)$ a complete ν-dimensional totally geodesic submanifold of $\bar{M}(k)$ orthogonal to the orbits of $A(s)$. The map $f : \mathbb{R}\times S\to\bar{M}(k)$ defined by $f(s,p) = A(s)p$ is a generalized helicoid iff the covariant derivative $\frac{d}{ds}a$ of the velocity vector of $a(s) = f(s,p)$ lies in the tangent of $A(s)S$ at $f(s,p)$ for each $s\in\mathbb{R}$ and $p\in S$.*

Note that a generalized helicoid in a sphere or a hyperbolic space is the restriction of a certain submanifold in Euclidean space. The classification of generalized helicoids in $\mathbb{R}^{N+1}$, up to the motion of the ambient space, is the following [BDJ]:

$$f(s,t_1,\dots,t_\nu) = \sum_{i=1}^{r} t_i e_i(s) + \sum_{i=1}^{\nu-r} t_{r+i}V_{2r+i} + sbV_{\nu+r+1},$$

where the vectors $V_1,\dots,V_{N+1}$ are any special orthonormal bases of $\mathbb{R}^{N+1}$, the fields $e_i(s)$ are given by $e_i(s) = \cos a_i s\ V_{2i-1} + \sin a_i s\ V_{2i}$, $(1\le i\le\nu)$, and $a_1,\dots,a_\nu,b,s,t_1,\dots,t_\nu$ are real numbers and variables. An analogous result is obtained for $\mathbb{H}^{N+1}$.

5.72 Theorem. [BDJ] *Let $M^{\nu+1}$ be a minimal ruled submanifold of $\bar{M}^N(k)$. Then there is a generalized helicoid $f : \mathbb{R} \times \bar{M}^\nu(k) \to \bar{M}^N(k)$ and an open set $U \subset \mathbb{R} \times \bar{M}^n(k)$ such that f restricted to U parameterizes M.*

Consider the generalization of helicoids, where instead of "screw lines" the "nicely curved submanifolds" in space forms are used.

The notion of a (locally) symmetric submanifold M of the Euclidean space $\mathbb{R}^n$ is given by D. Ferus [Fe 6] as a submanifold (locally) mapped into itself for each $x \in M$ by the reflection of $\mathbb{R}^n$ with respect to the (affine) normal space to M at x. W. Strübing [St] studied locally symmetric submanifolds of any Riemannian space.

5.73 Definition. [CMR] A submanifold $M \subset \bar{M}(c)$ is called *(locally) symmetric* if for each $x \in M$, M is (locally) mapped into itself by the geodesic reflection of $\bar{M}(c)$ with respect to the subspace $F_x = \exp_x(T_x M^\perp)$. F_x is called a *submanifold of symmetry*.

O. Kowalski and I. Külich [KoK] defined the notion of a k-symmetric submanifold of $\mathbb{R}^n$, $(k \geq 2)$, using suitable isometries of $\mathbb{R}^n$. They observed that a 2-symmetric submanifold M is invariant under the reflections of $\mathbb{R}^n$ with respect to subspaces of the normal spaces of M; so the 2-symmetric submanifolds appear as a generalization of the symmetric submanifolds.

5.74 Definition. [CMR] A submanifold $M \subset \bar{M}(c)$ is called *2-symmetric* if for each $x \in M$ there exists an involutive isometry σ_x of $\bar{M}(c)$ such that locally M is mapped by σ_x into itself and x is a fixed point of σ_x isolated on M.

It is easy to check that a totally geodesic submanifold of a symmetric submanifold is a 2-symmetric submanifold, but generally it is not a symmetric submanifold.

The vector space $\overset{1}{N}_x M$ generated by the values of the second fundamental form $h = \overset{0}{h}$ at a fixed point $x \in M$ is called the *first normal space* of M at x. Then the *k-th normal space* $\overset{k}{N}_x M$ at $x \in M$ is the vector subspace of $T_x\bar{M}$ generated by values of the $(k-1)$*-th fundamental form* $\overset{k-1}{h}$ of M (see [CMR])

$$\overset{k-1}{h} : T_x M \times \overset{k-1}{N}_x M \to (T_x M \oplus \overset{1}{N}_x M \oplus \cdots \oplus \overset{k-1}{N}_x M)^\perp.$$

Note that for $k \geq l$ we have $\overset{k}{h} = 0$, for $k \leq l$ we have $\overset{k}{N}_x M \neq 0$, l is the maximal value of k such that $\overset{k}{N}_x M \neq 0$ and that the vector spaces $\overset{k}{N}_x M$ are mutually orthogonal. For a submanifold M we set $V_x = \oplus_{i=0}^{q} \overset{2i+1}{N_x} M$, where $q = [\frac{l-1}{2}]$.

5.75 Theorem. [CMR] *If M is a 2-symmetric submanifold in $\bar{M}(c)$, then*

$$\overset{k}{\nabla} h = 0 \ \text{ for each } k. \tag{5.31}$$

5.76 Definition. [CMR] A submanifold $M \subset \bar{M}(c)$ is called *nicely curved* if $\dim \overset{k}{N_x} M$ does not depend on the points $x \in M$ for each $k \in \mathbb{N}$.

Recall that a submanifold M of a space form $\bar{M}(c)$ is called an *essential submanifold* if it is not contained in any proper totally geodesic submanifold of $\bar{M}(c)$.

5.77 Lemma. *(see* [CMR]*) If M is a connected nicely curved submanifold of a real space form $\bar{M}(c)$, then M is an essential submanifold of a totally geodesic submanifold of $\bar{M}(c)$ of dimension equal to* $\dim(\oplus_{i=0}^{l} \overset{i}{N_x} M)$. *In particular, the normal space $T_x M^{\perp}$ of M at x is given by $T_x M^{\perp} = \oplus_{i=1}^{l} \overset{i}{N_x} M$.*

5.78 Definition. [CMR] A submanifold $\tilde{M}$ of $\bar{M}(c)$, which is a tubular neighborhood of a nicely curved M satisfying condition (5.31) in the set $\{\exp_x v\}_{x \in M,\, v \in V_x}$, is foliated by totally geodesic submanifolds of $\bar{M}(c)$ and is called a *multihelicoid*.

If $\dim M = 1$, condition (5.31) implies that M is just a curve with constant curvature, and $\tilde{M}$ becomes the helicoid associated to a curve in the sense of Definition 5.66.

5.79 Theorem. [MR 1] *Any multihelicoid $\tilde{M}$ in $\bar{M}(c)$ over a nicely curved submanifold M satisfying condition* (5.31) *is minimal.*

For a nicely curved submanifold M in $\bar{M}(c)$, we can consider the map $\nu : M \to G(p, \bar{m} - p)$ into the Grassmannian manifold where $p = \dim(\oplus_{i=0}^{q} \overset{2i+1}{N}{}_x M)$, given by the formula

$$\nu(x) = \oplus_{i=0}^{q} \overset{2i+1}{N}{}_x M, \quad (x \in M,\ q = [\frac{l-1}{2}]). \tag{5.32}$$

Recall that in any Grassmannian manifold $G(p, q)$ there exist flat totally geodesic submanifolds for each dimension $m \leq \min(p, q - p)$. If H is a flat totally geodesic submanifold of $G(p,q)$, then the map $\exp_a : T_aH \to H$, $(a \in H)$, is a totally geodesic (local) isometry, see [MR 3]. If we identify $\mathbb{R}^m$ with T_aH, then $\exp_a$ gives a totally geodesic map of $\mathbb{R}^m$ into $G(p,q)$.

5.80 Definition. A map $f : M \to N$ between two manifolds is called *totally geodesic* if the covariant derivative $\nabla(df)$ of the differential df is zero, see [EL] and Example 2.6.

5.81 Theorem. [MR 3] *Let M be a complete connected nicely curved submanifold of dimension m in $\mathbb{R}^{\bar{m}}$ and let $\nu : M \to G(p, \bar{m} - p)$ be a totally geodesic map defined by* (5.32). *Then*

1) *if* $\dim \ker d\nu = 0$, *then $\nu(M)$ is a complete totally geodesic submanifold of $G(p, \bar{m} - p)$,*

2) *if* $\dim \ker d\nu \neq 0$, *then* $\ker d\nu$ *defines a foliation on M whose leaves are Euclidean spaces of dimension r; moreover, the leaf space B is a complete totally geodesic submanifold of $G(p, \bar{m} - p)$, the map ν factors into a Riemannian submersion with totally geodesic leaves, $\pi : M \to B$, followed by a totally geodesic immersion $j : B \to G(p, \bar{m} - p)$, ($B$ is a totally geodesic submanifold of $G(p, \bar{m} - p)$ and its connection coincides with the one induced by j) and the fiber space $M \to B$ has a flat connection with totally geodesic horizontal leaves,*

3) *M has nonnegative curvature and is locally symmetric.*

In [MR 3] an explicit construction of totally geodesic homotheties of $\mathbb{R}^m$ into a Grassmannian manifold $G(p,q)$ is given, which allows us to obtain examples of (minimal) symmetric ruled submanifolds of $\mathbb{R}^m \times \mathbb{R}^{p+q}$. Any such ruled submanifold M is the orbit $\mathcal{O}(F_{\bar{x}})$, $(\bar{x} \in M)$, of the generator $F_{\bar{x}}$ through any fixed point $\bar{x} \in M$, under the action of the subgroup $\mathcal{O}$ of $I(\mathbb{R}^m \times \mathbb{R}^{p+q})$, spanned by the reflections of $\mathbb{R}^m \times \mathbb{R}^{p+q}$ with respect to the generators. Moreover, the orbit of each point of M is a 2-symmetric submanifold of $\mathbb{R}^m \times \mathbb{R}^{p+q}$.

For a submanifold $M \subset \bar{M}$ with generator $L \subset M$, we can consider a geodesic reflection of $\bar{M}$ with respect to L, which maps a neighborhood of L in M into M.

5.82 Definition. [MR 1,2] A *ruled submanifold* $M^{\nu+n}$ is called *symmetric*, if the geodesic reflection of $\bar{M}$ with respect to every generator L maps some neighborhood of L in M into M maps and generators into generators.

Since a geodesic reflection of a space form $\bar{M}(c)$ with respect to a totally geodesic submanifold is an isometry (rigid motion), a sufficient condition for a ruled submanifold $M \subset \bar{M}(c)$ to be symmetric is that M be locally mapped into itself by a geodesic reflection with respect to each of its generators.

Any multihelicoid M, associated to a nicely curved 2-symmetric submanifold in $\bar{M}(c)$, is even a symmetric ruled submanifold in the sense of Definition 5.82 [MR 2]. The following theorem partially answers the question of how to classify symmetric ruled submanifolds in space forms.

5.83 Theorem. [MR 2] *A symmetric ruled submanifold $M^{\nu+n}$ in a space form $\bar{M}(c)$ is a minimal submanifold.*

On the other hand, minimal submanifolds of a Euclidean space can be considered as a special case of submanifolds of finite type, which were introduced by B.-Y. Chen in 1984 (for a recent survey see [Ch 4,5]).

5.84 Definition. [Ch 3] A submanifold $M \subset \mathbb{R}^N$ is said to be of *finite type* if each component of its position vector field f can be written as a finite sum of eigenfunctions of the Laplacian Δ of M, i.e., $f = f_0 + f_1 + \cdots + f_m$, where f_0 is a constant vector and $\Delta f_i = \lambda_i f_i$, $(1 \le i \le m)$. If, in particular, all eigenvalues $\{\lambda_1, \lambda_2, \dots, \lambda_m\}$ are mutually different, then M is said to be of *m-type*.

Note that every minimal submanifold of Euclidean space is of 1-type (with $\lambda_1 = 0$), since $\Delta f = 0$. Also note that a plane curve of finite type is part of a circle or a straight line. In [Di] the Catalan Theorem was extended to ruled submanifolds of finite type with $n = 1$ in $\mathbb{R}^N$ (see [CDVV] for surfaces).

5.85 Theorem. [Di] *A ruled submanifold $M^{\nu+1}$ in $\mathbb{R}^N$ is of finite type iff M is part of a cylinder on a curve of finite type or part of a generalized helicoid. In particular, a ruled submanifold $M^{\nu+1}$ in $\mathbb{R}^{\nu+2}$ is of finite type iff M is part of a hyperplane, a circular cylinder, a helicoid Γ^2 in $\mathbb{R}^3$, a cone K^3 with a top p on a minimal ruled surface in the sphere S^3 centered at p, or a cylinder on Γ^2 or K^3.*

Finite type ruled submanifolds modelled on spherical submanifolds are studied in [FGL]. Ruled surfaces of finite type in S^3 are studied in [HV]. The test for a finite type Gauss map for ruled submanifolds is given in [Bai].

5.86 Theorem. [Bai] *The only ruled submanifolds $M^{\nu+1}$ in Euclidean space*

$\mathbb{R}^{\nu+p}$ with finite type Gauss map are the cylinders over curves of finite type and the $(\nu+1)$-dimensional Euclidean spaces.

The only ruled submanifolds $M^{\nu+1}$ of $\mathbb{R}^N$ with constant nonzero mean curvature are *generalized cylinders*, that is, products of the type $\mathbb{R}^\nu \times c(t) \subset \mathbb{R}^\nu \times \mathbb{R}^{N-\nu}$, where $c(t) \subset \mathbb{R}^{N-\nu}$ is a curve with constant first curvature [BDJ]. For $k < 0$ there are no ruled submanifolds $M^{\nu+1}$ of $\bar{M}(k)$ with constant nonzero mean curvature [BDJ]. If $k > 0$ the situation is quite different.

5.87 Theorem. [BDJ] *a) If $M^{\nu+1}$ is a ruled submanifold in $\bar{M}^N(k)$, $(k > 0)$, with constant nonzero mean curvature, then $N \geq 2\nu + 1$ and M is locally isometric to the product $\mathbb{R} \times \bar{M}^\nu(k)$. b) If $f : M^{\nu+1} \to S^{2\nu+1}(k)$ is a ruled submanifold with a mean curvature of nonzero constant length, then $\nu + 1$ is even. Given $a > 0$, then for any complete smooth curve c in the symmetric space $O(2p)/U(p)$ there exists an isometric ruled immersion $f_{c,a} : \mathbb{R} \times S^{2p-1} \to S^{4p-1}$ with a mean curvature of constant length a, which is well determined up to a rigid motion of S^{4p-1}. Moreover, if c and c_1 differ by a rigid motion of $O(2p)/U(p)$ then $f_{c,a}$ and $f_{c_1,a}$ are the same up to a rigid motion of S^{4p-1}.*

Thus we have a complete classification of ruled submanifolds $M^{\nu+1}$ of codimension ν in space forms with mean curvature of constant length.

If in Theorem 5.87 the mean curvature vector is parallel or the normal bundle is trivial, then $\nu = 1$ and M^2 is the product of two circles in $S^3 \subset S^N$.

5.5. Submanifolds with spherical generators

Next we consider submanifolds with generators which are *totally umbilic* in an ambient space.

5.88 Definition. [Rec 2] A 1-form λ on $TM^\perp$ is called a *principal curvature* of a submanifold $M \subset \bar{M}$ if $T_m(\lambda) = \{x \in T_m M : A_\xi x = \lambda(\xi)x,\ \xi \in T_m M^\perp\}$ is at least a 1-dimensional subspace for all $m \in M$.

Let $G(\lambda) \subset M$ be an open *regularity domain*, where the dimension of $T_m(\lambda)$ is equal to the minimal value $\mu(\lambda, M)$. Then a principal curvature function λ is smooth on $G(\lambda)$.

5.89 Theorem. [Rec 2] *Suppose that the principal curvature function λ of a (complete) submanifold $M \subset \bar{M}(k)$ with $\mu(\lambda, M) > 1$ is given. Then the* *subspaces $T_m(\lambda)$ form an integrable distribution on $G(\lambda)$, whose leaves $\{L(\lambda)\}$ are (complete) totally umbilic submanifolds (spherical generators) in $\bar{M}$, and thus in M.*

5.90 Remark. There are isometric immersions without any principal curvature, for instance, the immersion of the Veronese surface into S^4. The principal curvatures for hypersurfaces in a conformally flat Riemannian space are studied in [Hw]. For recent investigations of the completeness of curvature surfaces of submanifolds in complex space forms and pseudo-Riemannian spaces, see [Bö 1,2].

5.91 Corollary. [Rec 2], [O'N 2] *Let $M(k) \subset \bar{M}(\bar{k})$ be an isometric immersion between space forms with $k > \bar{k}$ and* $\dim M < \dim \bar{M} \leq 2 \dim M - 2$. *Then at every $m \in M$ there exists exactly one principal curvature λ_m with* $\dim T(\lambda_m) \geq 2$*; moreover,* $\dim T(\lambda_m) \geq 2 \dim M - \dim \bar{M} + 1 \geq 3$, *(for the case* $\dim M + 2 = \dim \bar{M}$ *see* [Hen]*). To this principal curvature function all of Theorem* 5.89 *applies.*

The analogous property is true for conformally flat submanifolds of low codimension. These submanifolds have been extensively studied during the last 30 years. Conformally flat hypersurfaces have natural relations with other classes (defined extrinsically) such as special quasiumbilical hypersurfaces, loci of n-spheres and canal hypersurfaces [Ch 1]. The local structure of conformally flat hypersurfaces was discovered by E. Cartan [Car 1]: they are generically foliated by codimension-1 spheres.

5.92 Definition. [MZ] A submanifold M^m of $\mathbb{R}^{m+p}$, $(p \leq m)$, is said to be *generically foliated by spheres* if there exists a dense open set $U \subset M$ such that $U = \bigcup_{i=0}^{p} U_i$, where U_i (for every i) is an open set of M which is foliated by $(m-p+i)$-dimensional umbilical submanifolds of $\mathbb{R}^{m+p}$. We shall denote this class of submanifolds by $\mathcal{F}S_m^p$. The points of U are called *generic points*. A submanifold M is said to be *strongly generically foliated by spheres* if $M \in \mathcal{F}S_m^p$ and the restriction of $TM^{\perp}$ to each leaf L in U is parallel in $TL^{\perp}$. We will denote this class of submanifolds by $S\mathcal{F}S_m^p$.

This definition implies, in particular, that the leaves in U_i are open sets of standard spheres S^{m+p+i}. Let $C\mathcal{F}_m^p$ be the class of conformally flat submanifolds of $\mathbb{R}^{m+p}$. In view of Theorem 5.89 we have the following theorem.

5.93 Theorem. *(local)* [MZ] *If $1 \leq p \leq m-3$ then $C\mathcal{F}_m^p \subset S\mathcal{F}S_m^p$.*

5.94 Theorem. *(global)* [MZ] *Let M^m be an oriented conformally flat submanifold of $\mathbb{R}^{m+p}$, $(p \leq m-3)$. If M is k-regular (i.e., the leaves have dimension k everywhere and form a regular foliation on M, see Theorem 5.93), then M is a sphere bundle (the fibers are spheres) over an $(m-k)$-dimensional manifold.*

For classification of the hypersurfaces in Theorem 5.94 see [CDM].

5.95 Example. [MZ] Consider the standard immersion of the sphere $S^{m-2}(1)$ in $\mathbb{R}^{m-1}$ and any isometric immersion of a compact surface $\mathbb{H}^2(-1)$ of curvature -1 into $\mathbb{R}^{17}$ (by the Nash Theorem). The product $S^{m-2}(1) \times \mathbb{H}^2(-1)$ is then isometrically immersed in $\mathbb{R}^{m+16}$. If $m \geq 19$, we obtain the situation of Theorem 5.94.

The following theorem generalizes the classical investigations by B. Riemann [Rie] (*a minimal surface in $\mathbb{R}^3$ that is foliated by circles in parallel planes must be either a piece of a catenoid or the example now called the "Riemann staircase"*) and A. Enneper [Enn] (*if a minimal surface in $\mathbb{R}^3$ is foliated by circles or by circular arcs, then the planes containing these curves must be parallel*).

5.96 Theorem. [Jagy] *If $M^{\nu+1}$ is a complete and nonplanar minimal hypersurface in $\mathbb{R}^{\nu+2}$, $(\nu \geq 2)$, and an open subset of M is foliated by pieces of ν-dimensional spheres, then M is a higher-dimensional catenoid (a hypersurface of revolution).*

Appendix to Section 5.3.1

Corollary 5.51 can be improved in case of embeddings into a cylinder of Euclidean space.

5.97 Definition. A hypersurface $C(s,r)$ in $\mathbb{R}^{N+1}$, which is congruent to

$$\sum_{i=1}^{s+1}(x_i)^2 - r^2 = 0,$$

is called a *circular cylinder of radius r with s−dimensional parallel and $(N-s)$−dimensional ruling (generator).*

For $s = N$ we obtain a *hypersphere* of radius $r > 0$.

5.98 Theorem. *Let M^n be a compact manifold with condition $Ric_M^q \leq qc^2$ for some integer $q \in [1, n-1]$ and a constant $c > 0$. Then M^n cannot be isometrically embedded into $\mathbb{R}^{n+p}$, $(p \leq n-q)$ inside of a circular cylinder of radius $r = \frac{1}{c}$ with $(2p+q-1)$–dimensional parallel.*

Proof. Suppose on the contrary that M^n is isometrically embedded into $\mathbb{R}^{n+p}$, $(p \leq n-q)$ inside of a circular cylinder $C(2p+q-1, r)$ of radius $r = \frac{1}{c}$ with $(2p+q-1)$–dimensional parallel. Since the dimension n of M^n is greater than the dimension $n-p-q$ of the axes of a cylinder $C(2p+q-1, r)$, then, decreasing radius r and keeping the axes we obtain a smaller cylinder $C(2p+q-1, r_1)$ of radius $r_1 \in (0, r)$, which is tangent to M^n at some point m. See Figure 21.

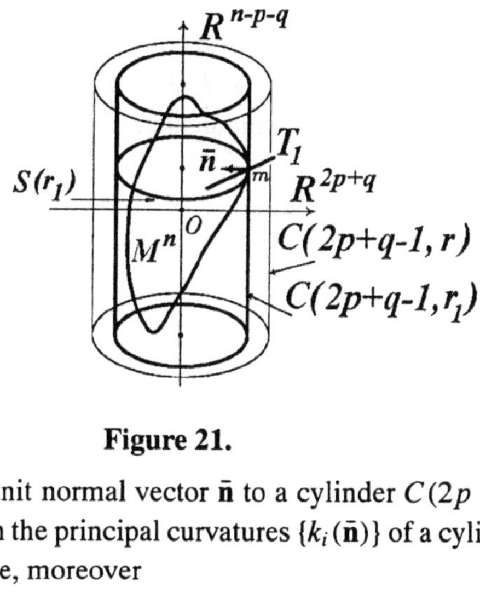

Figure 21.

Consider a field of unit normal vector $\bar{\mathbf{n}}$ to a cylinder $C(2p+q-1, r_1)$, directed inside of it. Then the principal curvatures $\{k_i(\bar{\mathbf{n}})\}$ of a cylinder $C(2p+q-1, r_1)$ are nonnegative, moreover

$$0 = k_1(\bar{\mathbf{n}}) = \cdots = k_{n-p-q}(\bar{\mathbf{n}}) < k_{n-p-q+1}(\bar{\mathbf{n}}) = \cdots = k_{n+p-1}(\bar{\mathbf{n}}) = \frac{1}{r_1}.$$

In view of the relation between dimensions of the cylinder, its parallel and the submanifold M^n: $n + (2p+q-1) - (n+p-1) = p+q$, there exists a $(p+q)$–dimensional subspace $T_1 \subset T_m M$ which is tangent to a parallel of the cylinder.

The principal curvatures of cylinder $C(2p+q-1, r_1)$ in directions $\boldsymbol{\lambda} \in T_1$ are constant and equal to the principal curvature $\frac{1}{r_1}$ of a parallel ($(2p+q-1)$–dimensional sphere) $S(r_1)$. Since the principal curvature $A_{\bar{\mathbf{n}}(m)}(\boldsymbol{\lambda}, \boldsymbol{\lambda})$ of a submanifold M^n for normal vector $\bar{\mathbf{n}}(m)$ in the direction $\boldsymbol{\lambda} \in T_1$ is not less than $\frac{1}{r_1}$, in view of the formula

$$A_{\bar{\mathbf{n}}(m)}(\boldsymbol{\lambda}, \boldsymbol{\lambda}) = (h(\boldsymbol{\lambda}, \boldsymbol{\lambda}), \bar{\mathbf{n}}(m)) \tag{5.33}$$

we have the following estimate for a second fundamental form h of a submanifold M^n in directions of T_1:

$$|h(\boldsymbol{\lambda}, \boldsymbol{\lambda})| \overset{(5.33)}{\geq} A_{\bar{\mathbf{n}}(m)}(\boldsymbol{\lambda}, \boldsymbol{\lambda}) \frac{1}{r} \boldsymbol{\lambda}^2, \qquad (\boldsymbol{\lambda} \in T_1).$$

Hence there exist $q+1$ orthonormal vectors $\{x_0; x_1, \ldots, x_q\}$ in T_1 with the property

$$Ric_M^q(x_0; x_1, \ldots, x_q) \overset{(5.26')}{=} Ric_h^q(x_0; x_1, \ldots, x_q) > \frac{q}{r^2},$$

because in the opposite case, by Lemma 5.48 for a restriction of the second fundamental form $h : T_1 \times T_1 \to T_m M^{\perp}$, we shall obtain the inequality $p > \dim T_1 - q = p$, which is a contradiction. □

CHAPTER VI

DECOMPOSITION OF RULED SUBMANIFOLDS

In **Chapter 6** we study ruled submanifolds of a Riemannian space.

Section 6.1 contains *cylinder theorems* in a Riemannian space of nonnegative sectional curvature. We begin with the fact that a complete strongly parabolic submanifold M in a space form $\bar{M}(0)$, satisfying the additional condition $Ri_M \geq 0$, is a cylinder with $\mu(M)$-dimensional *generator* $\{L\}$ ([Bor 4], [Har] for $K \geq 0$). Theorem 6.8 is based on the traditional method of the Riccati ODE from Section 4.3; Theorem 6.3 and Theorem 6.4 are based on the method developed in Section 4.2 using the *volume of L-parallel vector fields.*

The restrictions on the topology of a ruled submanifold $M^{n+\nu}$ with ν-dimensional generator in a sphere $S^N(k)$ appear only under additional conditions for codimension [Pet] or curvature. For example, from $K_{mix} > 0$ it follows that $\nu < \rho(n)$; it is possible to realize $S^\nu \times S^n$ as a double ruled *submanifold of Segre type* (with $K_{mix} = 0$) only when $\operatorname{codim} M \geq n\nu$. In **Section 6.2** we introduce and study the class of ruled submanifolds with $K_{mix} \geq 0$. Some *Segre type* decomposition theorems (with estimate of codimension) are proved for ruled submanifolds $M^{n+\nu}$ satisfying the inequality $\nu \geq \rho(n)$ in a Riemannian space of positive sectional curvature.

Theorem 6.14 is based on the extrinsic synthetic procedure in a Riemannian space whose curvature tensor is similar to that of CROSS, on the relationship between the condition $K_{mix} \geq 0$ and the concept of a *t-uniquely projectable submanifold along a generator*. Theorem 6.18 and Theorem 6.20 (for a submanifold of a Riemannian space with positive curvature) are based on a combination of the extrinsic procedure of Theorem 6.14 with a method of variation in Chapter IV for foliations.

Corollaries give the classical type of test for the existence of a Segre embedding in S^N or CP^N and for the optimal *Ferus type estimates of dimension of a generator* for the case of $K_{mix} > 0$.

6.1. Cylindricity of submanifolds in a Riemannian space of nonnegative curvature

For a submanifold $M \subset \bar{M}$ the term $\overline{Ric}^{\top}(x)$ with $x \in TM$ is defined by the same formula as $Ric\,(x)$ for a manifold M, combined with the help of the sectional curvature of $\bar{M}$; see (6.2) below.

6.1 Lemma. *Let $M \subset \bar{M}$ be a complete submanifold and suppose that for some point $m \in M$ with maximal rank $r(m) = r(M)$ we have*

$$Ric\,(x) \geq \overline{Ric}^{\top}(x), \quad (x \in T_m M). \tag{6.1}$$

Then $\mu(M) = \dim M - r(M)$ and the second quadratic form A_H for a mean curvature vector H at m is positive on the subspace $(\ker h)^{\perp}$.

Proof. Let $\{x_i\}$ be an orthonormal base of $T_m M$ and suppose that the vector $y \perp \ker h$. From the Gauss equation (5.4) it follows that

$$(h(y, y), h(x_i, x_i)) = (R(y, x_i)x_i, y) - (\bar{R}(y, x_i)x_i, y) + h^2(x_i, y).$$

We sum this equation over i and denote by $\eta = \sum_i h(x_i, x_i)$ the $(\dim M)$-times mean curvature vector of M, obtaining

$$(A_\eta y, y) = (h(y, y), \eta) = [Ric\,(y) - \overline{Ric}^{\top}(y)] + \sum_i h^2(x_i, y). \tag{6.2}$$

From (6.1) and (6.2) it follows that the restriction of the linear operator A_η on the subspace $(\ker h)^{\perp}$ is positive definite (see [Ab 4] for the case of $\bar{M} = S^N$) and the same holds for all normal vectors at m near η. Since the normal vectors with maximal rank $r(M)$ form an open dense subset of $T_m M^{\perp}$, we have $\mu(M) = \dim M - r(M)$. $\square$

6.2 Definition. Let $M \subset \bar{M}$ be a complete curvature invariant submanifold, that is, M satisfies the property

$$\bar{R}(x, y)z^{\perp} = 0, \ (x,\ y,\ z \in TM) \tag{6.3}$$

and $k \geq 0$. For any point $m \in M$ with $\mu(m) = \mu(M) > 0$, we consider the set $\Gamma(m, h)$, the sheaf of geodesics $\gamma : [0, \frac{\pi}{\sqrt{k}}] \to M,\ (\gamma(0) = m,\ \gamma'(0) \in \ker h$, where h is a second fundamental form of M), and define a nonnegative number

$$\alpha(m, h) = \sup\{(\nabla_y z, \gamma') : \ y,\ z \perp \ker h,\ y \perp z,\ |y| = |z| = 1,\ \gamma \in \Gamma(m, h)\}.$$

Note that $\alpha(m,h)$ is an analog of the turbulence $\alpha(L)$ for a relative nullity foliation on the regularity domain $\bar{G} \subset M$ and is well-defined in view of Theorem 5.6. Let $\delta(\bar{R})$ be, in analogy to $\delta(R)$ (see Chapter IV), the nonnegative number

$$\delta(\bar{R}) = \sup\{(\bar{R}(x,y)u,z) : x,\ u \in \ker h,\ y,\ z \in TM \cap (\ker h)^{\perp},$$
$$x \perp u,\ y \perp z,\ |x| = |y| = |u| = |z| = 1\}.$$

For example, $\delta(\bar{R}) = 0$ holds for a conformally flat metric on $\bar{M}$ along M.

6.3 Theorem. *Let $M \subset \bar{M}$ be a complete not-totally geodesic analytical curvature invariant submanifold with $r(M) < \dim M$, satisfying the condition*

$$k_2 \geq \bar{K}(x,y) \geq k_1 \geq 0,\ (x,\ y \in TM), \tag{6.4}$$

and suppose that there exists a point $m \in M$ of maximal rank satisfying (6.1) *and one of the following inequalities:*

$$(k_2 - k_1)\max\{a(m,h)^2, k\} \leq 0.3 k_2 k, \tag{6.5a}$$

$$(k_2 - k_1 + 2\delta(\bar{R}))a(m,h)\max\{a(m,h)^2,k\}^{\frac{3}{2}} \leq 0.004\, k^2 k_2,\ (\textit{with } k_1 \geq \frac{1}{2} k_2), \tag{6.5b}$$

where $k = \frac{k_1+k_2}{2}$. Then $k_1 = k_2 = 0$ and M is a cylindrical ruled submanifold with a flat $\mu(M)$-dimensional generator.

Proof. 1. By Lemma 6.1 we have $\mu(M) = \dim M - r(M)$. The regularity domain $\bar{G} \subset M$ is foliated in the case of (6.3) by $\mu(M)$-dimensional complete generators $\{L\}$ (see Theorem 5.6), i.e., $\bar{G}$ is a developable ruled submanifold. The structural tensor B for this relative nullity (totally geodesic) foliation is defined by (2.3).

2. If $\xi \in TM^{\perp}$ is a local unit vector field of normal vectors satisfying the conditions $(\nabla^{\perp}\xi)_m = 0$ and $\xi(m) = \eta$ (the direction of mean curvature vector), then we have

$$(\nabla^{\perp}_x A_{\xi})y = A_{\eta}B(x,y),\ (x \in T_m L,\ y \in T_m L^{\perp}) \tag{6.6}$$

in view of (6.3) and the Codazzi equation (5.5).

From (6.6) it follows that the linear operators $B(x,*)$, $(x \neq 0)$, are self-adjoint for the positive scalar product $\{y,y\} := (A_{\eta}y, y)$ in $T_m L^{\perp}$, whose definition is based on (6.2); thus these operators have $r(M)$ eigenvectors, which form an affine basis in $T_m L^\perp$. The same is true for points in a neighborhood of m.

3. For a relative nullity foliation $\{L\}$, as in the proof of Theorem 4.10 with $\varepsilon = 0$ (for (6.5 a)) or Theorem 4.23 (for (6.5 b)), we deduce $k_2 = k_1 = 0$; moreover the structural tensor B is zero (because its eigenvalues vanish) on a neighborhood U_m of m. Thus U_m is isometric to the product $\mathbb{R}^{\mu(M)} \times L^\perp$ and the leaves (generators) $\{L\}$ are flat (see the proof of Lemma 4.14). Since M is an analytic manifold, it is cylindrical and locally isometric to the product $\mathbb{R}^{\mu(M)} \times L^\perp$. □

Recall the sequence $\nu(n) = \max\{t : t < \rho(n - t)\}$, defined in Section 5.2.2.

6.4 Theorem. *Let $M \subset \bar{M}$ be a complete, 'not-totally' geodesic analytical curvature invariant submanifold and assume that either M or $\bar{M}$ is compact, $\mu(M) > \nu(\dim M)$, ($\mu(M) > 0$ for the Kählerian case),*

$$k_2 \geq \bar{K}(x, y) \geq k_1 \geq 0, \quad (x, y \in T\bar{M}), \tag{6.7}$$

and one of the inequalities (6.5a) or (6.5b) holds for some point m with $\mu(m) = \mu(M)$. Then $k_1 = k_2 = 0$ and M is a cylindrical ruled submanifold with a flat $\mu(M)$-dimensional generator.

Proof. 1. The regularity domain $\bar{G} \subset M$ is foliated in the case of (6.3) by $\mu(M)$-dimensional complete rulings $\{L\}$, which are tangent to the distribution $\ker h$ (see Theorem 5.6). The structural tensor $B : TL \times TL^\perp \to TL^\perp$ for this relative nullity (totally geodesic) foliation is defined by (2.3).

2. If $k_1 = 0$, then from (6.5a) or (6.5b) it follows that $k_2 = 0$. Since $\bar{G}$ is a developable ruled submanifold in a flat space form and the conditions of Lemma 6.5 below hold (either M is compact, or a closure of $\bar{G}$ is compact), $\bar{G}$ is a cylindrical ruled submanifold with a flat $\mu(M)$-dimensional generator. Since the submanifold M is analytic the proof of Theorem 6.4 for the case $k_1 = 0$ is complete.

3. Assume that $k_1 > 0$ for (6.5a). In view of the Gauss equation (5.4) the relative nullity foliation $\{L\}$ satisfies the conditions of Theorem 4.10, case a) with $\varepsilon = 0$. Hence, the structural tensor B is without eigenvectors (see Lemma 3.40), that is, $\mu(M) < \rho(\dim M - \mu(M))$ and hence $\mu(M) \leq \nu(\dim M)$, which is a contradiction in the real case. For the Kählerian case we also get a contradiction, because the structural tensor B satisfies the property $B(x, Jy) =$

$B(Jx, y) = JB(x, y)$, see [Ab 3], and by Lemma 3.41, for every point $m \in \bar{G}$, some linear operator $B(x, *)$, $(x \in T_m \setminus \{0\})$ has an eigenvector.

4. Assume that $k_1 > 0$ for (6.5b). Then the rulings are compact Riemannian manifolds with diameter not greater than $\frac{\pi}{\sqrt{k_1}}$. Remember that L-parallel fields are globally defined on rulings with trivial holonomy. By Lemma 1.16 almost every ruling has trivial holonomy; the field L_0 below is such a ruling.

In view of the given estimate for the index of relative nullity, for any point $m \in L_0$ there exist unit vectors $x_1 \in T_m L_0$, $y_1 \in T_m L_0^{\perp}$ and a real $\lambda \leq 0$ with the property $B(x_1, y_1) = \lambda y_1$, where B is structural tensor of the foliation $\{L\}$. (For the Kählerian case the property $B(x, Jy) = B(Jx, y) = JB(x, y)$ holds and by Lemma 3.41 for any point $m \in L_0$ and some vector $x \neq 0$ the linear operator $B(x, *)$ has an eigenvector).

Thus, to complete the proof, it suffices to repeat parts 3, 4, and 5 of the proof of Theorem 4.23, where $a(L)$ is replaced by $a(m, h)$, to obtain a contradiction to the assumption $k_1 > 0$. □

The following lemma, which was used in the proof of Theorem 6.4, completes the results in [Bor 4] about cylinders in $\mathbb{R}^N$.

6.5 Lemma. *Let $M^{n+\nu}$ be a developable ruled submanifold with ν-dimensional ruling $\{L\}$ in a flat space form $\bar{M}(0)$ and suppose that either $M^{n+\nu}$ is closed or $\bar{M}$ is compact. Then $M^{n+\nu}$ is a cylindrical ruled submanifold.*

Proof. The rulings $\{L\}$ are totally geodesic in $\bar{M}(0)$ and hence have a flat metric. From (5.3) and the Gauss equation (5.4), it follows that the mixed sectional curvature is zero. Assume that $m \in L_0$ and $\gamma : \mathbb{R} \to L_0$, $(\gamma(0) = m)$, is any geodesic. By compactness, a geodesic γ is either closed or ergodic in some compact flat submanifold $T^a \subset M$, which is covered by a torus; moreover, T^a is totally geodesic in $\bar{M}(0)$. Consequently, there exists an integer sequence $\{t_i\}_{i \in N} \to \infty$ with the properties $\gamma(t_i) \to m$ and $\gamma'(t_i) \to \gamma'(0)$. Since $K_{mix} = 0$, the L-parallel Jacobi fields on M along γ have the simple form $y(t) = y_0 + t y_1$ and the nondegenerate Jacobi tensor along γ has the form $Y(t) = E + tY_1$. For a structural tensor of the foliation $\{L\}$, we deduce from (3.19) the equation

$$Y_1 = B(t)(E + tY_1). \tag{6.8}$$

If λ is a real eigenvalue of the matrix Y_1, then from $\det Y(t) \neq 0$ it follows that $\lambda = 0$. Assume that $Y_1 \neq 0$, i.e., for some orthonormal base in $T_m L^{\perp}$, the matrix Y_1 can be written in the form $Y_1 = \begin{pmatrix} 0 & 0 \\ 0 & \bar{Y} \end{pmatrix}$, where $\bar{Y}$ is a nondegenerate matrix. Then

$$Y(t) = \begin{pmatrix} E & 0 \\ 0 & E + t\bar{Y} \end{pmatrix}, \quad B(t) = \begin{pmatrix} A & 0 \\ 0 & \bar{B} \end{pmatrix}.$$

Moreover, $\bar{Y} = \bar{B}(t)(E + t\bar{Y})$ and $\bar{B}(t)$ is a nondegenerate matrix. Thus $E + t\bar{Y}$ is also a nondegenerate matrix and in view of (6.8) we have

$$\bar{B}(t) = \frac{1}{t}\bar{Y}(\bar{Y} + \frac{1}{t}E)^{-1}. \tag{6.9}$$

Consequently, $\bar{B}(t) \to 0$ for $t \to \infty$ and hence $B(t_i) \to 0$ for $i \to \infty$. But $B(t_i) \to B(0) = Y_1 \neq 0$, which is a contradiction.

Thus $B(0) = 0$. Since the point m and the geodesic γ were arbitrary, the structural tensor B is zero; hence the orthogonal distribution $TL^{\perp}$ is totally geodesic and integrable. By the de Rham Decomposition Theorem, the submanifold M is locally isometric to the product $L \times L^{\perp}$, i.e., M is cylindrical. □

6.6 Corollary. *Let $M^n \subset \bar{M}^{n+p}$ be a complete curvature invariant submanifold satisfying condition* (6.4) *for $k_1 > 0$ and condition* (6.5*a*) *or* (6.5*b*) *for some point $m \in M$ with $\mu(m) = \mu(M)$. Also suppose that one of the following properties holds:*
a) $\mu(M) > \nu(n)$, ($\mu(M) > 0$ for the Kählerian case),
b) (6.1) *for the point $m \in M$,*
c) for some integer $q \in [1, n-2]$ the extrinsic partial curvature $Ric_h^q \leq 0$ (see Section 5.1) *and* $2p < \begin{cases} n - \nu(n) - q + \delta_{1q}, & \\ n - q + \delta_{1q}, & \text{for the Kählerian case.} \end{cases}$
Then M is a totally geodesic submanifold.

6.7 Remark. Theorem 6.3, Theorem 6.4 and Corollary 6.6 generalize certain results in [Bor 4] and [Ab 3-5] (see also Section 5.3); moreover, for the Kählerian case, the holomorphic bisectional curvature (see [GK]) of $\bar{M}$ in (6.4), or (6.7) can be considered. Case c) of Corollary 6.6 follows from a), because from $Ric_h^q \leq 0$ we get the inequality for $\mu(M)$ given in a), see (5.29). The analytic condition in Theorem 6.3 and Theorem 6.4 may be changed by the requirement that the function $\mu(m)$ be constant. Note that the sectional curvature in Theorem 6.3 and Theorem 6.4 for (6.5b) is 0.5-pinched, in contrast to (6.5a) with less requirement on the curvature tensor, but with greater (equal to 0.7) pinching of curvature.

6.8 Theorem. *Let $M \subset \bar{M}$ be a complete nontotally geodesic analytical curvature invariant submanifold with $r(M) < n$ satisfying the following conditions:*

$$Ric(x) \geq \overline{Ric}^{\top}(x), \quad (x \in TM), \tag{6.10}$$

$$\bar{K}_{|M} \geq 0. \tag{6.11}$$

Then M is a cylindrical ruled submanifold with an $(\dim M - r(M))$*-dimensional generator.*

6.9 Corollary. *Let $M \subset \bar{M}$ be a complete curvature invariant submanifold satisfying the conditions in* (6.10) *and suppose that $\bar{K}_{|M} > 0$ and $r(M) <$* $\dim M$. *Then M is a totally geodesic submanifold.*

6.10 Remark. Theorem 6.8 and its Corollary 6.9 complete the results in [Ab 4] (where the cases $\bar{M} = \mathbb{R}^N$ and $\bar{M} = S^N$ are studied) and also the results in [Bor 4].

Proof of Theorem 6.8 and its Corollary. Let $\{L^{\mu(M)}\}$ be a relative nullity (totally geodesic) foliation on the domain of regularity $\bar{G}$, which is defined by a relative nullity distribution by (6.3) and Theorem 5.6. Let η be the $\dim M$–times the mean curvature vector of M. As in the proof of Lemma 6.1, for $y \subset TL^{\perp}$ it follows from the Gauss equation (5.4) that the operator $A^{\perp}_{\eta} : TL^{\perp} \to TL^{\perp}$ is positive definite and that $\mu(M) = \dim M - r(M)$. From (6.3) it follows that

$$(\nabla_x A^{\perp}_{\xi})y = A_{\xi} B(x, y), \ (x \in TL, \ y \in TL^{\perp}) \tag{6.12}$$

and, in view of the symmetry of $A^{\perp}_{\xi}$ (the restriction of A_{ξ}), we have the following property:

$$A^{\perp}_{\xi} B(x, *) = B^{*}(x, A^{\perp}_{\xi} *), \ (x \in TL), \tag{6.13}$$

where $B^{*}(x, *)$ is the operator adjoint to $B(x, *)$ in $TL^{\perp}$. Consider the covariant derivative of (6.13) with respect to x at a point $m \in \bar{G}$. By (6.12) and $(\nabla_x x)_m =$

$(\nabla_x^{\perp}\xi)_m = 0$ we obtain

$$\begin{aligned}
\nabla_x(A_\xi^{\perp}B(x,*)) &= (\nabla_x A_\xi^{\perp})B(x,*) + A_\xi^{\perp}(\nabla_x B)(x,*) \\
&= A_\xi^{\perp}B(x, B(x,*)) + A_\xi^{\perp}[-B(x,(B(x,*)) - R(*,x)x] \\
&= -A_\xi^{\perp}R(*,x)x; \\
\nabla_x(B^*(x, A_\xi^{\perp}*)) &= (\nabla_x B^*)(x, A_\xi^{\perp}*) + B^*(x, (\nabla_x A_\xi^{\perp})*) \\
&= -B^*(x, B^*(x, A_\xi^{\perp}*)) - R(A_\xi^{\perp}*, x)x \\
&\quad + B^*(x, B^*(x, A_\xi^{\perp}*)) \\
&= -R(A_\xi^{\perp}*, x)x.
\end{aligned}$$

Consequently,

$$A_\xi^{\perp}R(*,x)x = R(A_\xi^{\perp}*, x)x, \ (x \in TL). \tag{6.14}$$

The equality (6.13) means that the operator $B(x,*)$ is self-adjoint with respect to the scalar product defined by $A_\eta^{\perp}$, that is, $B(x,*)$ has $r(M)$ linearly independent eigenvectors $\{y_i\}$

$$B(x, y_i) = \lambda_i y_i, \ \ (A_\eta^{\perp}y_i, y_j) = \delta_{ij}. \tag{6.15}$$

The covariant derivative of the first equation in (6.15) along a geodesic γ, $(\gamma'(0) = x)$, after the scalar product with $A_\eta^{\perp}y_i$, can be written in the form

$$((B(x,*)-\lambda_i E)y_i', A_\eta^{\perp}y_i) = (\lambda_i'+\lambda_i^2)(A_\eta^{\perp}y_i, y_i)+(R(y_i,x)x, A_\eta^{\perp}y_i). \tag{6.16}$$

Since

$$((B(x,*) - \lambda_i E)y_i', A_\eta^{\perp}y_i) = (A_\eta^{\perp}y_i', B(x,y_i) - \lambda_i y_i) = 0, \tag{6.17}$$

Given the symmetry of $A_\eta^{\perp}$ and the equalities (6.14) and (6.15), we have

$$\lambda_i' + \lambda_i^2 + (A_\eta^{\perp}R(y_i,x)x, y_i) = 0. \tag{6.18}$$

Equality (6.14) means that the operator $A_\eta^{\perp}R(*,x)x$ is symmetric and, by (6.11), it is nonnegative definite. Thus summing (6.18) over i leads to the following inequality along γ :

$$f'(t) + \frac{1}{r}f^2(t) \le 0, \ (-\infty < t < \infty), \tag{6.19}$$

where the function $f(t) = \sum_i \lambda_i(t)$ is well-defined for all t and $r = r(M)$. By comparing (6.19) to the Riccati equation, we obtain $f(t) \equiv 0$ and $\lambda_1 = \cdots = \lambda_{r(M)} = 0$. (In the case of inequality with sectional curvature assumed in Corollary 6.9 and $r(M) \neq 0$, we have a strict inequality in (6.19), which leads to a contradiction and completes the proof of Corollary 6.9). Since $r(M)$ linear independent vectors $\{y_i\}$ belong to the kernel of $B(x, *)$, this operator is zero. Thus the distribution $TL^\perp$ is integrable and the integral manifolds are totally geodesic in M. By de Rham's decomposition theorem, the domain $\bar{G}$ is locally isometric to the product $L^{\dim M - r(M)} \times D^{r(M)}$, and hence is a cylindrical ruled submanifold. Since the embedding is analytical, the same is true for M in general. □

6.11 Remark. From Theorem 6.3, Theorem 6.4, Theorem 6.8 and results in Section 5.3.2, we can deduce cylinder theorems for submanifolds with *nonpositive extrinsic curvature* with "small" codimension and certain additional conditions as well as for k-saddle submanifolds.

6.2. Ruled submanifolds in CROSS and the Segre embedding

The dual nature of ruled submanifolds allows us to apply the methods of submanifold theory to totally geodesic foliations and vice-versa. The curvature tensor of ambient CROSS along the ruled submanifolds of classes 1–3 (see Section 5.2.2) has the following properties:

$$\bar{R}(x, y)z \begin{cases} = 0, & (x \in TL,\ y \in TM,\ z \perp (x \wedge y)), \\ \| Jz, & (x \in TL,\ y \in TM,\ z \perp (x \wedge y \wedge Jx \wedge Jy)) \\ & \text{in the Kählerian case)} \end{cases} \tag{6.20}$$

$$\bar{R}(y, x)x = kx^2 y, \quad (x \in TL,\ y \in TL^\perp), \tag{6.21}$$

where $k > 0$. In this case the nonnegativity of the mixed sectional curvature

$$K(x, y) \geq 0, \ (x \in TL,\ y \in TL^\perp) \tag{6.22}$$

is not sufficient to decompose a ruled submanifold $M^{n+\nu}$ even when $\nu \geq n$.

We can see this with the help of the Hopf fibration $S^3 \subset S^7 \xrightarrow{\pi} HP^1$. Namely, let $B^2 \subset HP^1 \approx S^4$ be a smooth surface (a small sphere, for example) and let $M^5 = \pi^{-1}(B^2)$ be its lift into S^7. Obviously, M^5 is a ruled submanifold with 3-dimensional generators and $\nu = 3 > 2 = n$. Condition (6.22) holds, because the map $\pi_{|M^5}$ is a Riemannian submersion, but M^5 is not a metric product.

We shall analyze the following condition for the case (6.21):

$$(A_\xi x)^2 + (A_\xi u)^2 \leq k, \quad (x,\, u \in TL,\; x \perp u,\; |x| = |u| = |\xi| = 1). \tag{6.23}$$

Note that (6.23) follows from the "clearer" inequality

$$||A_\xi||^2 = trace\,(A_\xi^2) \leq 2k, \quad (|\xi| = 1). \tag{6.24}$$

From (6.23) it follows that $|A_\xi x|^2 \leq k, \quad (x \in TL,\; |x| = 1)$, and hence

$$|h(x, y)|^2 \leq k, \quad (x \in TL,\; y \in TL^\perp,\; |x| = |y| = 1).$$

But then, in view of the Gauss equation (5.4) and (6.21), we obtain inequality (6.22).

Let $x_1, \dots, x_\nu, y_1, \dots, y_n$ be an orthonormal base in $T_m M$ such that $x_i \in T_m L$, $y_i \in T_m L^\perp$. Let A_ξ denote the symmetric matrix of the second quadratic form for a normal vector $\xi \in T_m M^\perp$ ($|\xi| = 1$) for this basis. If, for example, $A_\xi x_i = \sqrt{k} y_j$ for some numbers i, j (extremal situation), then the matrix A_ξ satisfying the condition (6.23) has only two nonzero elements; its rank is equal to two.

We define the nondecreasing sequence $\bar\rho(n) = \max\{\rho(t) : t \leq n\}$ of natural numbers. From the definition of $\rho(n)$ (see Section 1.1) we obtain the estimates

$$2[\log_2 n] \leq \bar\rho(n) \leq 2[\log_2 n] + 2, \quad ([x] \text{ denotes the integer part of } x).$$

The main result of this section is the following theorem.

6.12 Theorem. *Let $M^{n+\nu}$ be a ruled (Kähler) submanifold with complete rulings $\{L^\nu\}$ in a real (or Kähler) space form $\bar M(k)$, $(k > 0)$, satisfying condition (6.23) and $\nu \geq \bar\rho(n)$ (or $\nu > 2$ for the Kählerian case). Then $M^{n+\nu}$ is locally isometric to the product $L^\nu \times M_2^n$ and* codim $M \geq \nu n$. *Moreover, if* codim $M = \nu n$*, then M is congruent to the domain of the Segre submanifold.*

From Theorem 6.12 it follows (see Theorem 5.34) that a ruled (Kähler) submanifold $M^{n+\nu}$ in S^N (or CP^N) with complete rulings $\{L^\nu\}$ and $K_{mix} > 0$ has the property $\nu < \rho(n)$ (or $\nu = 2$ and n is divisible by 4 for the Kählerian case).

Theorem 6.12 is given below in a more general form. See Theorem 6.14 and its Corollary 6.17.

An orthogonal projection in Euclidean space $\mathbb{R}^m$ onto the subspace $\mathbb{R}^l$ can be described with the help of a family of $(m-l)$-dimensional planes which are orthogonal to $\mathbb{R}^l$ and coincide with inverse images of points from $\mathbb{R}^l$ under projection. If, for example, a surface M^2 in $\mathbb{R}^3$ has a one-valued orthogonal projection onto some plane $\mathbb{R}^2$, then it is in general a graph of some function $z = f(x, y)$.

Borisenko [Bor 2] studied orthogonal projections in the Riemannian symmetric spaces S^m, CP^m and HP^m onto the totally geodesic submanifolds S^l, $CP^{\frac{l}{2}}$ and $HP^{\frac{l}{4}}$, respectively; submanifolds with regular projection are called uniquely projectible. Moreover, a submanifold M^l having regular projection onto some totally geodesic subspace S^{l+t}, $CP^{\frac{1}{2}(l+t)}$ or $HP^{\frac{1}{4}(l+t)}$ is called t-uniquely projectible, see Section 5.2. We shall transport this notion for submanifolds with generators to an arbitrary ambient Riemannian space. The following synthetic concept is effective for local (i.e., in a neighborhood of a generator) research on ruled and parabolic (see Chapter 7) submanifolds.

6.13 Definition. A smooth submanifold $M \subset \bar{M}$ is said to be *t-uniquely projectable along a generator $L \subset M$*, if t is a minimal integer such that for some point $m \in L$ there exists a subspace $V \subset T_m M^{\perp}$ of dimension $\operatorname{codim} M - t$, that stays transversal to M under $\bar{\nabla}$-parallel displacement along any path in L. For $t = 0$ such an M is called *uniquely projectable along L*.

In the case of a space form $\bar{M}(c)$, Definition 6.13 is equivalent to the classical one. Note that an arbitrary Riemannian manifold $\bar{M}$ does not contain totally geodesic submanifolds such that orthogonal projection onto a given L has the property that (even locally) inverse images of points from L under projection are totally geodesic submanifolds. An example is a ruled submanifold in CP^m of class 2, see Definition 5.30.

6.14 Theorem. *Let $M^{n+\nu} \subset \bar{M}^N$ be a ruled submanifold with complete rulings $\{L^\nu\}$ and satisfying conditions* (6.20), (6.21) *and* (6.23) *with $k > 0$. Suppose there exists an integer $m \in [1, n]$ such that*

$$\frac{1}{k} Ric_M \geq (n+\nu) - (m+1). \tag{6.25}$$

If $M^{n+\nu}$ is t-uniquely projectable along the generators with $t \geq \nu m$, then $M^{n+\nu}$ is locally isometric to the product $M_1^m \times M_2^{n-m+\nu}$, where M_2 is a ruled *submanifold developed along* $\{L^\nu\}$.

Proof. We divide the proof into steps. Assume that $k = 1$.

Step 1. Since M is not uniquely projectable along L, it follows that for any point $a \in L$ there exist unit vectors $x \in T_a L$, $v \in T_a M^\perp$ and a positive number t_0 such that under $\bar\nabla$-parallel displacement along a geodesic $\gamma(t) \subset L$, $(\gamma'(0) = x)$, at the point $t = t_0$ the vector $\bar v$ satisfies $\bar v \in T_{\gamma(t_0)} M$. The Jacobi vector field $y(t)$, $(y(t_0) = \bar v)$ in $\bar M$ that is induced by the rulings $\{L\}$ along γ and hence is tangent to M, is given (in view of (6.21)) by the formula

$$y(t) = (\cos t)\bar y + (\sin t)[\overline{B(x,y)} + \overline{h(x,y)}],$$

where "$-$" denotes $\bar\nabla$-parallel fields. From the equality

$$y(t_0) = (\cos t_0)\bar y + (\sin t_0)[\overline{B(x,y)} + \overline{h(x,y)}] = \bar v$$

and the conditions $y \perp v$ and $B(x,y) \perp v$, we obtain $h(x,y) \parallel v$ and $B(x,y) = \lambda y$. Since $y(t) \in T_\gamma M$ for all t and $\cot(t)$ is a π-periodic function, we may assume, by taking the opposite direction of γ if necessary, that $\lambda = -\cot(t_0) \le 0$ and $t_0 \in (0, \frac{\pi}{2}]$.

Step 2. We will show that $\lambda = 0$ and that under $\nabla^\perp$-parallel displacement along $\gamma(t)$, $(0 \le t \le \frac{\pi}{2})$, a normal vector v rotates uniformly in $\bar M$ by the angle $\frac{\pi}{2}$. Let $v(t)$ be $\nabla^\perp$-parallel along a γ vector field of normal vectors to M, which contains $v(0) = v$. We identify the Euclidean spaces $(T_\gamma L)^\perp$ (all normal vectors to L in $\bar M$) with $\mathbb{R}^N$ by $\bar\nabla$-parallel displacement along γ. Then a vector field $v(t)$ corresponds to a smooth curve on the unit sphere $S^{N-1} \subset \mathbb{R}^N$. The distance between the points $v(0)$ and $v(t_0)$ on S^{N-1} is equal to $\frac{\pi}{2}$, because the vectors $v(0)$ and $v(t_0)$ are orthogonal. Then the length l of a curve $v(t)$ is not less than $\frac{\pi}{2}$:

$$l = \int_0^{t_0} |\bar\nabla_{\gamma'} v(t)| dt \ge \frac{\pi}{2}.$$

From the definition of $v(t)$ it follows that

$$\bar\nabla_{\gamma'} v(t) = -A_{v(t)}\gamma'$$

and by (6.23) we obtain

$$|\bar\nabla_{\gamma'} v(t)| \le 1.$$

This gives us the inequality $l \le t_0$ and leads (by the condition $0 < t_0 \le \frac{\pi}{2}$) to an extremal situation: $t_0 = \frac{\pi}{2}$, $\lambda = 0$, $|A_{v(t)}\gamma'| \equiv 1$. Moreover, the curve

$v(t)$ is a quarter of a great circle on the sphere S^{N-1}, i.e.,

$$v(t) = (\cos t)\bar{v} + (\sin t)\bar{z}_1, \ (0 \le t \le \frac{\pi}{2}),$$

where the $\bar{\nabla}$-parallel vector field $\bar{z}_1$ along γ contains the unit vector $z_1 = -A_v x$.

Step 3. We will "chip off" of a tangent vector field

$$z_1(t) = \bar{\nabla}_{\gamma'} v(t) = (\cos t)\bar{z}_1 - (\sin t)\bar{v}$$

along a geodesic γ. For convenience, we will give without proof the following simple proposition.

6.15 Proposition. *Let $w(t)$ be a Jacobi vector field in $\bar{M}$ satisfying condition* (6.21) *along a geodesic $\gamma \subset L$ and suppose that a vector field $\bar{v}$ is $\bar{\nabla}$-parallel along γ. If the vectors $w(0)$ and $\bar{\nabla}_{\gamma'(0)} w(0)$ are orthogonal to $\bar{v}$, then $w(t)$ is also orthogonal to $\bar{v}$ for all t.*

The derivative

$$\bar{\nabla}_{\gamma'} z_1(t) = \nabla_{\gamma'} z_1(t) + h(\gamma', z_1(t)) = -v(t)$$

is orthogonal to $T_{\gamma(t)}M$; this implies $\nabla_{\gamma'} z_1(t) = 0, \ h(\gamma', z_1(t)) = -v(t)$. Thus from (6.21), (6.23) and Gauss equation (5.4) it follows that $R(z_1(t), \gamma')\gamma' = 0$. If a vector $w \in T_a L^{\perp}$ is orthogonal to z_1, then $h(x, w)$ is orthogonal to v, because

$$(h(x, w), v) = (A_v x, w) = -(z_1, w) = 0.$$

So, by Proposition 6.15, any L-parallel Jacobi vector field $w(t)$, $(w(0) \perp z_1)$, along γ is orthogonal to $\bar{v}$ for all t. In particular, the vectors $w(\frac{\pi}{2})$ and $\bar{\nabla}_{\gamma'(\frac{\pi}{2})} w(t)$ at the point $\gamma(\frac{\pi}{2})$ are orthogonal to $\bar{v} = z_1(\frac{\pi}{2})$. Thus, the orthogonal complement to $z_1(\frac{\pi}{2})$ in $T_{\gamma(\frac{\pi}{2})} L^{\perp}$ is an invariant subspace of the linear operator $B(\gamma'(\frac{\pi}{2}), *)$. Since a field of normal vectors $v(t)$ on any interval of $\gamma(t)$ with length $\frac{\pi}{2}$ rotates by an angle of $\frac{\pi}{2}$ in $\bar{M}$, it follows that for any t, the orthogonal complement to $z_1(t)$ in $T_{\gamma(t)} L^{\perp}$ forms an invariant subspace of the operator $B(\gamma'(t), *)$.

Step 4. We will show that $v(\frac{\pi}{2})$ stays orthogonal to M under $\bar{\nabla}$-parallel displacement along the geodesic $\gamma \subset L$, $(\gamma_1(0) = \gamma(\frac{\pi}{2}), \ \gamma_1'(0) \perp \gamma'(\frac{\pi}{2}))$.

Let $u \in T_{\gamma(\frac{\pi}{2})} L$ be any unit vector orthogonal to $\gamma'(\frac{\pi}{2})$. Since $|A_{v(\frac{\pi}{2})} \gamma'(\frac{\pi}{2})| = 1$, in view of (6.20), the equality $A_{v(\frac{\pi}{2})} u = 0$ holds. Thus for any $w \in$

$T_{\gamma(\frac{\pi}{2})}L^{\perp}$ the vector $h(u, w)$ is orthogonal to $v(\frac{\pi}{2})$:

$$(h(u, w), v(\frac{\pi}{2})) = (A_{v(\frac{\pi}{2})}u, w) = 0.$$

From this and Proposition 6.15 it follows that the normal vector $v(\frac{\pi}{2})$ is stationary along γ_1.

Step 5. We will show that the tangent vector z_1 becomes orthogonal to M at "time" $t = \frac{\pi}{2}$ under $\bar{\nabla}$-parallel displacement along any geodesic in L with the initial point a.

For any unit vector $u \in T_aL$, $(u \perp x)$, and real $\varepsilon \approx 0$, consider the geodesic

$$\gamma_\varepsilon \subset L, \quad \gamma_\varepsilon(0) = a, \ \gamma_\varepsilon'(0) = (\cos\varepsilon)x + (\sin\varepsilon)u.$$

For some $t_1 \approx \frac{\pi}{2}$ the point $\gamma_\varepsilon(t_1)$ belongs to a hypersurface in L of the form $\exp(U)$, where U consists of vectors $T_{\gamma(\frac{\pi}{2})}L$ orthogonal to $\gamma'(\frac{\pi}{2})$. As was shown before, the normal vector $v(\frac{\pi}{2})$ is stationary along a geodesic between the points $\gamma(\frac{\pi}{2})$ and $\gamma_\varepsilon(t_1)$. By (6.20) the result of $\bar{\nabla}$-parallel displacement of z_1 along the boundary of the triangle $\overline{a, \gamma(\frac{\pi}{2}), \gamma_\varepsilon(t_1)}$ is equal to z_1. Thus z_1 under $\bar{\nabla}$-parallel displacement along γ_ε becomes a normal vector at "time" $t = t_1$.

We will show that $t_1 = \frac{\pi}{2}$. There exists an L-parallel Jacobi vector field $y_\varepsilon(t)$ along γ_ε with the initial value $y_\varepsilon(0) = z_1$. The derivative of this vector field at the point t_1 has the form

$$\bar{\nabla}_{\gamma_\varepsilon'(t_1)}y_\varepsilon(t)) = u_1 + v_1, \ (u_1 \in T_{\gamma_{\varepsilon(t_1)}}L^{\perp}, \quad v_1 \in T_{\gamma_{\varepsilon(t_1)}}M^{\perp}).$$

By (6.21) the vector field $y_\varepsilon(t)$ is given by the formula

$$y_\varepsilon(t) = \cos(t_1 - t)\bar{y}_0 + \sin(t_1 - t)(\bar{u}_1 + \bar{v}_1),$$

where $\bar{y}_0$, $\bar{u}_1$ and $\bar{v}_1$ are $\bar{\nabla}$-parallel vector fields along γ_ε, which contain, respectively, the vectors $y_0 = y_\varepsilon(t_1)$, u_1 and v_1. For $t = 0$ we get

$$(\cos t_1)\bar{y}_0 + (\sin t_1)(\bar{u}_1 + \bar{v}_1) = z_1. \tag{6.26}$$

From (6.26) and the fact that $\bar{v}_1$ and $\bar{z}_1$ are orthogonal to the vectors $\bar{y}_0$ and $\bar{u}_1$, we get the following equations at the point $\gamma_\varepsilon(t_1)$:

$$(\sin t_1)\bar{v}_1 = z_1, \ (\cos t_1)y_0 + (\sin t_1)u_1 = 0. \tag{6.27}$$

Thus the vector u_1 is parallel to nonzero y_0, that is, y_0 is an eigenvector for the linear operator $B(\gamma_\varepsilon'(t_1), *)$. As was shown in Step 2, the eigenvalue is equal to zero:

$$u_1 = B(\gamma_\varepsilon'(t_1, y_0) = 0.$$

But then from (6.27) it follows that $\cos t_1 = 0$ and $t_1 = \frac{\pi}{2}$.

Step 6. We will "chip off" a tangent vector z_1.

Consider along γ_ε, as for γ above, a ∇-parallel vector field $z_\varepsilon(t)$, $(z_\varepsilon(0) = z_1)$, for which the curvature $R(z_\varepsilon(t), \gamma_\varepsilon')\gamma_\varepsilon'$ is equal to zero. Since u was arbitrary (in the construction of γ_ε) we obtain the property $R(z_1, u)u = 0$, $(u \in T_aL)$. As was shown in Step 3 above, the $(n-1)$-dimensional subspace V_1 containing the vectors from $T_aL^\perp$ which are orthogonal to z_1 is invariant under the linear operators $B(u, *)$, $(u \in T_aL)$. Note that z_1 is not orthogonal to y:

$$(z_1, y) = (-A_v x, y) = -(h(x, y), v) = -(v, v) \neq 0.$$

Step 7. We will "chip off" an m-dimensional distribution $Z \subset TL^\perp$.

Since the subspace V_1, consisting of the vectors orthogonal to z_1, is invariant under the operators $B(u, *)$, we may repeat the above procedure until we get $s \geq 1$ linearly independent unit vectors $z_i \in T_aL^\perp$, $(1 \leq i \leq s)$, satisfying the conditions $R(z_i, u)u = 0$; moreover, the orthogonal complement $V_a \subset T_aL^\perp$ and the subspace $Z_a = z_1 \wedge \cdots \wedge z_s$ are invariant under the operators $B(u, *)$ and do not contain their real eigenvectors. We may assume that $(z_i, z_j) = \delta_{ij}$.

Let $\{\xi_{i\alpha} = h(x_\alpha, z_i)\}$ be unit normal vectors at a point a, where $\{x_\alpha\}_{1 \leq a \leq \nu}$ is an orthonormal base of T_aL. $A_{\xi_{i\alpha}} x_\alpha = z_i$, so by (6.23) we obtain $A_{\xi_{i\alpha}} x_\beta = 0$, $(\alpha \neq \beta)$. From this it follows that

$$(h(x_\alpha, z_i), h(x_\beta, z_j)) = 0, \quad (\alpha \neq \beta \text{ or } i \neq j),$$

i.e., the normal vectors $\{\xi_{i\alpha}\}$ form an orthonormal system and generate a νs-dimensional subspace $N_1 \subset T_aM^\perp$.

Since under $\bar{\nabla}$-parallel displacement along a geodesic $\gamma_\alpha \subset L$, $(\gamma_{\alpha'(0)} = x_\alpha)$, the normal vector ξ becomes tangent to M at "time" $t = \frac{\pi}{2}$, and since M is t-unique projectable along L with $t \geq \nu m$, we obtain the inequality $s \geq m$. By the Gauss equation (5.4) and (6.21), the mixed sectional curvature of a ruled submanifold is not greater than one, and so the Ricci curvature satisfies

$$Ric_M \leq (n + \nu) - (s + 1) \leq \dim M - (m + 1).$$

Comparing the last inequality with (6.25), we obtain $s = m$ and

$$R(w, u)u = u^2 w, \quad R(z, u)u = 0, \quad (u \in T_aL \; w \in V_a, \; z \in Z_a).$$

Since we can freely choose the point a, there exist regular sub-bundles

$$Z \subset TL^{\perp}, \ (\dim Z = m); \quad V \subset TL^{\perp}, \ (\dim V = n - m).$$

Moreover, V is $\bar{\nabla}$-parallel along $\{L\}$, and the orthogonal complement N_2 to N_1 in $TM^{\perp}$ is $\bar{\nabla}$-parallel along the generators $\{L\}$.

Step 8. We will show that the distributions Z and $V \oplus TL$ are integrable and totally geodesic in M. From this and de Rham's decomposition theorem, it will follow that M is locally isometric to the product $M_1^m \times M_2^{n-m+\nu}$ and that integral manifolds for $V \oplus TL$ are developed along $\{L\}$ ruled submanifolds.

We use the Codazzi equation (5.5), which in the case of (6.20) has the form

$$(\bar{\nabla}_x h)(z, y) = (\bar{\nabla}_z h)(x, y), \quad (x \in TL, \ z \in TM, \ y \perp x \wedge z). \tag{6.28}$$

Let $y_i \subset V, \ z_j \subset Z$ be local vector fields. We calculate (6.28) for $z = z_j, \ y = y_i$:

$$\begin{aligned} (\bar{\nabla}_x h)(z_j, y_i) &= \nabla_x^{\perp}(h(z_j, y_i)), \\ (\bar{\nabla}_{z_j} h)(x, y_i) &= -h(\nabla_{z_j} x, y_i) - h(x, \nabla_{z_j} y_i), \end{aligned} \tag{6.29}$$

and find the projections of the vectors in (6.29) onto the normal vectors $\xi_s = h(x, z_s)$:

$$\begin{aligned} (h(x, \nabla_{z_j} y_i), \xi_s) &= (A_{\xi_s} x, \nabla_{z_j} y_i) = -(\nabla_{z_j} z_s, y_i), \\ (\nabla_x^{\perp}(h(z_j, y_i)), \xi_s) &= \nabla_x (A_{\xi_s} z_j, y_i) = 0, \\ (h(\nabla_{z_j} x, y_i), \xi_s) &= (A_{\xi_s} y_i, \nabla_{z_j} x) = 0. \end{aligned}$$

Thus

$$((\bar{\nabla}_x h)(z_j, y_i), \xi_s) = 0, \ ((\bar{\nabla}_{z_j} h)(x, y_i), \xi_s) = (\nabla_{z_j} z_s, y_i),$$

and since y_i was arbitrary we obtain $\nabla_{z_j} z_s \perp V$. Analogously with the help of (6.28) we will prove that $\nabla_{z_j} z_s \perp TL$. Assume that $w \subset TL, \ w \perp x$ and $\nabla_x w = 0$. Then it is not difficult to calculate the following:

$$\begin{aligned} (\bar{\nabla}_{z_j} h)(x, w) &= h(\nabla_{z_j} x, w) - h(x, \nabla_{z_j}, w), \\ ((\bar{\nabla}_x h)(z_j, w) &= \nabla_x^{\perp}(h(z_j, w)), \\ (h(x, \nabla_{z_j} w), \xi_s) &= (z_s, \nabla_{z_j} w)) - (\nabla_{z_j} z_s, w), \\ (\nabla_x^{\perp}(h(z_j, w)), \xi_s) &= 0, \\ (h(w, \nabla_{z_j} x), \xi_s) &= 0. \end{aligned}$$

Then $(\nabla_{z_j} z_s, w) = 0$ and hence $\nabla_{z_j} z_s \perp TL$.

Finally we obtain $\nabla_{z_j} z_s \subset Z$; this means that the distribution Z is integrable and certain integral manifolds are totally geodesic in M.

Analogous considerations for (6.28) with $z = v_j$ and $y = v_i$ allow us to obtain the inclusion $\nabla_{v_i} v_j \subset V \oplus TL$. As was shown before, $\nabla_x v_i \subset V$ and a field v_i along $\{L\}$ is tangent to a submanifold of the one-parameter variation $\{L\}$, that is, $[x, v_i] \subset V \oplus TL$. From this it follows that

$$\nabla_{v_i} x = \nabla_x v_i - [x, v_i] \subset V \oplus TL .$$

Consequently, the distribution $V \oplus TL$ is integrable and its integral submanifolds $\{M_2\}$ are totally geodesic in M. The submanifolds $\{M_2\}$ are ruled with generators $\{L\}$ because V is invariant under the operators $B(x, *)$; the submanifolds $\{M_2\}$ are developed along $\{L\}$, the distribution N_2 is stationary along $\{L\}$ and the distributions Z and N_1 "change" places under $\bar{\nabla}$-derivation along $\{L\}$. □

6.16 Corollary. *A ruled submanifold $M \subset \bar{M}$ satisfying the properties* (6.20), (6.21) *and $K_{mix} > 0$ along a complete generator L is uniquely projectible.*

Proof. Let $k = 1$. Assume on the contrary that there exists a point $p \in L$ and unit vectors $x \in T_pL$, $v \in T_pM^{\perp}$ such that along a geodesic $\gamma \subset L$, $(\gamma'(0) = x)$, the $\bar{\nabla}$-parallel displacement $\bar{v}$ of a vector v is tangent to M at some moment t_0. As in Step 1 of the proof of Theorem 6.14, we consider a Jacobi vector field $y(t)$, $(y(t_0) = \bar{v})$, in $\bar{M}$ that is induced by the rulings $\{L\}$ along γ. We then obtain

$$h(x, y(0)) \,||\, v, \;\; B(x, y(0)) = \lambda y(0),$$

where $\lambda = -\cot(t_0) \leq 0$ and $t_0 \in (0, \frac{\pi}{2}]$.

A $\nabla^{\perp}$-parallel field $v(t)$, $(v(0) = v)$, of normal vectors to M along the geodesic γ corresponds to a smooth curve on the unit sphere $S^{N-1} \subset \mathbb{R}^N$, see Step 2 in the proof of Theorem 6.14. From the definition of $v(t)$ and in view of (5.3), (5.4) and $K_{mix} > 0$ we obtain $|\bar{\nabla}_{\gamma'} v(t)| < 1$. Then the length

$$l = \int_0^{t_0} |\bar{\nabla}_{\gamma'} v(t)| \, dt$$

of the curve $v(t)$ is smaller than $\frac{\pi}{2}$. But the vectors $v(0)$ and $v(t_0)$ are orthogonal in $\mathbb{R}^N$ and the distance between $v(0)$ and $v(t_0)$ on the unit sphere S^{N-1} is $\frac{\pi}{2}$, which is a contradiction. □

6.17 Corollary. *Let $M^{n+\nu} \subset \bar{M}$ be a ruled submanifold with complete rulings $\{L^\nu\}$, and suppose that conditions* (6.20), (6.21) *and* (6.23) *with $k > 0$ are satisfied. If $\nu \geq \bar{\rho}(n)$ (or $\nu > 2$ for the Kählerian case), then $M^{n+\nu}$ is locally isometric to the product $L^\nu \times M_2^n$ and* $\operatorname{codim} M \geq \nu n$. *In the case of $\bar{M} = S^N$ or CP^N and* $\operatorname{codim} M = \nu n$ *a submanifold M is congruent to the domain of the Segre submanifold.*

Proof. The structural tensor $B(x, y)$ has eigenvectors, by Lemma 3.40 (real case) and Lemma 3.42 (Kählerian case). By repeating the proof of Theorem 6.14, we see that the distribution Z (which is geodesic and integrable) coincides with $TL^\perp$ and that a submanifold M is isometric to the product $L \times M_2^n$. Let $\{x_\alpha\}$ and $\{z_i\}$ denote orthonormal bases of T_aL and $T_aL^\perp$, respectively. As in the proof of Theorem 6.14, we get that the vectors $\{h(x_\alpha, z_i)\}$ form an orthonormal system in $T_aM^\perp$, i.e., $\operatorname{codim} M \geq \nu n$.

Let $\operatorname{codim} M = \nu n$. If $\bar{M} = S^{n+\nu+\nu n}(k)$, then $\{h(x_\alpha, z_i)\}$ is an orthonormal base of $T_aM^\perp$, and in view of the above, the vectors from $TL^\perp$ are asymptotic, i.e., the leaves $\{a\} \times M_2$ are domains of great spheres in $S^{n+\nu+\nu n}$. Thus in view of Theorem 5.3, M is congruent to the domain of the Segre submanifold. The case $\bar{M} = CP^{n+\nu+\nu n}(k)$ is analogous. □

It follows from the proof of Theorem 6.14 that (6.23) may be changed (without affecting Theorem 6.14) by the integral inequality for the second quadratic forms as follows:

Along any geodesic $\gamma : [0, \frac{\pi}{2\sqrt{k}}] \to M$, $(\gamma' \in TL)$,

$$\int_\gamma ||A_{\xi(t)}|| dt \leq \pi\sqrt{k},$$

where $\xi(t)$ is $\nabla^\perp$-parallel along γ unit normal field.

6.3. Ruled submanifolds in a Riemannian space of positive curvature and Segre type embeddings

Combining the area of L-parallel fields method for foliations (see Theorem 4.10) with the synthetic method of Theorem 6.14 on ruled submanifolds in space forms, we obtain a Segre type decomposition in a Riemannian space with positive sectional curvature.

For a ruled submanifold M in a Riemannian space $\bar{M}$ with nonconstant mixed sectional curvature along the generators we assume

$$k_2 \geq \bar{K}(x, y) \geq k_1 > 0, \ (x \in TL, \ y \in TL^{\perp}). \tag{6.30}$$

Below we consider the following inequality (where $k = \frac{k_2+k_1}{2}$) that is analogous to (6.23):

$$\sqrt{(A_\xi x)^2 + (A_\xi u)^2} \leq \sqrt{k}\{1 - 5\left(1 - \frac{k_1}{k_2}\right)\left(\frac{a(L_0)^2}{k} + 1\right)\}, \quad (x, u \in TL, \ x \perp u, \ |x| = |u| = |\xi| = 1). \tag{6.31}$$

6.18 Theorem. *Let $M^{n+\nu} \subset \bar{M}$ be a ruled submanifold with a complete ν-dimensional ruling $\{L\}$ and suppose that conditions* (6.20), (6.30) *and* (6.31) *hold. If $\nu \geq \bar{\rho}(n)$ ($\nu > 2$ for the Kählerian case), then $k_2 = k_1$, $M^{n+\nu}$ is locally isometric to the product $L^\nu \times M_2^n$ and* codim $M \geq \nu n$.

Proof. The proof is broken down into steps. We will obtain the equality $k_2 = k_1$ and then the proof of the theorem will be completed with the help of Corollary 6.17.

Step 1. Let L_0 be a generator and let $p \in L_0$ be any point. In view of the given inequality for the dimensions of L_0 and M (see Lemma 3.40 and Lemma 3.42) there exist unit vectors $x_1 \in T_pL$, $y_1 \in T_pL^{\perp}$ and a real $\lambda \leq 0$ with the property $B(x_1, y_1) = \lambda y_1$ (a bilinear operator $B : TL \times TL^{\perp} \to TL^{\perp}$ is defined for a totally geodesic foliation $\{L\}$ on M by rule (2.3), for the Kählerian case the condition $B(Jx, *) = JB(x, *)$ holds). Let $y(t)$, $(y(0) = y_1)$, be an L-parallel Jacobi vector field in M along a geodesic $\gamma : [0, \frac{\pi}{\sqrt{k}}] \to L_0$, $(\gamma'(0) = x_1)$, i.e., it satisfies the equation $\nabla_{\gamma'} y(t) = B(\gamma', y(t))$. Since $\{L\}$ are rulings, $y(t)$ is also a Jacobi vector field in $\bar{M}$ and

$$\bar{\nabla}_{x_1} y(t) = \lambda y + v, \quad v = h(x_1, y_1). \tag{6.32}$$

From (6.31) and the method of Theorem 4.10, it follows that $v \neq 0$ and also that $|v| \leq \sqrt{k}$. In view of (6.32), (6.20) and (6.30) we have $y(t) = Y(t) + u(t)$, where

$$\begin{aligned} Y(t) =& y_1[\cos(\sqrt{k}t) + \frac{\lambda}{\sqrt{k}} \sin(\sqrt{k}t)] + \frac{v}{\sqrt{k}} \sin(\sqrt{k}t) = \\ &- y_1[\sin(\sqrt{k}(t - t_0))/\sin(\sqrt{k}t_0)] + v \sin(\sqrt{k}t)/\sqrt{k}, \end{aligned}$$

with $t_0 = arccot\,(-\lambda/\sqrt{k})/\sqrt{k} \in (0, \frac{\pi}{2\sqrt{k}}]$, $u(0) = u'(0) = 0$.

Step 2. In view of $|y(0)|' = \lambda \le 0$ (and for $\lambda = 0$ since $|y|''(0) < 0$) the inequality $|y(t_0)| \le |y(2t_0)|$, that is,

$$|u(t_0)| + |u(2t_0)| + \frac{|v|}{\sqrt{k}} \sin(\sqrt{k}t_0) \le |Y(2t_0)|, \tag{6.33}$$

is sufficient to ensure the existence of a local minimum for the function $|y(t)|$ on the interval $(0, 2t_0]$. From Lemma 4.7 for $\varepsilon_1 = \frac{k_2-k_1}{2}$ and the inequality $|Y(t)| \le \sqrt{2}$ it follows that

$$|u(t)| \le \sqrt{2}\frac{k_2 - k_1}{3k_1 - k_2}(1 - \cos(\sqrt{k}t)), \ (0 \le t \le 2t_0). \tag{6.34}$$

Since

$$Y(2t_0) = -y_1 + \frac{v}{\sqrt{k}} \sin(2\sqrt{k}t_0)$$

and $v \perp y_1$, it follows that $|Y(2t_0)| \ge 1$ and $|y(2t_0)| \ge 1 - |u(2t_0)|$. In view of the above estimates, condition (6.33) follows from the inequality

$$\sqrt{2}\frac{k_2 - k_1}{3k_1 - k_2}[2 - \cos(\sqrt{k}t_0) - \cos(2\sqrt{k}t_0)] + \omega \sin(2\sqrt{k}t_0) \le 1, \tag{6.33'}$$

where $\omega = \frac{|v|}{\sqrt{k}} \in [0, 1]$. Since the function $f(t) = \cos(t) + \cos(2t)$ is monotonically decreasing for $0 \le t \le \frac{\pi}{2}$ $(f' < 0)$,

$$\omega \le 1 - 3\sqrt{2}\frac{1 - \delta}{3\delta - 1} \tag{6.35}$$

is sufficient to prove (6.33'), and consequently (6.33). Note that the right hand side of (6.35) is positive for $\delta = \frac{k_1}{k_2} \ge 0.724$.

Step 3. We will show that assuming

$$\omega < \sqrt{\frac{2\delta}{1+\delta}\left(\frac{3\delta - 1}{\delta + 1}\right)^2 \Big/ \left(\frac{a(L_0)^2}{k} + 1\right)} - \pi\frac{1 - \delta}{1 + \delta} \tag{6.36}$$

leads to a contradiction. Inequality (6.35) follows from (6.36) for $\delta \ge 0.7$ because for such a δ we have

$$\sqrt{\frac{2\delta}{1+\delta}\left(\frac{3\delta - 1}{\delta + 1}\right)^2} - \pi\frac{1 - \delta}{1 + \delta} \le 1 - 3\sqrt{2}\frac{1 - \delta}{3\delta - 1}.$$

The last inequality divided by $1 - \delta > 0$ is equivalent to

$$1 \Big/ \left(1 + \sqrt{\frac{2\delta}{1+\delta}}\right) + 4\left(\frac{2\delta}{1+\delta}\right)^{\frac{3}{2}} \ge 3\sqrt{2}\frac{1 + \delta}{3\delta - 1} - \pi$$

which holds in the case of $\delta \geq 0.7$. Note that the last inequality (after dividing by $1-\delta > 0$) is equivalent to

$$f(\delta) := 1\Big/\left(1+\sqrt{\frac{2\delta}{1+\delta}}\right) + 4\left(\frac{2\delta}{1+\delta}\right)^{\frac{3}{2}} \geq 3\sqrt{2}\frac{1+\delta}{3\delta-1} + \pi,$$

which, after the change of parameter

$$x = \sqrt{\frac{2\delta}{1+\delta}}, \quad \left(\delta \in (\frac{1}{3}, 1],\ x \in (\frac{1}{\sqrt{2}}, 1]\right),$$

may be written in the form

$$\tilde{f}(x) := \frac{1}{1+x} + 4x^3 - \frac{3\sqrt{2}}{2x^2-1} + \pi \geq 0.$$

The function $\tilde{f}(x)$ is increasing for $x \in (\frac{1}{\sqrt{2}}, 1]$, because its derivative is positive:

$$\tilde{f}'(x) = -\frac{1}{(1+x)^2} + 12x^2 + \frac{12\sqrt{2}x}{(2x^2-1)^2} \geq 12x^2 - 1 \geq 5 > 0.$$

Thus the function $f(\delta)$ is also increasing for $\delta \in (\frac{1}{3}, 1]$. With the help of a computer we get an upper bound for the unique root: $f(0.7) > 0$. Thus in the case of (6.36) (see Step 2) the function $|y(t)|$ has its first local minimum $t_m \in [0, \frac{\pi}{\sqrt{k}}]$. In particular, $|y(t)| \leq 1$ for $0 \leq t \leq t_m$.

Assume that $q = \gamma(t_m)$. Consider the opposite parameterization of γ with parameter $\tau = t - t_m$, i.e., $\gamma(0) = q,\ \gamma(t_m) = p$ and, in particular, $y(t_m) = y_1$. The vector field $y(\tau)$ may be written in the form $y(\tau) = Y_1(\tau) + u_1(\tau)$, where

$$Y_1(t) = y(q)\cos(\sqrt{k}\tau) + y'(q)\sin(\sqrt{k}\tau)/\sqrt{k} + w\sin(\sqrt{k}\tau)/\sqrt{k}$$

and $w = h(y(q), \gamma')$. Since (in view of (6.31)) the inequality $|w| \leq \sqrt{k}|y(q)|$ holds, we have

$$|y(q)\cos(\sqrt{k}\tau) + w\sin(\sqrt{k}\tau)/\sqrt{k}| \leq |y(q)|$$

(all three vectors $y(q),\ y'(q)$ and w are mutually orthogonal). Next we deduce $|y'(q)| \leq a(L_0)$ (see Step 3 of the proof of Theorem 4.10). Consequently

$$|Y_1(t_m)| \leq |y(q)|\left(\frac{a(L_0)^2}{k} + 1\right)^{\frac{1}{2}}.$$

From Lemma 4.7 with $\varepsilon_1 = \frac{k_2-k_1}{2}$ it follows that

$$|u_1(t_m)| \leq 2\frac{1-\delta}{3\delta-1}|y(q)|\left(\frac{a(L_0)^2}{k}+1\right)^{\frac{1}{2}}.$$

Finally we obtain

$$|y(p)| \leq |Y_1(t_m)| + |u_1(t_m)| \leq |y(q)|\left(1+2\frac{1-\delta}{3\delta-1}\right)\left(\frac{a(L_0)^2}{k}+1\right)^{\frac{1}{2}} =$$
$$|y(q)|\left(\frac{a(L_0)^2}{k}+1\right)^{\frac{1}{2}}\frac{1+\delta}{3\delta-1},$$

i.e., (in view of $|y(p)| = |y_1| = 1$) we have

$$|y(q)| \geq \frac{3\delta-1}{1+\delta}\Big/\left(\frac{a(L_0)^2}{k}+1\right)^{\frac{1}{2}}. \tag{6.37}$$

Assume that γ has its initial parameterization again. Let $\bar{V}_2(t)$ denote the area of the parallelogram in $T\bar{M}$ with sides the vectors $y(t)$ and $\bar{\nabla}_{\gamma'}y(t)$. Note that $\bar{V}_2(0) = |v| = \omega\sqrt{k}$. As in the proof of Theorem 4.10 we deduce the inequality

$$\bar{V}_2(t_m) \geq \sqrt{k_1}|y(t_m)|^2$$

at the point t_m, which is the local minimum of the function $|y(t)|$. Thus

$$\bar{V}_2(t_m) - \bar{V}_2(0) \geq \sqrt{k_1}|y(t_m)|^2 - \omega\sqrt{k}.$$

From this and

$$\bar{V}_2(t_m) - \bar{V}_2(0) \leq \int_0^{t_m} |\bar{V}_2'(t)|\,dt \leq \left(\int_0^{t_m} |y(s)|^2\,ds\right)\frac{k_2-k_1}{2} \leq t_m\frac{k_2-k_1}{2},$$

using (6.37) and $t_m \leq \frac{\pi}{\sqrt{k}}$, the inequality opposite to (6.36) follows, whence we obtain a contradiction.

Step 4. We temporarily put into (6.31) the parameter $\frac{1}{c} > 0$ instead of the constant 5. From (6.31) and the inequality opposite to (6.36)

$$\omega \geq \sqrt{2\frac{\delta}{1+\delta}\left(\frac{3\delta-1}{1+\delta}\right)^2\Big/\left(\frac{a(L_0)^2}{k}+1\right)} - \pi\frac{1-\delta}{1+\delta} \tag{6.38}$$

it follows that

$$\omega \geq \left(\sqrt{\frac{2\delta}{1+\delta}}\left(\frac{3\delta-1}{1+\delta}\right)^2 - \frac{c\pi}{1+\delta}\right)\Big/\left(\frac{a(L_0)^2}{k}+1\right). \tag{6.39}$$

We will show that $k_2 = k_1$ and that under $\nabla^\perp$-parallel displacement along $\gamma(t)$, $(0 \le t \le \frac{\pi}{2\sqrt{k}})$, the unit normal vector $v(0) = \frac{v}{|v|}$ rotates uniformly in $\bar{M}$ by the angle $\frac{\pi}{2}$ (this generalizes the method of proof of Theorem 6.14, Step 2, and condition (6.31) is partially used).

We will estimate the deviation of the normal vector $v(0)$ from the normal space $T_{\gamma(t_0)}M^\perp$ under $\bar{\nabla}$-parallel displacement along γ at the point $t = t_0$.

Since

$$y(t_0) = v \sin(\sqrt{k}t_0)/\sqrt{k} + u(t_0) \in T_{\gamma(t_0)}M,$$

the cosine of angle $\varphi \in [0, \frac{\pi}{2}]$ (between the normal vector v and the space $T_{\gamma(t_0)}M^\perp$) is not greater than $|u(t_0)|\sqrt{k}/(\sin(\sqrt{k}t_0)|v|)$. From (6.34) and (6.39) it follows that

$$\cos\varphi \le \frac{|u(t_0)|\sqrt{k}}{\sin(\sqrt{k}t_0)|v|} \le \frac{\sqrt{2}k(k_2 - k_1)(q - \cos(\sqrt{k}t_0))}{\omega(3k_1 - k_2)\sin(\sqrt{k}t_0)} \le \frac{\sqrt{2}(1-\delta)/(3\delta-1)[a(L_0)^2/k+1]}{\sqrt{2\delta/(1+\delta)[(3\delta-1)/(1+\delta)]^2 - c\pi/(1+\delta)}}. \tag{6.40}$$

Let $v(t)$ be a $\nabla^\perp$-parallel field of normal vectors to M along γ, containing the vector $v(0) = \frac{v}{|v|}$. We identify the Euclidean space $(T_\gamma L)^\perp$ (all normal vectors to L_0 in $\bar{M}$) with $\mathbb{R}^{n+p}$ by the $\bar{\nabla}$-parallel displacement along γ. Then the field $v(t)$ corresponds to a smooth curve on the unit sphere $S^{n+p-1} \subset \mathbb{R}^{n+p}$.

We will estimate the distance l between the points $v(0)$ and $v(t)$ in S^{n+p-1}. From the definition of the field $v(t)$ it follows that $\bar{\nabla}_{\gamma'} v(t) = -A_{v(t)}\gamma'$ and with the help of (6.31) we obtain

$$|\bar{\nabla}_{\gamma'} v(t)| \le \sqrt{k}\{1 - \frac{1}{c}(1-\delta)\left(\frac{a(L_0)^2}{k} + 1\right)\}.$$

Thus

$$l \le \int_0^{t_0} |\bar{\nabla}_{\gamma'} v(t)|dt \le t_0\sqrt{k}\{1 - \frac{1}{c}(1-\delta)\left(\frac{a(L_0)^2}{k} + 1\right)\} \le \frac{\pi}{2}\{1 - \frac{1}{c}(1-\delta)\left(\frac{a(L_0)^2}{k} + 1\right)\}.$$

Note that $\varphi \le l$, that is, in the case of (6.32) the inequality

$$\cos\varphi \ge \cos l \ge \sin\{\frac{\pi}{2c}(1-\delta)\left(\frac{a(L_0)^2}{k} + 1\right)\} \ge \frac{1}{c}(1-\delta)\left(\frac{a(L_0)^2}{k} + 1\right) \tag{6.41}$$

holds. We use the well-known inequality

$$\sin x \geq \frac{2}{\pi}x, \quad (0 \leq x \leq \frac{\pi}{2}).$$

Let some real number c satisfy the inequality

$$\frac{c\sqrt{2}}{3\delta - 1} \leq \sqrt{\frac{2\delta}{1+\delta}\left(\frac{3\delta - 1}{1+\delta}\right)^2 - \frac{c\pi}{1+\delta}},$$

which is equivalent to

$$c \leq \sqrt{\frac{2\delta}{1+\delta}} \cdot \frac{(3\delta - 1)^3}{(3\pi + \sqrt{2})\delta - (\pi - \sqrt{2})} \cdot \frac{1}{1+\delta}. \tag{6.42}$$

(Note that the denominator in (6.42) is greater than 0 for $\delta > 0.16$). Then the inequalities (6.40) and (6.41) are compatible only for the extremal situation $k_2 = k_1$, $\lambda = 0$ and $t_0 = \frac{\pi}{2\sqrt{k}}$.

We now explain how to choose the parameter c. Note that from (6.31) it follows that $c \geq 1 - \delta$, which agrees with (6.42) for $\delta \geq 0.8$ (Actually we can consider the inequality

$$f(\delta) = \sqrt{\frac{2\delta}{1+\delta}} \cdot \frac{3\delta - 1}{(2\pi + 1)\delta + 1} \cdot \frac{3\delta - 1}{1 - \delta} \cdot \frac{2\delta}{1+\delta} \geq 1,$$

in which the left hand side is increasing for $\delta \in (0.34, 1]$, because it is a product of four increasing factors). With the help of a computer we find an upper bound for the unique root: $f(0.8) > 0$.

Since we want to get a minimum value for δ, we must choose $c = 0.2$, so $\frac{1}{c} = 5$. This completes the proof of the theorem. □

6.19 Corollary. *Let $M^{n+\nu} \subset \bar{M}$ be a ruled submanifold with complete ν-dimensional ruling $\{L\}$ and suppose that there exists a point $m \in M$ such that along any geodesic $\gamma : [0, \frac{\pi}{\sqrt{k}}] \to L_0$, $(\gamma(0) = m)$, the following conditions hold:* (6.20), (6.30) *and*

$$\frac{|h(\gamma', y)|}{|y|} < \sqrt{k}\{1 - 5\left(1 - \frac{k_1}{k_2}\right)\left(\frac{a(L_0)^2}{k} + 1\right)\}, \quad (y \in T_\gamma L^\perp), \tag{6.43}$$

with $x = \gamma'$ and $k = \frac{k_2+k_1}{2}$. Then $\nu < \rho(n)$, or $\nu = 2$ and n is divisible by 4 for the Kählerian case.

In the case of $k_2 = k_1$, condition (6.43) means (in view of the Gauss equation (5.4)) that the sectional curvature $K(\gamma', y)$ is positive.

Proof. Actually, we repeat the proof of Theorem 6.18.

If inequality (6.43) is given instead of (6.31), then the first inequality of (6.41) is exact and we obtain a contradiction. Thus for the case of (6.43) the operator $B(x,*)$, $(x \neq 0)$ has no eigenvectors and by Lemma 3.40 (or Lemma 3.42 for the Kählerian case) the proof of the corollary is complete. □

Combining the 3-dimensional volume method of L-parallel fields, (see Theorem 4.23) with the synthetic method of Theorem 6.14 also leads us to a Segre type decomposition (Theorem 6.20, given below with idea of the proof) analogous to Theorem 6.18, but this time we cannot obtain more precise curvature pinching.

6.20 Theorem. *Let $M \subset \bar{M}$ be a ruled submanifold with rulings $\{L\}$ and suppose that the following conditions hold:*

$$k_2 \geq \bar{K}(x,y) \geq k_1 > 0, \ (x \in TL, \ y \in TM), \tag{6.44}$$

$$\bar{R}(x,y)z = 0, \ \ (x \in TL, \ y \in TM, \ z \perp x \wedge y), \tag{6.45}$$

or for the Kählerian case

$$\bar{R}(x,y)z \,||\, Jz, \ (x \in TL, \ y \in TM, \ z \perp x \wedge Jx \wedge y \wedge Jy), \tag{6.45'}$$

$$\sqrt{(A_\xi x)^2 + (A_\xi u)^2} \leq \sqrt{k}\{1 - 5.5\left(1 - \frac{k_1}{k_2}\right)\left(\frac{a(L)^2}{k} + 1\right)^2\}, \tag{6.46}$$

where $x, u \in TL$, $x \perp u$, $|x| = |u| = |\xi| = 1$ and $k_1 \geq 0.82 k_2$. If

$$\dim L \geq \bar{p}(\operatorname{codim} L), \ (\dim L > 2 \textit{ for the Kählerian case}),$$

then $k_1 = k_2$, and M is locally isometric to the product $L \times L^\perp$, $(K_L = const > 0)$; moreover $\operatorname{codim} M \geq (\dim L)(\operatorname{codim} L)$.

6.21 Remark. If $k_1 = k_2$ then from (6.46) and (6.44), in view of the Gauss equation (5.4), it follows that $K_{mix} \geq 0$. From (6.44) it also follows that $k_2 \geq K_L \geq k_1$. If in Theorem 6.20 the space $\bar{M}$ coincides with S^N or CP^N (i.e., $k_1 = k_2$ and (6.45) or (6.45') holds) and $\operatorname{codim} M = (\dim L)(\operatorname{codim} L)$, then the submanifold M is congruent to a domain of the Segre embedding.

Idea of Proof. By (6.44), the sectional curvature $K_L \geq k_1 > 0$, that is, the rulings are compact Riemannian manifolds with diameter not greater than $\frac{\pi}{\sqrt{k_1}}$.

Recall that L-parallel fields are globally defined on a ruling with trivial holonomy. By Lemma 1.16, almost every ruling has trivial holonomy; below L_0 is such a ruling.

Assume that $m \in L_0$. In view of the given estimate for dim L there exist unit vectors $x_1 \in T_mL$, $y_1 \in T_mL^\perp$ and a real $\lambda \leq 0$ with the property $B(x_1, y_1) = \lambda y_1$, where B is a structural tensor of the foliation $\{L\}$. Consider the L-parallel Jacobi field $Y : L_0 \to TL_0^\perp$, $(Y(m) = y_1)$, along the geodesics on L_0. Let $\gamma_1 : [0, \frac{\pi}{\sqrt{k}}] \to L_0$, $(\gamma_1(0) = m)$, be a geodesic with initial value $\gamma_1'(0) = x_1$.

For the restriction $y_1(t) = Y(\gamma_1(t))$ of the field Y along γ_1 we have

$$\bar{\nabla}_{x_1} y_1(t) = \lambda y_1 + v_1, \tag{6.47}$$

where $v_1 = h(x_1, y_1)$. From (6.47) and the proof of Theorem 4.10 it follows that $v_1 \neq 0$ and $|v_1| \leq \sqrt{k}$. Let

$$\gamma_2 : [0, l] \to L_0, \ (\gamma_2(0) = m', \ \gamma_2(l) = m, \ l \leq diam\, L \leq \frac{\pi}{\sqrt{k_1}})$$

be the shortest geodesic between some minimum point $m' \in L_0$ for the length function $|Y| : L_0 \to R_+$ (m' exists because L_0 is compact) and the point m, and let $y_2(t) = Y(\gamma_2(t))$ be the restriction of Y along γ_2.

Our plan for the rest of the proof is as follows (for the complete proof see [Rov 19]):

Step 1. By modifying Step 2 of the proof of Theorem 4.22 along γ_2 we will deduce in step 2 below for $\omega_1 = \frac{|v_1|}{\sqrt{k}}$ the lower bound

$$\omega_1 \geq \left(\frac{3.02k_1 - 1.02k_2}{k_1 + k_2}\right)^3 \frac{k_1}{k} / \left(\frac{a(L)^2}{k} + 1\right)^2 - \frac{\pi(k_2 - k_1)}{2\sqrt{k_1 k}}. \tag{6.48}$$

Step 2. By modifying in Step 4 the exterior-geometric part of the proof of Theorem 6.18 we will estimate along γ_1 the velocity of rotation of a normal subspace and obtain the extremal situation $k_2 = k_1$. Thus, Theorem 6.20 will be reduced to Corollary 6.17. □

CHAPTER VII

DECOMPOSITION OF PARABOLIC SUBMANIFOLDS

In **Chapter VII** we apply the ideas of Chapter VI to the decomposition of parabolic submanifolds (with degenerate second quadratic forms, see Section 5.1) of a Riemannian space with positive curvature.

In Lemma 7.2 we show that the classical Segre embedding is also a parabolic submanifold and then investigate the relationship of the condition $K_M > 0$ with the synthetic concept of a *t-uniquely projectable submanifold.* Theorem 7.3 gives a criterion for a parabolic submanifold M in CROSS with $K_M \geq 0$ to be a Segre embedding. Theorem 7.8 contains a criterion for a parabolic submanifold M of a Riemannian space with positive curvature to be a Segre type embedding; its proof combines the extrinsic (synthetic) procedure of Section 7.1 with the variational method of Theorem 7.3 for foliations.

7.1. Parabolic submanifolds in CROSS

The curvature tensor of a submanifold M in $S^N(k)$ or a Kählerian submanifold M in $CP^N(k)$ has the following properties:

$$\bar{R}(x,y)z\begin{cases} = 0, & (x,\ y \in TM,\ z \perp (x \wedge y)), \\ \| Jz, & (x,\ y \in TM,\ z \perp (x \wedge y \wedge Jx \wedge Jy)), \text{ for the Kählerian case} \end{cases} \tag{7.1}$$

and

$$\bar{R}(y,x)x = kx^2 y,\ (x,\ y \in TM;\ \text{for the Kählerian case}, y \perp (x \wedge Jx)). \tag{7.2}$$

We consider in Section 7.1 the inequality for second quadratic forms (compare with (6.23))

$$A_\xi^2 x + A_\xi^2 u \leq k,\ (x, u \in TM,\ x \perp u,\ |x| = |u| = |\xi| = 1), \tag{7.3}$$

or the more clear but stronger inequality (compare with (6.24))

$$\|A_\xi\|^2 \leq 2k,\ (|\xi| = 1). \tag{7.4}$$

For a submanifold $M \subset \bar{M}$ with property (7.1) or (7.2), it follows from assumption (7.4) $K_M \geq 0$ in view of the Gauss equation (5.4).

7.1 Definition. For a submanifold $M \subset \bar{M}$ and any s-dimensional subspace $U_s \subset T_qM^\perp$, let $h(U_s) = Ph : T_qM \times T_qM \to U_s$ be a bilinear form where h is a second fundamental form of M and $P : T_qM^\perp \to U_s$ is an orthoprojector. The *s-null-index* of M, where $s \leq \operatorname{codim} M$, is defined by the formula

$$\mu_s(M) = \min\{\dim N(U_s) : U_s \subset T_qM^\perp,\ q \in M\},$$

where $N(U_s)$ is the null-space of the bilinear form $h(U_s)$. Note that this is different from [CDa 1]).

The following inequalities hold, where $p = \operatorname{codim} M$ and $r(M)$ is the maximal rank of the second quadratic forms of a submanifold M:

$$\mu(M) = \mu_p(M) \leq \cdots \leq \mu_2(M) \leq \mu_1(M) = \dim M - r(M). \tag{7.5}$$

7.2 Lemma. *The Segre submanifold $M^{n+\nu} = f(S^\nu \times S^n) \subset S^N$ in the case of $\nu \neq n$ is parabolic and satisfies the conditions*

$$r(M) = 2\min\{\nu, n\}, \quad \mu_1(M) = \mu_2(M) + 1. \tag{7.6}$$

Proof. Assume, for instance, that $\nu > n$. Let $q \in M^{n+\nu}$ and $\{x_1, \dots, x_\nu, y_1, \dots, y_n\}$ be an orthonormal base of T_qM such that $x_i \in T_qS^\nu$ and $y_i \in T_qS^n$. In view of the Gauss equation (5.4) and the condition

$$K(x, y) = 0, \ (x \in T_qS^\nu,\ y \in T_qS^n)$$

we have

$$h^2(x, y) = k, \ (x \in T_qS^\nu,\ y \in T_qS^n).$$

Then from the linearity in x, it follows that

$$(h(x_i, y), h(x_s, y)) = 0, \ (i \neq s,\ y \in T_qS^n).$$

Again by the linear property for y, it follows that

$$(h(x_i, y_j), h(x_s, y_t)) + (h(x_i, y_t), h(x_s, y_j)) = 0, \ (i \neq s,\ j \neq t).$$

Both terms in the last equation are the same and are thus equal to zero (in view of the Gauss equation (5.4), we have $R(x_i, y)x_s = R(x_s, y)x_i$, which follows from $R(x_i, x_s)y = 0$ and the first Bianchi identity for a curvature tensor). Thus the νn vectors $\{\xi_{ij} = h(x_i, y_j)\}$ form an orthonormal base of $T_q M^\perp$. The rank of the matrix $A_{\xi_{ij}}$ is equal to two; only two of its elements are nonzero, see Section 6.2. For any unit normal vector ξ the matrix A_ξ may be written in the form

$$A_\xi = \begin{pmatrix} 0 & A \\ A^* & 0 \end{pmatrix}, \quad A = \{a_{ik} = (A_\xi x_i, y_k)\}, \ A^* \text{ is the transpose matrix of } A.$$

Thus the rank of A_ξ is not greater than $(n + \nu) - (\nu - n) = 2n$. To see this in another way, note that a normal vector $\xi_\varepsilon = \varepsilon_1 \xi_{11} + \cdots + \varepsilon_n \xi_{nn}$ for $\varepsilon = (\varepsilon_1, \ldots, \varepsilon_n)$ with small $\varepsilon_i \neq 0$ gives the matrix of the maximal possible rank $2n$, which completes the proof of the first equality in (7.6).

Note that $\mu_1(q) = \nu - n$ and

$$\ker A_{\xi_\varepsilon} = \{x_{n+1}, \ldots, x_\nu\}, \quad \ker A_{\xi'} = \{\varepsilon_2 x_{n+1} - s x_{n+2}, x_{n+2}, \ldots, x_\nu\},$$

where $\xi' = \xi_\varepsilon + s\xi_{1,1+n}$ and $s \neq 0$, $\varepsilon_i \neq 0$. Hence $\ker A_{\xi_\varepsilon} \cap \ker A_{\xi'} = \{x_{n+2}, \ldots, x_\nu\}$ is a $(\nu - n - 1)$-dimensional subspace. Since the last property holds for an open set of pairs of normal vectors (bivectors in $T_q M^\perp$), we conclude that $\mu_2(q) = \nu - n - 1$. □

The numerical sequence $\bar{\rho}(n)$ was defined in Section 6.2.

7.3 Theorem. *Let $M \subset \bar{M}$ be a complete submanifold satisfying properties (7.1a), (7.2), $0 < r(M) \leq \dim M - \bar{\rho}(\dim M)$ and one of the following inequalities:*

$$a)\ (7.3), \quad b)\ K_M \geq 0, \quad c)\ \frac{1}{k} Ric_M \geq \dim M - 2,$$

where $k > 0$ and $\mu_1(M) \leq \mu_2(M) + 1$ for b) and c). Then M is locally isometric to the product

$$S^{\dim M - \frac{1}{2}r(M)} \times M_2^{\frac{1}{2}r(M)} \quad \text{for a) and b)},$$
$$\text{or } S^{\dim M - 1} \times M_2^1, \ r(M) = 2 \ \text{for c)}$$

and $\operatorname{codim} M \geq \frac{1}{2}r(M)(\dim M - \frac{1}{2}r(M))$. *Moreover, if $\bar{M} = S^m(k)$ and* $\operatorname{codim} M = \frac{1}{2}r(M)(\dim M - \frac{1}{2}r(M))$, *then the submanifold M is congruent to the Segre embedding.*

Proof. The proof of a) is divided into steps.

Step 1. In view of (7.1) and Theorem 5.13, for any unit normal $\xi_0 \in T_q M^\perp$ with maximal rank $r(\xi_0) = r(M)$ at a point q, there exists a ruling $L_0 \subset$

M, $(T_qL_0 = \ker A_{\xi_0})$ which is a complete totally geodesic submanifold in $\bar{M}$, with a normal vector ξ_0 that is stationary along L and has constant rank. Let us extend ξ_0 onto a transversal neighborhood of the point q in M by $\nabla^\perp$-parallel displacement along geodesics orthogonal to L_0. Since the function $r(\xi)$ is semi-continuous, such normal vectors $\{\xi\}$ may be assumed to be of rank $r(M)$ and as such, they define an $r(M)$-parameter family of rulings $\{L\}$ on M having the same property that normal vectors $\{\xi\}$ are stationary along $\{L\}$. On some neighborhood U_q, the rulings $\{L\}$ define a totally geodesic foliation and the normal vectors $\{\xi\}$ form a vector field on U_q. Then by rule (2.3), the structural tensor $B : T_qL \times T_qL^\perp \to T_qL^\perp$ is defined. In view of the inequality for $r(M)$ and Lemma 3.40, there exist unit vectors $x_1 \in T_qL$, $y_1 \in T_qL^\perp$ and a real $\lambda \leq 0$ having the property $B(x_1, y_1) = \lambda y_1$. The Jacobi vector field $y(t)$, $(y(0) = y_1)$, induced by the family of rulings $\{L\}$ along the geodesic $\gamma : [0, \frac{\pi}{\sqrt{k}}] \to L$, $(\gamma'(0) = x_1)$, has the initial value

$$\bar{\nabla}_{x_1} y(t) = \lambda y_1 + v, \quad v = h(x_1, y_1) \tag{7.7}$$

and in view of (7.2) may be written in the form

$$\begin{aligned} y(t) &= y_1 \left(\cos(\sqrt{k}t) + \frac{\lambda}{\sqrt{k}} \sin(\sqrt{k}t) \right) + \frac{v}{\sqrt{k}} \sin(\sqrt{k}t) \\ &= -y_1 \frac{\sin(\sqrt{k}(t - t_0))}{\sin(\sqrt{k}t_0)} + \frac{v}{\sqrt{k}} \sin(\sqrt{k}t), \end{aligned}$$

where $t_0 = arccot\,(-\frac{\lambda}{\sqrt{k}})/\sqrt{k} \in (0, \frac{\pi}{2\sqrt{k}}]$.

From (7.3) or (7.4) it follows that $|v| \leq \sqrt{k}$, but in contrast to Theorem 6.18 on ruled submanifolds, the vector v in (7.7) may be zero. Assume that $k = 1$.

Step 2. We shall "chip off" the unit vector z_1, which is parallel to $A_{\xi_0} y_1$, in the case of $v = 0$.

Let L_s be a generator on M, which corresponds to a normal vector $\xi_s \in \{\xi\}$ through the point $\gamma_1(s) = \exp(sy_1)$ for a small $s > 0$. The formula $\xi_s = \bar{\xi}_0 + s\bar{w} + o(s)$ holds, where $\bar{\xi}_0$ and $\bar{w}$ are $\bar{\nabla}$-parallel vector fields along a geodesic γ_1, which contain a normal vector ξ_0 and a nonzero vector $w = -A_{\xi_0} y_1$. The Jacobi vector field $y(t)$ along γ is induced by the geodesic variation $Q(s, t)$, $0 < t \leq t_0$, $|s| \leq \varepsilon$ [Dom] having properties $\gamma_s = Q(s, *) \subset L_s$, $(\gamma_s(0) = \gamma_1(s))$, that is, $(\frac{\partial}{\partial s} Q)(0, t) = y(t)$. From $y(t_0) = 0$ it follows that the point $p_s = Q(s, t_0)$ belongs to an $o(s)$-neighborhood of the point $\gamma(t_0)$. Let $\tilde{\xi}_s$ be a unit normal vector to M at the point p_s, which is obtained by $\bar{\nabla}$-parallel displacement of ξ_s along γ_s. Consider the extension of ξ_0 to a field of normal vectors to M in a neighborhood of γ, where $y(t) \neq 0$. Since the vector field $\bar{\nabla}_{y(t)}\xi$ along γ is $\bar{\nabla}$-parallel, the result of $\bar{\nabla}$-parallel displacement of $\tilde{\xi}_s(t)$ along $Q(*, t)$ to the point $\gamma(t)$ differs from a vector $\eta_t = \bar{\xi}_0 + s\bar{w}$ by $O(s)$-error. For $t = t_0$ the resulting vector η_t differs from a normal vector to M by $O(s)$-error. Since $\bar{\xi}_0$ is also orthogonal to M at a point $\gamma(t_0)$, the vector $\bar{w} = (\eta_{t_0} - \bar{\xi}_0)/s$ differs from a unit normal vector to M at the point $\gamma(t_0)$ by $O(s)$-error. Passing to the limit with $s \to 0$ leads to a vector $\bar{w} \in T_{\gamma(t_0)}M^{\perp}$. Consequently, a normal vector $\bar{w}$ to M at the point $\gamma(t_0)$ under $\nabla^{\perp}$-parallel displacement along γ to the point $q = \gamma(0)$ for "time" t_0 rotates in $\bar{M}$ by an angle of $\frac{\pi}{2}$.

Then, as in the proof of Theorem 6.14, the extremal situation is obtained: $t_0 = \frac{\pi}{2}$, $\lambda = 0$ and the unit normal vector $v_1 \parallel \bar{w}$ at the point $P = \gamma_0(\frac{\pi}{2})$ is included in the $\nabla^{\perp}$-parallel field of normal vectors along γ_0, which are of the form $v_1(t) = (\cos t)\bar{v}_1(0) + (\sin t)\bar{v}_1$. Then for the unit tangent ∇-parallel field along γ_0 we have

$$z_1(t) = \bar{\nabla}_{\gamma_0'} v_1(t) = (\cos t)\bar{z}_1 - (\sin t)\bar{v}_1, \quad z_1 \in \bar{v}_1$$

and the following conditions hold:

$$R(z_1(t), \gamma_0')\gamma_0' = 0, \quad h(\gamma_0', z_1(t)) = -v_1(t), \quad z_1 \parallel A_{\xi_0} y_1.$$

As in the proof of Theorem 6.14 we obtain that the orthogonal complement to z_1 in $T_q L^{\perp}$ is an invariant subspace for the operator $B(x, *)$.

The operator of a second quadratic form $A_{\xi(s)}$ of the unit normal vector

$$\xi(s) = (\cos s)\bar{\xi}_0 + (\sin s)v_1, \quad (s \approx 0)$$

at the point p has rank $r(M)$. By (7.1) and the equation

$$|A_{v_1}\gamma_0'(\frac{\pi}{2})| = 1,$$

the conditions $A_{v_1}u = 0$ and $u \perp \gamma_0'(\frac{\pi}{2})$ hold. Thus the kernel of the operator $A_{\xi(s)}$, $(s \approx 0)$, contains the vectors from $T_P L$ which are orthogonal to $\gamma_0'(\frac{\pi}{2})$. In this case the vector $\gamma_0'(\frac{\pi}{2})$ does not belong to the kernel of $A_{\xi(s)}$, because the normal vector v_1 is not stationary along γ_0. Consequently, v_1 remains a normal vector under $\bar{\nabla}$-parallel displacement along any geodesic $\gamma_1 \subset L_0$ with initial value $\gamma_1'(0) \perp \gamma_0'(\frac{\pi}{2})$. Let $\{L(s)\}$ be a one-parameter family of rulings on M which correspond to the normal vectors $\{\xi(s)\}$ at the point p. In view of the above considerations, every $L(s)$, $(s \neq 0)$, intersects L_0 in a $(\nu - 1)$-dimensional totally geodesic submanifold, which is orthogonal to the vector $\gamma_0'(\frac{\pi}{2})$ at the point P. Thus the Jacobi vector field $y_1(t) \in T_{\gamma_0(t)}L^\perp$, $(y_1(\frac{\pi}{2}) = 0)$, which is induced by the family $\{L(s)\}$, is given by the formula $y_1(t) = (\cos t)\bar{y}_1$, where $\bar{y}$ is a $\bar{\nabla}$-parallel vector field along γ_0 containing y_1. From the above we obtain the equations

$$R(y_1(t), \gamma_0')\gamma_0' = y_1(t), \quad h(\gamma_0', y_1(t)) = 0.$$

Vectors y_1 and z_1 are orthogonal because they are eigenvectors for a Jacobi operator $R(*, x_1)x_1 : T_qL^\perp \to T_qL^\perp$ with different eigenvalues (1 and 0). Note that the unit normal vector $\tilde{\xi}_s/|\tilde{\xi}_s|$ at the point p_s has an s^2-order difference from the normal vector

$$\xi(s) = (\cos s)\bar{\xi}_0 + (\sin s)v_1$$

at the point $p = \gamma_0(\frac{\pi}{2})$, and that the distance between p_s and p has s^2-order. Thus the Jacobi vector field $y_1(t)$ is induced by the rulings $\{L\}$. So $B(x_1, y_1) = 0$.

By similar considerations for other geodesics $\gamma \subset L_0$, $(\gamma(0) = q)$, we obtain that vectors from $T_qL^\perp$, which are orthogonal to z_1 form an invariant subspace for the operators $B(u, *)$, $(u \in T_qL)$, with the following properties:

$$R(y_1, u)u = h(u, y_1) = B(u, y_1) = 0, \; R(z_1, u)u = z_1, \; (u \in T_qL, \; |u| = 1). \tag{7.8}$$

Step 3. We shall "chip off" the unit vector z_1, which is parallel to $A_v x$, in the case of $v \neq 0$. As in the proof of Theorem 6.14, we obtain that a normal vector $v_1(0) = \frac{v}{|v|}$ under $\nabla^\perp$-parallel displacement along $\gamma_0(t)$, $(0 \leq t \leq \frac{\pi}{2})$, rotates uniformly in $\bar{M}$ by the angle $\frac{\pi}{2}$. Thus $\lambda = 0$ and the $\nabla^\perp$-parallel vector field $v_1(t)$ is given by the formula

$$v_1(t) = (\cos t)\bar{v}_1(0) + (\sin t)\bar{z}_1, \; (z_1 \in T_qL^\perp),$$

where $\bar{v}_1(0)$ and $\bar{z}_q$ are $\bar{\nabla}$-parallel vector fields along γ_0, containing the vectors $v_1(0)$ and $z_1 \,||\, A_v x$. Thus the unit tangent vector field

$$z_1(t) = \bar{\nabla}_{\gamma_0'} v_1(t) = (\cos t)\bar{z}_1 - (\sin t)\bar{v}_1(0)$$

is ∇-parallel along γ_0, and the following equations hold:

$$R(z_1(t), \gamma_0')\gamma_0' = 0, \; h(\gamma_0', z_1(t)) = -v_1(t).$$

Since A_{ξ_0} is nondegenerate on $T_qL^\perp$, there exists a unit vector $y_1 \in T_qL^\perp$ with the property $A_{\xi_0}y_1 \,||\, z_1$. Let $y_1(t)$, $(y_1(0) = y_1)$, be a Jacobi vector field along γ_0, which is induced by $\{L\}$. As for the case of $v = 0$ (see Step 2) we consider the rectangle $Q(s,t)$ of geodesics $\gamma_s \subset L_s$, which induces the vector field $y_1(t)$. After $\nabla^\perp$-parallel displacement of ξ_0 along the boundary of $Q(s,t)$, we obtain the normal vector $\bar{\xi}_0 + s\bar{v}_1(\frac{\pi}{2}) + o(s)$ at the point p_s. Thus $A_{\bar{\xi}_0}(y_1(\frac{\pi}{2})) = 0$, and in view of the nondegeneracy of $A_{\bar{\xi}_0}$ on $TL^\perp$, the vector $y_1(\frac{\pi}{2})$ is zero; this means that $y_1(t) = (\cos t)\bar{y}_1$. As above we deduce that the orthogonal complement to z_1 in $T_qL^\perp$ is an invariant subspace for the operator $B(x, *)$.

By investigating other geodesics $\gamma \subset L_0$, $(\gamma(0) = q)$, as in the case of $v = 0$ (see Step 2), we obtain the properties in (7.8). We also get that the orthogonal complement to z_1 in $T_qL^\perp$ is an invariant subspace for the operators $B(u, *)$, $(u \in T_qL^\perp)$.

Step 4. We will repeat the "chipping off" the unit vectors $\{z_i\}$, $\frac{1}{2}r(M)$ times.

Let V_2 be the orthogonal complement to the plane $y_1 \wedge z_1$ in $T_qL^\perp$ and let $B_2 = P_2B : T_qL \times V_2 \to V_2$ be a bilinear operator, where $P_2 : T_qL^\perp \to V_2$ is an orthoprojector and B is the structural tensor (2.3) for a totally geodesic foliation $\{L\}$. In view of the given inequality for the rank $r(M)$ and Lemma 3.40, the structural tensor B_2 has an eigenvector: $B_2(x, y) = \lambda y$ for some unit vectors $x \in T_qL$, $y = V_2$ and a real $\lambda \le 0$.

Since the subspace $V_2 \oplus \{y_1\}$ is invariant for B, we have $B(x, y) = \lambda y + \mu y_1$. From this it follows that

$$y(t) = (\cos t + \lambda \sin t)\bar{y} + (\sin t)(\mu\bar{y}_1 + \bar{v}), \ (v = h(x, y)),$$

for a Jacobi vector field $y(t)$, $(y(0) = y)$, (where $\bar{y}, \bar{y}_1, \bar{v}$ are $\bar{\nabla}$-parallel vector fields along γ_0), which is induced by $\{L\}$ along the geodesic $\gamma_0 \subset L_0$, $(\gamma_0'(0) = x)$. The two cases $v = 0$ and $v \neq 0$ are both possible and we will consider each of these.

If $v = 0$, then $t_0 = arccot\,(-\lambda) \in (0, \frac{\pi}{2}]$ implies that $y(t_0) = (\sin t_0)\mu y_1$. Since the L-parallel Jacobi vector field $y(t)$ is not parallel to $\bar{y}_1$ for any t, it follows that $\mu = 0$ and $y(t) = (\cos t + \lambda \sin t)\bar{y}$. As for "chipping off" z_1 in Step 3, we obtain for $y_2 = y$ and $z_2 \,||\, A_{\xi_0}y$ the following equalities:

$$R(y_2, u)u = B(u, y_2) = 0, \ R(z_2, u)u = z_2, \ (u \in T_qL, \ |u| = 1). \quad (7.9)$$

Also note that the orthogonal complement of the plane $z_1 \wedge z_2$ in $T_qL^\perp$ is an invariant subspace for the operators $B(u, *)$. Note that the vectors y_2 and

y_1 are linearly independent. Since the symmetric operator A_{ξ_0} on $T_qL^\perp$ is nondegenerate, the vectors $A_{\xi_0}y_2$ and $A_{\xi_0}y_1$ are also linearly independent, which means that the same property holds for the vectors z_1 and z_2.

If $v \neq 0$, then for $t_0 = arccot\,(-\lambda) \in (0, \frac{\pi}{2})$ the vector $y(t_0) = (\sin t_0)(\mu \bar{y}_1 + \bar{v})$ is tangent to M. Since the vector field $\bar{y}_1$ along γ_0 is tangent to M, the vector $\bar{v}$ is tangent to M at the point $\gamma_0(t_0)$. By using (7.3), as above, we see that $t_0 = \frac{\pi}{2}$, $\lambda = 0$ and that the unit $\nabla^\perp$-parallel field of normal vectors $v_2(t)$, $(v_2(0) = \frac{v}{|v|})$ along $\gamma_0(t)$, $(0 \leq t \leq \frac{\pi}{2})$, rotates uniformly in $\bar{M}$ by the angle $\frac{\pi}{2}$. Assume $z_2 = -A_{v_2(0)}x$ and find a unit vector y_2 such that $A_{\xi_0}y_2 \,||\, z_2$. As above, we obtain for z_1, that y_2 stays tangent under $\bar{\nabla}$-parallel displacement along L_0, and the identities in (7.8) hold. From the conditions

$$h(x, y) \,||\, v_2(0), \quad h(x, z_1) \,||\, v_1(0), \quad y \perp z_1$$

it follows that the vectors $v_2(0)$ and $v_1(0)$ are linearly independent. Thus the vectors $z_2 = -A_{v_2(0)}x$ and $z_1 = -A_{v_1(0)}x$ are also linearly independent. Since the operator A_{ξ_0} is nondegenerate on $T_qL^\perp$, the vectors y_2 and y_1 are linearly independent.

After $\frac{1}{2}r(M)$ analogous steps we find the unit vectors $\{z_i\}$, $\{y_i\}$, $(1 \leq i \leq \frac{r(M)}{2})$ in $T_qL^\perp$ with the following properties: – the $\{z_i\}$ span the $\frac{1}{2}r(M)$-dimensional subspace $Z_q \subset T_qL^\perp$ with the condition

$$R(z, u)u = 0, \quad (u \in T_qL,\ z \in Z_q),$$

– the $\{y_i\}$ span the $\frac{1}{2}r(M)$-dimensional subspace $Y_q \subset T_qL^\perp$ that is orthogonal to Z_q, invariant for the structural tensor B, and satisfies the property

$$R(y, u)u = y, \quad (u \in T_qL,\ y \in Y_q,\ |u| = 1).$$

Moreover, the condition $A_{\xi_0}y_i \,||\, z_i$, holds for all i. Analogous vectors $\{z_i\}$, $\{y_i\}$ exist for all points near q, and thus there exist $\frac{1}{2}r(M)$-dimensional distributions Z and Y from $TL^\perp$ on the neighborhood of q with certain conditions for the curvature. Furthermore, the distribution Y is orthogonal to Z and $\bar{\nabla}$-parallel along the rulings $\{L\}$.

Step 5. We will prove that the distributions Z and $W = Y \oplus TL$ define a metric decomposition of M into $M_1^{\frac{1}{2}r(M)} \times S^{\dim M - \frac{1}{2}r(M)}$.

As in the proof of Theorem 6.14, we see that distributions Z, W on some neighborhood of q are integrable and that certain integral submanifolds are totally geodesic in M. By de Rham's decomposition theorem the neighborhood of a point q is isometric to the product $M_1^{r(M)} \times M_2^{\dim M - \frac{1}{2}r(M)}$.

The curvature tensor of a totally geodesic submanifold M_1 in M is given by the restriction of the curvature tensor of M. As above, we obtain the properties

$$R(y,x)x = 0, \ R(z,x)x = z, \quad (x \in TL, \ y \in Y, \ z \in Z),$$

and thus (in view of (7.2)) a normal subspace to M_1 is stationary along $\{L\}$. We also have the following property for the curvature:

$$0 < K(x, w) < 1, \quad (x \in TL, \ w \in TL^{\perp} \setminus (Y \cup Z)).$$

From the discussion above, it follows that a metric decomposition of a neighborhood of the point q is independent of the choice of the initial normal vector ξ with rank $r(\xi) = r(M)$. We see that the metric decomposition $M_1^{\frac{1}{2}r(M)} \times M_2^{\dim M - \frac{1}{2}r(M)}$ is well-defined on an open subset $G \subset M$ of points with maximal rank $r(M)$. Additionally, from the above, it follows that $(\dim M - \frac{1}{2}r(M))$-dimensional submanifolds $\{M_1\}$ in M have the relative null-index $\mu(M_1) \geq l - r(M)$; moreover, the kernel of any normal vector $\xi \in TM^{\perp}$ with rank $r(\xi) = r(M)$ belongs to the relative null-space of M_1 and is tangent to a complete $(\dim M - r(M))$-dimensional totally geodesic submanifold in M_1. From the estimate for rank we get

$$\dim M - r(M) \geq \rho(\dim M_1 - \mu(M_1)),$$

and hence, by Theorem 1.38, $\mu(M_1) = \dim M_1$. Therefore M_1 is a totally geodesic submanifold in $\bar{M}$. In view of (7.2) the submanifolds $\{M_1\}$ have constant sectional curvature and are locally isometric to the sphere $S^{\dim M - \frac{1}{2}r(M)}$. As in the proof of Theorem 6.14, we obtain the following lower bound for codimension:

$$\operatorname{codim} M \geq \frac{r(M)}{2}(\dim M - \frac{r(M)}{2}).$$

Since the curvature tensor and the second fundamental form of M are continuous, we may extend such the metric decomposition onto a boundary $\partial\bar{G}$ of the regularity domain $\bar{G}$. By using special base points of $\partial\bar{G}$ (see the proof of Theorem 6.14), we obtain for small $\varepsilon = (\varepsilon_1, \dots, \varepsilon_{\frac{1}{2}r(M)})$ the normal vector $\xi_\varepsilon = \varepsilon_1\xi_{1,1} + \cdots + \varepsilon_{\frac{1}{2}r(M)}\xi_{\frac{1}{2}r(M),\frac{1}{2}r(M)}$ with rank $r(M)$. Thus the rank of the points of $\partial\bar{G}$ is equal to $r(M)$. Since $\bar{G}$ is open and closed in M, we have $\bar{G} = M$. We see that M is locally isometric to the product $M_1^{\frac{1}{2}r(M)} \times S^{\dim M - \frac{1}{2}r(M)}$.

In the case of $\operatorname{codim} M = \frac{1}{2}r(M)(\dim M - \frac{1}{2}r(M))$, as in the proof of Theorem 6.14, we deduce that $M \subset S^m(k)$ is congruent to the Segre submanifold. The proof for parts b) and c) is similar. □

The decomposition of Kählerian parabolic submanifolds can be deduced from weaker assumptions on rank $r(M)$ and on curvature.

7.4 Theorem. *Let $M \subset \bar{M}$ be a complete Kählerian submanifold satisfying properties* (7.1*b*), (7.2), $0 < r(M) < \dim M$ *and suppose that any of the following conditions holds:*

a) $|h(x, y)|/(|x| \cdot |y|) \le \sqrt{k}$,

b) $\frac{1}{k} Ric_M \ge \dim M$,

c) the (bi)sectional curvature of M is nonnegative.

Then M is locally isometric to the product, respectively,

$$a),\ c)\ CP^{\frac{1}{2}\dim M - \frac{1}{4}r(M)} \times M_2^{\frac{1}{2}r(M)},\quad b)\ CP^{\frac{1}{2}\dim M - 1} \times M_2^2,\ r(M) = 4,$$

and $\operatorname{codim} M \ge \frac{1}{2}r(M)(\dim M - \frac{1}{2}r(M))$. *If, moreover,* $\bar{M} = CP^m(k)$ *and* $\operatorname{codim} M = \frac{1}{2}r(M)(\dim M - \frac{1}{2}r(M))$, *then M is congruent to a complex Segre submanifold.*

Proof. As for the real case in Theorem 7.3, we find an $r(M)$-dimensional family of rulings $\{L\}$, which form a totally geodesic foliation on some neighborhood of a point $q \in M$. We will show that the structural tensor (2.3) of this foliation at the point q has the following property:

$$B(x, Jy) = B(Jx, y) = JB(x, y), \quad (x \in T_qL,\ y \in T_qL^\perp). \tag{7.10}$$

From the Codazzi equation (5.5) and (7.1b), with the help of the $\nabla^\perp$-parallel normal vector field $\{\xi\}$ at the point q, it follows that

$$(\bar{\nabla}_x A_\xi) y = (\bar{\nabla}_y A_\xi) x, \quad (x \in T_qL,\ \xi \in T_qL^\perp).$$

Let $\tilde{x} \ni x$ be a local vector field on TL, and let $\tilde{y} \ni y$ be a local vector field on $TL^\perp$. In view of the equation $A_\xi J = -JA_\xi$, which is true for Kählerian submanifolds, we obtain

$$\begin{aligned} A_{\xi_0} B(x, Jy) &= -(\bar{\nabla}_{Jy} A_\xi) x = -(\bar{\nabla}_x A_\xi) Jy \\ &= -\nabla_x (A_\xi J y) + A_{\xi_0}(\nabla_x J\tilde{y}) = J[\nabla_x (A_\xi \tilde{y}) - A_{\xi_0}(\nabla_x \tilde{y})] \\ &= J(\bar{\nabla}_x A_\xi) y = J(\bar{\nabla}_x A_\xi) x = -JA_{\xi_0}(\nabla_y \tilde{x}) = A_{\xi_0} JB(x, y). \end{aligned} \tag{7.11}$$

Since the operator A_{ξ_0} is nondegenerate on $T_qL^{\perp}$, it follows from (7.11) that $B(x, Jy) = JB(x, y)$. Since M and $\{L\}$ are Kählerian submanifolds, we also have the equation $B(Jx, y) = JB(x, y)$ from (2.3), and equation (7.10) holds. By Lemma 3.41, a bilinear operator $B(x, y)$ with the property (7.10) has an eigenvector. In view of this property, the proof for the Kählerian case is analogous to the real case in Theorem 7.3. □

7.5 Definition. Let $d(\xi)$ be the dimension of a maximal plane on an asymptotical cone for a normal vector $\xi \in TM^{\perp}$. The integer $d(M) = \min\{d(\xi) : |\xi| = 1\}$ is called the *planar order* of a submanifold M (see Definition 5.21).

The rank and the planar order of a submanifold M are related in the following way, see [Ab 5]:

$$r(M) \leq 2(\dim M - d(M)).$$

Thus from Theorem 7.3 we get the following corollary.

7.6 Corollary. *Let M be a complete nontotally geodesic submanifold satisfying condition* (7.1), *a) in S^N, b) Kählerian in CP^N, and suppose that*

$$a)\ d(M) \geq \frac{1}{2}\{\dim M + \bar{\rho}(\dim M)\},\quad b)\ d(M) > \frac{1}{2}\dim M.$$

Then M is isometric to a product and $\operatorname{codim} M \geq \dim M - 1$.

7.2. Parabolic submanifolds in a Riemannian space of positive curvature

For the nonconstant sectional curvature of $\bar{M}$ along a submanifold M

$$k_2 \geq \bar{K}(x, y) \geq k_1 > 0,\ (x,\ y \in TM), \tag{7.12}$$

and $k = \frac{k_2+k_1}{2}$, we assume in Section 7.2, instead of (7.3) or (7.4), the analogous inequality

$$||A_\xi|| \leq \sqrt{2k}\{1 - 5\left(1 - \frac{k_1}{k_2}\right)\left(\frac{a(\xi_0)^2}{k} + 1\right)\},\ (|\xi| = 1), \tag{7.13}$$

or the slightly weaker inequality

$$|h(x, y)|/(|x| \cdot |y|) \leq \sqrt{k}\{1 - 5\left(1 - \frac{k_1}{k_2}\right)\left(\frac{a(\xi_0)^2}{k} + 1\right)\}. \tag{7.13'}$$

The following real $a(\xi)$ is related to the turbulence from Section 4.2.

7.7 Definition. Let $M \subset \bar{M}$ be a complete submanifold with property (7.1) and $k > 0$. For a unit normal vector $\xi \in T_m M^{\perp}$ with the condition $r(\xi) = r(M) < \dim M$ we consider $\Gamma(m, \xi)$, the sheaf of geodesics $\gamma : [0, \frac{\pi}{\sqrt{k}}] \to M$, $(\gamma(0) = m)$, with the initial directions $\gamma'(0) \in \ker A_\xi$, and define a real

$$a(\xi) = \sup \frac{|(\nabla_{\gamma'} A_{\xi(t)})y| + k}{|A_{\xi(t)} z|},$$

where $y, z \perp \ker A_{\xi(t)}$, $y \perp z$, $|y| = |z| = 1$, $\xi(0) = \xi$, $\nabla^{\perp}_{\gamma'} \xi(t) = 0$ and $\gamma \in \Gamma(m, \xi)$.

We will see that $a(\xi)$ is an analog of the turbulence $\alpha(L)$ for some foliation by generators on a domain of M. In view of Corollary 5.14, $a(\xi)$ is well-defined.

Combining the method of variation of Theorem 4.10 for foliations with the synthetic method of Theorem 7.3 and Theorem 7.4 for parabolic submanifolds in CROSS leads to the next theorem.

7.8 Theorem. *Let $M \subset \bar{M}$ be a complete submanifold satisfying conditions* (7.1), (7.12) *and* (7.13) *or* (7.13′) *with $\mu_1(M) \le \mu_2(M) + 1$ (in the Kählerian case without the inequalities for μ_i), where $k = \frac{k_1 + k_2}{2}$ and ξ is any unit normal vector having rank $r(\xi) = r(M)$. If*

$$0 < r(M) \le \begin{cases} \dim M - \bar{\rho}(\dim M), & \\ \dim M - 1, & \textit{for the Kählerian case,} \end{cases}$$

then $k_2 = k_1$, M is locally isometric to the product $L^{\dim M - \frac{1}{2}r(M)} \times M_2^{\frac{1}{2}r(M)}$ and $\operatorname{codim} M \ge \frac{r(M)}{2}(\dim M - \frac{r(M)}{2})$.

If (7.13′) in Theorem 7.8 is given with strict inequality, then tests of totally geodesic submanifolds may be obtained.

7.8′ Corollary. *Let $M \subset \bar{M}$ be a complete submanifold with properties* (7.1), (7.12), (7.13′) *with strict inequality ($x, y \in TM$, k and ξ as in Theorem* 7.8*) and $r(M) \le \dim M - \nu(\dim M)$ (for the Kählerian case $r(M) < \dim M$). Then M is a totally geodesic submanifold.*

Proof of Theorem 7.8 and its Corollary. The proof is divided into steps. The purpose is to deduce the equality $k_2 = k_1$, i.e., Theorem 7.8 will be reduced to Theorems 7.3 and 7.4.

Step 1. From (7.1), Theorem 5.13 and Corollary 5.14, it follows that for any unit normal vector $\xi_0 \in T_q M^{\perp}$ with maximal rank $r(\xi_0) = r(M)$ at a point

q there exists a ruling $L_0 \subset M$, $(T_q L_0 = \ker A_{\xi_0})$, which is a complete totally geodesic submanifold in $\bar{M}$ with a normal vector ξ_0 which is stationary along L_0 and has constant rank.

We extend ξ_0 onto a neighborhood of the point q by $\nabla^\perp$-parallel displacement along the geodesics orthogonal to L_0 in M. Since $r(\xi)$ is a semi-continuous function, such normal vectors $\{\xi\}$ may be assumed to be of rank $r(M)$, and so they define an $r(M)$-parameter family of rulings $\{L\}$ on M with the same property of normals $\{\xi\}$ that are stationary along $\{L\}$. The neighborhood U_q may be chosen so that $\{L\}$ defines a totally geodesic foliation on U_q and the normal vectors $\{\xi\}$ form a vector field on U_q. Then its structural tensor $B : T_q L \times T_q L^\perp \to T_q L^\perp$ is defined by rule (2.3). In view of the given estimate for $r(M)$ and Lemma 3.40 (or Lemma 3.41 for the Kählerian case) there exist unit vectors $x_1 \in T_q L$, $y_1 \in T_q L^\perp$ and a real $\lambda \le 0$ with the property $B(x_1, y_1) = \lambda y_1$.

The Jacobi vector field $y(t)$, $(y(0) = y_1)$, that is induced by the family of rulings $\{L\}$ along the geodesic $\gamma : [0, \frac{\pi}{\sqrt{k}}] \to L$, $(\gamma'(0) = x_1)$, has initial value

$$\bar{\nabla}_{x_1} y(t) = \lambda y_1 + v, \quad v = h(x_1, y_1) \tag{7.14}$$

and may be written in the form $y(t) = Y(t) + u(t)$, where

$$\begin{aligned} Y(t) &= y_1 \left(\cos(\sqrt{k}t) + \frac{\lambda}{\sqrt{k}} \sin(\sqrt{k}t) \right) + \frac{v}{\sqrt{k}} \sin(\sqrt{k}t) \\ &= y_1 \frac{\sin(\sqrt{k}(t - t_0))}{\sin(\sqrt{k}t_0)} + v \frac{\sin(\sqrt{k}t)}{\sqrt{k}}, \end{aligned}$$

with $t_0 = arccot\,(-\lambda/\sqrt{k})/\sqrt{k} \in (0, \frac{\pi}{2\sqrt{k}}]$, $u(0) = u'(0) = 0$. From (7.12) or (7.13′) it follows that $|v| \le \sqrt{k}$, but in contrast to Theorem 6.18 with ruled submanifolds, the vector v in (7.14) may be zero.

Step 2. Assume in Steps 2–4 that there exists a real $t_1 \in [0, 2t_0]$ with the property $y(t_1) = 0$ (where t_1 is a minimal such number). We will show that

$$\sqrt{k}|t_1 - t_0| \le \frac{\pi}{2}\beta(1 - \delta), \quad (\beta = 3.2 \text{ for } \delta = \frac{k_1}{k_2} \ge 0.8), \tag{7.15}$$

from which it follows that $t_1 \to t_0$ when $\delta \to 1$.

From Lemma 4.7 for $\varepsilon_1 = \frac{k_2 - k_1}{2}$ (with the help of the inequality $|Y(t)| \le \sqrt{2}$), we get the following upper bound (see the proof of Theorem 4.10):

$|u(t)| \leq 2\sqrt{2}\frac{1-\delta}{3\delta-1}$. Since $|Y(t)| \geq \sin(\sqrt{k}|t-t_0|)$,

$$0 = |y(t_1)| \geq |Y(t_1)| - |u(t_1)| \geq \sin(\sqrt{k}|t_1 - t_0|) - 2\sqrt{2}\frac{1-\delta}{3\delta - 1},$$

that is, $\sin(\sqrt{k}|t_1 - t_0|) \leq \beta(1-\delta)$. In view of the well-known inequality

$$\sin x \geq \frac{2}{\pi}x, \quad (0 \leq x \leq \frac{\pi}{2})$$

(7.15) holds.

Step 3. Let $\bar{w}$ be a $\bar{\nabla}$-parallel vector field along a geodesic γ which contains a nonzero vector $w = -A_{\xi_0}y_1 \in T_qL^{\perp}$. Repeating the beginning of Step 2 of the proof of Theorem 7.3 allows us to deduce that $\bar{w}$ is a normal vector to M at the point $\gamma(t_1)$, which rotates in $\bar{M}$ by the angle $\frac{\pi}{2}$ for the values $t = t_1$ upon $\nabla^{\perp}$-parallel displacement along γ to the point $q = \gamma(0)$.

Step 4. Let $\tilde{w}(t) \subset T_\gamma M^{\perp}$, $(\tilde{w}(t_1) \,||\, \bar{w})$, be a unit $\nabla^{\perp}$-parallel field of normal vectors along γ . From Step 3 it follows that $\tilde{w}(0) \perp \bar{w}$. In view of (7.12) or (7.13′) the velocity of a rotation of the normal vector $\tilde{w}(t)$ in $\bar{M}$ along γ is bounded from above. That leads, with the help of (7.15), to the inequalities

$$\frac{\pi}{2} \leq \int_0^t |\bar{\nabla}_{\gamma'}\tilde{w}(t)|dt = \int_0^t |A_{\tilde{w}(t)}\gamma'|dt \leq t_1\sqrt{k}\{1 - 5(1-\delta)\left(\frac{a(\xi_0)^2}{k} + 1\right)\}$$

$$\leq \left(\frac{\pi}{2} + \frac{\pi}{2}\beta(1-\delta)\right)\{1 - 5(1-\delta)\left(\frac{a(\xi_0)^2}{k} + 1\right)\} \leq \frac{\pi}{2}[1 - (5-\beta)(1-\delta)],$$

which are compatible only for the extremal situation $\delta = 1$, i.e., $k_2 = k_1$. In the case of strict inequality in (7.13′) the proof of Corollary 7.8′ under the assumption $y(t_1) = 0$ is complete.

Step 5. Assume in Steps 5 and 6 that the vector field $y(t)$ is nonzero for $t \in [0, 2t_0]$. In a neighborhood of γ, where $y(t) \neq 0$ we obtain from (7.1) and the Codazzi equation (5.5)

$$(\nabla_x A_{\xi_0})y + A_{\xi_0}(\nabla_y x) = -A_{\nabla_y^{\perp}\xi}x, \quad (x \in \ker A_\xi, \ y \perp \ker A_\xi), \tag{7.16}$$

where $\xi \subset TM^{\perp}$ is defined as the extension of ξ_0 onto such a neighborhood. Since the Jacobi vector field $y(t)$ induces a one-parameter variation of geodesics (i.e., a 2-dimensional submanifold in $\bar{M}$), $[x, y] = 0$ holds; hence in view of (7.1) and $\bar{\nabla}_x\xi = 0$ we obtain

$$0 = \bar{R}(x, y)\xi_0 = \bar{\nabla}_x\bar{\nabla}_y\xi - \bar{\nabla}_y\bar{\nabla}_x\xi = \bar{\nabla}_x(\bar{\nabla}_y\xi).$$

Thus $\bar{\nabla}_y \xi = const$ along γ, that is,

$$|\nabla_y^\perp \xi| \le |\bar{\nabla}_y \xi| = |A_{\xi_0} y|(q) \le \sqrt{k}.$$

Consequently,

$$|(A_{\nabla_y^\perp \xi} x, z)| = |(h(x,z), \nabla_y^\perp \xi)| \le |h(x,z)| \cdot |A_{\xi_0} y| \le k.$$

From the last equation and (7.16) it follows that

$$\sup |\nabla_y x^\perp| \le \sup \frac{|(\nabla_x A_\xi) y| + k}{|A_\xi z|} = a(\xi_0).$$

Step 6. From the method of Theorem 4.10 (see also Step 1 of the proof of Theorem 6.18) it follows that $v \neq 0$. Repeating Steps 2–4 in the proof of Theorem 6.18 leads to the equality $k_2 = k_1$; in the case of the strict inequality in (7.13′), this completes the proof of Corollary 7.8′. □

7.3. Remarks on pseudo-Riemannian isometric immersions

The universal covering of a complete pseudo-Riemannian manifold M_p^n (with dimension n and a metric of signature $(p, n-p)$) of constant curvature k can be represented by (a hypersurface in the first and second cases)

$$\tilde{M}_p^n = \begin{cases} \sum_{i=1}^{p+1} x_i^2 - \sum_{j=1}^{q} x_{p+1+j}^2 = \sqrt{k}, & in\ \mathbb{R}_{p+1}^{n+1},\ if\ p \neq 1,\ k > 0, \\ \sum_{i=1}^{p} x_i^2 - \sum_{j=1}^{q+1} x_{p+j}^2 = -\sqrt{-k}, & in\ \mathbb{R}_{p+1}^{n+1},\ if\ q \neq 1,\ k < 0, \\ a\ pseudo\text{-}Euclidean\ space\ \mathbb{R}_p^{n+1}, & if\ k = 0. \end{cases}$$

Hence the case $k > 0$ is dual to the case $k < 0$.

In [AM 1] the relative nullity foliation of an indefinite Riemannian manifold isometrically immersed into an indefinite space form is studied. This foliation is totally geodesic and gives rise to a Riccati-type matrix ODE along a geodesic in a leaf. The differential equation is applied in several ways to estimate the index of relative nullity for geodesically complete Lorentz submanifolds in a Lorentz sphere.

In [AM 2] pseudo-Riemannian Kählerian strongly parabolic embeddings are studied. In [INR] products of complex submanifolds of indefinite complex space forms are studied and the criteria for a submanifold corresponding to the Segre embedding are given.

The decomposition results (analogous to Corollary 6.17 and Theorem 7.3) for submanifolds with generators of a pseudo-Riemannian manifold, which generalize certain results in [AM 1], are true. We will give them in their simplest forms. Let M_p^m ($S_p^m(1)$) be a pseudo-Riemannian manifold (a sphere with radius one) of dimension m with a metric of signature $(p, m-p)$.

7.9 Theorem. [Rov 12] *Let $M_p^{n+\nu} \subset S_q^m(1)$ be a ruled submanifold with geodesically complete rulings and the following condition on its second quadratic forms:*

$$(A_\xi x)^2 \geq 0,\ (A_\xi x)^2+(A_\xi u)^2 \leq 1,\ \ (x, u \in TL,\ \xi^2 = x^2 = u^2 = 1,\ (x, u) = 0).$$

If $\nu \geq p + \bar{\rho}(n)$, then $M_p^{n+\nu}$ is locally isometric to the product $S_p^\nu \times B_0^n$ and $m \geq n + \nu + \nu p$, $q \geq np$ hold, if moreover $m = n + \nu + \nu p$ and $q = np$, then the submanifold $M_p^{n+\nu}$ is congruent to a domain of the pseudo-Riemannian Segre embedding $S_p^\nu \times S_0^n \subset S_{np}^{n+\nu+\nu n}$.

7.10 Theorem. [Rov 12] *Let $M_p^l \subset S_q^{l+p}(1)$ be a geodesically complete submanifold satisfying one of the following conditions:*

a) $(A_\xi x)^2 \geq 0,\ (A_\xi x)^2 + (A_\xi y)^2 \leq 1,\ \ (\xi^2 = x^2 = y^2 = 1,\ (x, y) = 0)$,
b) $0 \leq K(x, y) \leq 1$,
c) $K(x, y) \leq 1,\ Ric\,(x) \geq l - 2,\ for\ b)\ c)\ :\ \mu_1(M) \leq \mu_2(M) + 1$,

where $x^2 = y^2 = 1$, $(x, y) = 0$ and the restriction of the metric on the planes $\ker A_\xi$, *$(\xi^2 = 1,\ r(\xi) = r(M))$ has signature $(p, l - r(M) - p)$. If $r(M) \leq l - p - \bar{\rho}(l - p)$, then M_p^l is locally isometric to the product*

$$a),\ b)\ S_p^{l-\frac{r(M)}{2}} \times B_0^{\frac{r(M)}{2}},\quad c)\ S_p^{l-1} \times B_0^1,\ r(M) = 2$$

and $p \geq \frac{r(M)}{2}(l - \frac{r(M)}{2})$, $q \geq p\frac{r(M)}{2}$. If, moreover, $p = \frac{r(M)}{2}(l - \frac{r(M)}{2})$ and $q = p\frac{r(M)}{2}$, then the submanifold M is congruent to the pseudo-Riemannian Segre embedding.

The existence of a ruled structure on a parabolic submanifold M^n with a definite metric in a pseudo-Riemannian space form is proved in [Bor 10].

Note that a pseudo-Riemannian manifold whose sectional curvature keeps its sign actually has constant curvature, see [Bor 10].

Indefinite isometric immersions of constant curvature were studied by many others.

7.11 Theorem. [Bor 10] *a) A pseudo-Riemannian manifold M_p^n of constant negative sectional curvature can not be locally isometrically embedded into $\mathbb{R}_{p'}^{2n-2}$, where $n - p = 2n - 2 - p'$.*

b) A complete pseudo-Riemannian manifold M_p^n of constant negative sectional curvature can not be locally isometrically embedded into $\mathbb{R}_{p'}^{2n-1}$, where $n - p = 2n - 1 - p' \neq 0, 1, 3, 7$.

7.12 Example. The local isometric immersion of M_p^n with constant negative curvature into $\mathbb{R}_{p'}^{2n-1}$, where $n - p = 2n - 2 - p'$, is given by the formulas [Bor 10]

$$x_{2i-1} = \begin{cases} a_i \exp(u_n) \cos(\frac{u_i}{a_i}), & if\ 1 \le i \le p-1, \\ a_i \exp(u_n) \cosh(\frac{u_i}{a_i}), & if\ p \le i \le n-1, \end{cases}$$
$$x_{2i} = \begin{cases} a_i \exp(u_n) \sin(\frac{u_i}{a_i}), & if\ 1 \le i \le p-1, \\ a_i \exp(u_n) \sinh(\frac{u_i}{a_i}), & if\ p \le i \le n-1, \end{cases}$$
$$x_{2i-1} = \int_0^{u_n} \sqrt{1 - \exp(2t)}\, dt,$$

where the x_i are coordinates in $\mathbb{R}_{p'}^{2n-1}$, and the a_i are real and satisfy the condition $\sum_i a_i^2 = 1$.

7.13 Remark. The proof of Theorem 7.11, which can be generalized for immersions into indefinite space forms, is based on Otsuki's lemma, see Section 5.3.1. For $n - p = 0$, in case 1, we have the known result for the local nonembedding of a Lobachevsky space, see Section 5.3.1. Perhaps Theorem 7.11 is also true for $n - p = 0, 1, 3, 7$ [Bor 10]; this is related to the fact that the parallelizable spheres have dimensions 1, 3, and 7.

Appendix A

GREAT SPHERE FOLIATIONS AND MANIFOLDS WITH CURVATURE BOUNDED ABOVE

V. Rovenskii and V. Toponogov

A.1. Great circle foliations

The interest in fibrations of the n-sphere by great ν-spheres is propelled by the Blaschke problem and by extremal theorems in Riemannian geometry, see [Bi], [Es 1-3,7], [Ga], [GW], [GWY], [GWZ 1,2], [GZ], [GG 1-3], [GrH], [Hop 1,2], [Häh], [Kam], [NT], [Li], [LV], [Lu], [RAT], [Rov 3,10], [Sat 1,2], [Vai], [Wilk] and [Ya 1-3].

On a round sphere, a foliation by geodesics is like a great circle fibration. The simplest of these, a *Hopf fibration* with fibers $\{S^1\}$, can be given as the collection of intersections of S^{2n-1} with all the holomorphic 2-planes $\{\sigma = (x \wedge Jx)\}$, where J is an *complex structure*, i.e., J is a linear operator in $\mathbb{R}^{2n}$, given for some orthonormal basis $\{e_i\}$ by the rule

$$Je_{2i-1} = e_{2i}, \ Je_{2i} = -e_{2i-1}, \ (1 \le i \le n). \qquad (A.1)$$

Let $\mathcal{F}_0(S^{2n-1})$ denote the *space of all Hopf fibrations* of the sphere S^{2n-1} (for $n = 2$ see Chapter 1). Each fiber spans a corresponding oriented 2-plane through the origin in $\mathbb{R}^{2n}$ and hence determines a point in the Grassmannian manifold $G(2, 2n)$.

A.1 Definition. ([GW] for n=2) The *skew-Hopf fibration* is given by the intersections of S^{2n-1} with all the holomorphic 2-planes $\{\sigma = (x \wedge Jx)\}$, where J is an *almost complex structure*, i.e., a linear operator in $\mathbb{R}^{2n}$, given for some affine basis $\{e_i\}$ by rule (A.1).

In other words, the skew-Hopf fibration is obtained from the Hopf fibration by applying a nondegenerate linear transformation of $\mathbb{R}^{2n}$ and then projecting the images of the fibers back to S^{2n-1}.

Let $\mathcal{F}_1(S^{2n-1})$ denote the *space of all skew-Hopf fibrations* of the sphere S^{2n-1}. Let $\mathcal{F}(S^{2n-1})$ denote the *space of all oriented great circle fibrations* of the sphere S^{2n-1}.

For simplicity, below we consider geodesic foliations on the 3-sphere (for $n > 2$ see [Rov 10]). The space $\mathcal{F}_0(S^3)$ is 2-dimensional and homeomorphic to a pair of disjoint 2-spheres. The 8-dimensional space $\mathcal{F}_1(S^3)$ is the disjoint union of two copies of $S^2 \times \mathbb{R}^6$. Both of these spaces are homogeneous [GW]:

$$\mathcal{F}_0(S^3) = O(4)/U(2), \ \mathcal{F}_1(S^3) = GL(4, \mathbb{R})/GL(2, \mathbb{C}). \qquad (A.2)$$

The space $\mathcal{F}(S^3)$ is infinite dimensional. Let V be a Hopf unit vector field and let D^2 be a small ball transverse to V at point p. As was shown in [GW], small C^1-perturbations of V on D^2, which are the identity on the neighborhood of the boundary ∂D^2, lead to different great circle foliations of S^3. For example, $G(2,4) \equiv S^2 \times S^2$ [GW]; a Hopf fibration $h \in \mathcal{F}_0(S^3)$ can be recognized by the fact that its orbit space M_h appears inside the Grassmannian as $\{point\} \times S^2$ or $S^2 \times \{point\}$.

For two 2-planes P and Q in $\mathbb{R}^4$, the smallest and largest angles α_{min} and α_{max} in the interval $[0, \frac{\pi}{2}]$, which any line in P makes with the plane Q, are called *the principal angles* between these planes. The same angles result upon interchanging the roles of P and Q. The relative positions of P and Q in $\mathbb{R}^4$ are completely determined by these principal angles. An orthonormal basis e_1, e_2, e_3, e_4 for $\mathbb{R}^4$ can always be chosen so that e_1, e_2 is an orthonormal basis for P and $\cos\alpha_{min} e_1 + \sin\alpha_{min} e_3$, $\cos\alpha_{max} e_2 + \sin\alpha_{max} e_4$ is an orthonormal basis for Q [GW]. Any two great circles equidistant (i.e., parallel, see Definition 4.3) in S^3 (in particular, two leaves of a Hopf fibration) determine two 2-planes from $\mathbb{R}^4$ with equal principal angles.

Not all submanifolds of $G(2,4)$ can appear in the role of M_f for $f \in \mathcal{F}(S^3)$.

A.2 Theorem. [GW] *A submanifold in $G(2,4) \equiv S^2 \times S^2$ corresponds to a fibration $f \in \mathcal{F}(S^3)$ if and only if it is the graph of a distance decreasing map $\tilde{f}$ from either S^2 factor to the other. The fibration f is differentiable if and only if the corresponding map $\tilde{f}$ is differentiable with $|df| \le 1$.*

Hence, the space $\mathcal{F}(S^3)$ is a deformation retract of the space $\mathcal{F}_0(S^3)$.

The catalogue with great circle fibrations of the 3-sphere (in Theorem A.2) is one of the first nontrivial examples in which one has a clear overview of all possible geodesic foliations of a fixed Riemannian manifold. There are plenty of nondifferentiable great circle fibrations of S^3. There also exist discontinuous fillings of S^3 by great circles: one can fill the closed solid torus $x_1^2 + x_2^2 \geq x_3^2 + x_4^2$ on S^3 as for a Hopf fibration and then fill the remaining open solid torus $x_1^2 + x_2^2 < x_3^2 + x_4^2$ as for a Hopf fibration defined by the second continuous partition of S^3 by great circles [GW].

A.3 Corollary. [GW] *Each $f \in \mathcal{F}(S^3)$ contains a pair of orthogonal fibers.*

A.4 Theorem. [Ga] *Each skew-Hopf fibration $f \in \mathcal{F}_1(S^3)$ corresponds to a distance decreasing map $\tilde{f} : S^2 \to S^2$ with convex image in a semi-sphere, which can be decomposed as follows: a) an orthogonal projection of S^2 to a plane through the center of the sphere, followed by b) a distance decreasing linear map from one 2-plane to another, and finally c) an inverse projection onto S^2.*

Thus the space $\mathcal{F}_1(S^3)$ is stratified with respect to the values of rank r of the linear map from part b): each fibration with $r = 2$ has exactly one pair of antipodes; in the degenerate case $r = 1$ each fibration has a one-parameter family (Hopf torus) of pairs of antipodes. The case $r = 0$ is the same as Hopf fibrations.

By Corollary A.3 for each $f \in \mathcal{F}(S^3)$ there exists an orthonormal *special basis* $\{e_i\}$ in $\mathbb{R}^4$ such that the sphere S^3 intersects the 2-planes $e_1 \wedge e_2$ and $e_3 \wedge e_4$ in a pair of orthogonal fibers (this is unique when $r = 2$). It is not difficult to see that, in the special basis, for each vector $\lambda = (1, 0, \lambda_3, \lambda_4)$ there exists a unique vector $h = (0, 1, h_3, h_4)$ such that the intersection of the plane $\lambda \wedge h$ with the sphere S^3 is a fiber of f. Hence, to each foliation $f \in \mathcal{F}(S^3)$ there corresponds a diffeomorphism of the 2-plane

$$\varphi = (\varphi_1, \varphi_2) : \mathbb{R}^2 \to \mathbb{R}^2, \quad h_3 = \varphi_1(\lambda_3, \lambda_4), \; h_4 = \varphi_2(\lambda_3, \lambda_4). \tag{A.3}$$

If $f \in \mathcal{F}_0(S^3)$, then the corresponding $\varphi : \mathbb{R}^2 \to \mathbb{R}^2$ is a linear orthogonal map.

A.5 Lemma. [Rov 3] *The space $\mathcal{F}_1(S^3)$ is characterized in $\mathcal{F}(S^3)$ by the property that $\varphi : \mathbb{R}^2 \to \mathbb{R}^2$ is a linear operator without real eigenvalues.*

Proof. Let $f \in \mathcal{F}_1(S^3)$ and suppose that an almost complex structure J in $\mathbb{R}^4$

is given, with respect to the special basis, by the following block matrix:

$$J = \begin{pmatrix} A & 0 \\ 0 & B \end{pmatrix}, \; A = \begin{pmatrix} a_{11} & a_{12} \\ a_{21} & a_{22} \end{pmatrix}, \; B = \begin{pmatrix} a_{33} & a_{34} \\ a_{43} & a_{44} \end{pmatrix}, \; A^2 = B^2 = -E. \tag{A.4}$$

Moreover, the matrices A and B have no real eigenvalues. It is easy to see that φ is a linear operator on $\mathbb{R}^2$ given by the following matrix without real eigenvalues:

$$F = \begin{pmatrix} a & b \\ c & d \end{pmatrix}, \; a = \frac{a_{33} - a_{11}}{a_{21}}, \; b = \frac{a_{34}}{a_{21}}, \; c = \frac{a_{43}}{a_{21}}, \; d = \frac{a_{44} - a_{11}}{a_{21}}. \tag{A.5}$$

Conversely, let φ be a linear transformation of $\mathbb{R}^2$ with matrix $F = \left(\begin{smallmatrix} a & b \\ c & d \end{smallmatrix}\right)$ without real eigenvalues, i.e., the discriminant of the characteristic quadratic equation is negative: $D = (a - d)^2 + 4bc < 0$. Then the matrix of form (A.4) with coefficients

$$a_{11} = -a_{22} = -\frac{a+d}{\sqrt{-D}}, \; a_{33} = -a_{44} = \frac{a-d}{\sqrt{-D}},$$

$$a_{12} = \frac{2(bc - ad)}{\sqrt{-D}}, \; a_{21} = \frac{2}{\sqrt{-D}}, \; a_{34} = \frac{2b}{\sqrt{-D}}, \; a_{43} = \frac{2c}{\sqrt{-D}} \tag{A.6}$$

defines an almost complex structure in $\mathbb{R}^4$. It is easy to see that the operator $\tilde{\varphi} : \mathbb{R}^2 \to \mathbb{R}^2$ corresponding to the induced foliation $\tilde{f} \in \mathcal{F}_1(S^3)$ has the matrix $F = \left(\begin{smallmatrix} a & b \\ c & d \end{smallmatrix}\right)$. □

We study the skew-Hopf fibrations in relation to the interesting space $\mathcal{F}_R(S^{2n-1})$ of analytic geodesic foliations of the sphere S^{2n-1}; see [Top 2-4] for the more general case.

A.6 Definition. [Top 2] The subspace $\mathcal{F}_R(S^{2n-1}) \subset \mathcal{F}(S^{2n-1})$ consists of fibrations for which there exists a tensor (multilinear function) $R : \mathbb{R}^4 \times \mathbb{R}^4 \times \mathbb{R}^4 \times \mathbb{R}^4 \to \mathbb{R}^1$ with the following properties:

R_1) curvature symmetries

$$R(x, y, z, w) = R(z, w, x, y) = -R(x, y, w, z),$$

$$R(x, y, z, w) + R(z, x, y, w) + R(y, z, x, w) = 0$$

R_2) for almost every unit vector $x \in \mathbb{R}^{2n}$ there exists a unique 2-plane $\sigma \ni x$ with

$$R(x, y, x, y) = 1, \; (y \in \sigma, \; y \perp x, \; |y| = 1),$$

R_3) if a 2-plane σ contains a fiber of f, then

$$R(x, y, x, y) = 1, \ (x, y \in \sigma, \ y \perp x, \ |x| = |y| = 1),$$

i.e., the *sectional curvature* of R is extremal on σ.

Note that the curvature tensor on the tangent space of CP^n with complex structure J, standard metric and corresponding Hopf fibration has the properties $R_1) - R_3)$.

A.7 Theorem. [Rov 3] $\mathcal{F}_R(S^3) = \mathcal{F}_1(S^3)$.

A.8 Remark. The analogous fact is true for $\mathcal{F}_R(S^{2n-1})$ with $n > 2$, see [Rov 10]. From Theorem A.7 follows the negative answer to the conjecture by Toponogov [Lemma 6, Top 2] that $\mathcal{F}_R(S^3)$ coincides with $\mathcal{F}_0(S^3)$, see also Section A.2.

Proof. For each $f \in \mathcal{F}(S^3)$ there exists a special orthonormal basis such that the 2-planes $e_1 \wedge e_2$ and $e_3 \wedge e_4$ intersect the sphere S^3 along fibers. Let $\varphi : \mathbb{R}^2 \to \mathbb{R}^2$ be a diffeomorphism that (by Lemma A.5) corresponds to the fibration $f \in \mathcal{F}(S^3)$.

1. First we will prove the inclusion $\mathcal{F}_R(S^3) \subset \mathcal{F}_1(S^3)$.

Let $f \in \mathcal{F}_R(S^3)$. For a special basis, in view of $R_1) - R_3)$, we have

$$R_{121i} = \delta_{i2}, \ R_{212i} = \delta_{i1}, \ R_{343i} = \delta_{i4}, \ R_{434i} = \delta_{i3},$$

where $R_{ijkp} = R(e_i, e_j, e_k, e_p)$ are components of the tensor R. We calculate the *sectional curvature* of R:

$$K(\lambda, h) = \frac{R(\lambda, h, \lambda, h)}{\lambda^2 h^2 - (\lambda, h)^2} = \frac{Q}{\lambda^2 h^2 - (\lambda, h)^2} + 1.$$

The quadratic form Q of the variables $\lambda_3, \lambda_4, h_3, h_4$ is given by the formula

$$Q = \sum_{i,k}(R_{2i2k} - \delta_{ik})\lambda_i \lambda_k + \sum_{j,p}(R_{1j1p} - \delta_{jp})h_j h_p + 2\sum_{i,p}(R_{12ij} + R_{2ij1})\lambda_i h_j. \tag{A.7}$$

The two-parameter family of vectors $(\lambda_3, \lambda_4, h_3, h_4)$, which correspond to the fibers of f, in view of $R_2) - R_3)$, lies in the kernel of the quadratic form Q. In view of $R_3)$, the subspace $\ker Q$ is 2-dimensional; hence the functions $h_3(\lambda_3, \lambda_4)$ and $h_4(\lambda_3, \lambda_4)$ of the variables λ_3 and λ_4 are linear. Since the fibers do not intersect in S^3, the linear operator φ has no real eigenvalues. By Lemma A.5, we have $f \in \mathcal{F}_1(S^3)$.

2. Now we will prove the inverse inclusion: $\mathcal{F}_1(S^3) \subset \mathcal{F}_R(S^3)$.

Let $f \in \mathcal{F}_1(S^3)$. Denote by $F = \begin{pmatrix} a & b \\ c & d \end{pmatrix}$ the matrix of the linear operator φ, which corresponds to f with respect to the special ortho-basis, see Lemma A.5. In view of multilinearity and the symmetries of the tensor R, it is sufficient to define only some of the components $\{R_{ijkp}\}$. It is easy to see that the properties $R_1) - R_3)$ for the tensor R and f hold, when we assume for arbitrary $\gamma < 0$ and $\beta < 0$ that

$$R_{121i} = \delta_{i2},\ R_{2i2i} = \delta_{i1},\ R_{343i} = \delta_{i4},\ R_{434i} = \delta_{i3},\ R_{1313} - 1 = \gamma,$$

$$R_{1414} - 1 = \beta,\ R_{2424} - 1 = b^2\gamma + d^2\beta,\ R_{1323} = a\gamma,\ R_{2324} = ab\gamma + cd\beta,$$

$$R_{1234} = \frac{1}{3}(-c\beta + b\gamma),\ R_{2323} - 1 = a^2\gamma + c^2\beta,\ R_{1424} = d\beta,\ R_{1314} = 0,$$

$$R_{2431} = -\frac{1}{3}(2b\gamma + c\beta),\ R_{2341} = -\frac{1}{3}(2c\beta + b\gamma). \qquad (A.8)$$

Thus $f \in \mathcal{F}_R(S^3)$. □

A.2. Extremal theorem for manifolds with curvature bounded above

A.2.1 Manifolds with curvature bounded above

A complete even-dimensional simply connected Riemannian manifold M with sectional curvature $0 < K_M \leq 4$ has injectivity radius $inj\ (M) \geq \frac{\pi}{2}$ and, hence, diameter $diam\ (M) \geq \frac{\pi}{2}$. For a complete odd-dimensional simply connected M the same inequalities for $inj\ (M)$ and $diam\ (M)$ are true under the stronger curvature restriction $1 \leq K_M \leq 4$ [GKM]. The optimum value for the curvature pinching constant δ in the last case is unknown, but for $\delta < \frac{1}{9}$ the proposition about injectivity radius and diameter of odd-dimensional manifold is wrong in view of Berger's example; see Remark A.9 [Be 2].

A.9 Remark. M. Berger considered a family of Riemannian metrics g_s, $(0 < s \leq 1)$, on the odd-dimensional spheres S^{2n+1}, which are defined by shrinking the standard metric in the direction of the Hopf circles in such a way that their lengths with respect to g_s become $2\pi s$. The range of the sectional curvature of such a metric g_s is in the interval $[s^2, 4 - 3s^2]$. Clearly, $\pi s < \pi/\sqrt{4 - 3s^2}$ for $s^2 < \frac{1}{3}$. This means that for any $\delta \in (0, \frac{1}{9})$ there exists the Berger metric g_s whose sectional curvature K is δ-pinched and whose injectivity radius is strictly smaller than $\pi/\sqrt{\max K}$. By considering the curvature along horizontal geodesics, we obtain that for any $0 < s < 1$ the conjugate radius of g_s is strictly greater than $\pi/\sqrt{\max K}$.

Note that $inj\,(M)$ is never greater than $diam\,(M)$. The manifolds with $inj\,(M) = diam\,(M)$ are precisely the Blaschke manifolds [Bes 1].

In situations when the extremal value for the curvature, the diameter or the volume of the manifold is considered (under other given conditions), one often obtains that the manifold is isometric to a model space from a finite list. A beautiful example of such an *extremal theorem* is the following:

A.10 Theorem. *(minimal diameter)* [Be 1] *Let M be a complete, connected, simply connected Riemannian manifold with sectional curvature $1 \leq K_M \leq 4$ and diameter $\frac{\pi}{2}$. Then M is isometric to CROSS: a round sphere of curvature four or one of the projective spaces CP^n, HP^n, or CaP^2, with its canonical metric.*

Berger uses direct geometric arguments to examine the curvature tensor in sufficient detail to prove that such a manifold must be locally symmetric and hence (since it is simply connected) it must be a symmetric space. Turning to the known classification of these spaces finishes the proof. J. Cheeger and D. Ebin [CE] also give a geometric proof of this result, which is even more direct. H. Gluck and coauthors [GWZ 2] reprove this theorem constructively, using Berger's geometric arguments to show that the exponential map from the tangent cut locus to the cut locus is a fibration of a round sphere by parallel great spheres; hence it is a Hopf fibration. Then they check how this fibration encodes the curvature tensor and use this to construct an isometry between M and a round sphere or a projective space. The latter theorem of Berger was generalized in the following ways:

1. Stability results: *there exists a constant $\delta_n < \frac{1}{4}$ such that any n-dimensional, complete, simply connected Riemannian manifold M^n with $\delta_n \leq K_M \leq 1$ is homeomorphic to CROSS* [Be 3]; *for a compact odd-dimensional manifold this δ is universal and smaller than $\frac{1}{4}(1+10^{-6})^{-2}$* [AM].

2. The upper curvature bound is replaced by the corresponding lower bound for the diameter or radius: *a complete, simply connected Riemannian manifold M^n with $K_M \geq 1$ and radius $rad\,(M) \geq \frac{\pi}{2}$ is either homeomorphic to the sphere or the universal covering $\tilde{M}^n$ is isometric to CROSS* [Wil].

An analogous result with the diameter is obtained in [GG 3], where only the case of $\mathbb{C}aP^2$ is unknown. Recall that the *radius, $rad\,(M)$*, of a compact connected Riemannian manifold is defined as the infimum of the function $p \to$

$rad_p(M) := \max_{q \in M} dist(p, q)$. Clearly, $inj(M) \leq rad(M) \leq diam(M)$. One of the key points in these results comes from studying Riemannian foliations on the round sphere. For 1- and 3-dimensional leaves they are always Hopf fibrations [GG 1,2]; a partial classification of Riemannian foliations on S^{15} with 7-dimensional leaves is given in [Wil], [Ra 3-4], and [Lu].

Thus it is natural to investigate $V^m(-\infty, 4)$, a complete simply connected Riemannian manifold, whose sectional curvature $K_V \leq 4$ and whose injectivity radius $inj(V) \geq \frac{\pi}{2}$. Since the diameter of this manifold is bounded from below, the case of the extremal value $\frac{\pi}{2}$ of the diameter is especially interesting. Note that the inequality $inj(V) \geq \frac{\pi}{2}$ is equivalent to the condition that *the perimeter of every nondegenerate geodesic bi-angle in $V^m(-\infty, 4)$ is not smaller than π.*

A.11 Theorem. [Top 3,4] *The manifold $M = V^{2n+1}(-\infty, 4)$ with diameter $\frac{\pi}{2}$ is isometric to a sphere of curvature four. The manifold $M = V^{2n}(-\infty, 4)$ with diameter $\frac{\pi}{2}$ is either isometric to a sphere of curvature four or its geodesics are grouped into the following families (as for projective spaces):*

F_1) for every point $p \in M$ and any vector $\lambda \in T_pM$ there exists an a-dimensional ($a = 2, 4$ or 8 and if $a = 8$, then $\dim M = 16$) subspace $d(\lambda) \subset T_pM$ such that all the geodesics $\gamma \subset M$, $(\gamma(0) = p,\ \gamma'(0) \in d(\lambda))$, form a totally geodesic submanifold $F(p, \lambda)$, which is isometric to the round sphere $S^a(4)$;

F_2) for all nonzero vectors $\lambda_1, \lambda_2 \in T_pM$ the submanifolds $F(p, \lambda_1)$ and $F(p, \lambda_2)$ either coincide or their intersection consists of only one point p.

The key point of Theorem A.11 is the following result.

A.12 Theorem. [Top 3,4] *If a Riemannian manifold $M = V^n(-\infty, 4)$ has a closed geodesic γ with length π and index $a - 1$, then there is an a-dimensional totally geodesic submanifold containing γ, which is isometric to the a-dimensional sphere of curvature four.*

We outline the idea of the proof of Theorem A.12.

A.13 Lemma. [Top 3,4] *Under the conditions of Theorem A.12 (and using the same notation), any two points P and Q of γ whose mutual distance on γ is $\frac{\pi}{2}$ are conjugate of multiplicity $a - 1$.*

The proof of Lemma A.13 uses the condition $inj(V) \geq \frac{\pi}{2}$ and the well-known lemma of the calculus of variations.

From Lemma A.13 we easily obtain the following lemma.

A.14 Lemma. [Top 3,4] *Under the conditions of Theorem A.12 (and using the same notation), there exists an arc σ of length $> \frac{\pi}{2}$ on the geodesic γ and an $(a-1)$-parameter family of parallel vector fields ν along σ such that the Riemann curvature for the 2-dimensional directions in ν along γ is equal to four.*

Using Lemma A.14 for all the fields ν, we can construct a sequence of triangles $\Delta_n(\nu)$, each of which has perimeter strictly smaller than π, which converges to γ. It follows from the condition $inj\ (V) \geq \frac{\pi}{2}$ that in every triangle $\Delta_n(\nu)$ we can span a cone $K_n(\nu)$ obtained as the set of the shortest lines between the vertices of $\Delta_n(\nu)$ and the opposite edges. For these cones the following lemma holds.

A.15 Lemma. [Top 3,4] *The Gauss curvature at any point of $K_n(\nu)$ does not exceed four.*

Lemma A.15 follows from Synge's Theorem [GKM].

A.16 Lemma. [Top 3,4] *The area of the cone $K_n(\nu)$ is not greater than the area of the triangle $\Delta_n^L(\nu)$ on the sphere of curvature four whose sides have the same length as that of the corresponding sides of $\Delta_n(\nu)$.*

Lemma A.16 follows from a theorem of A. D. Alexandrov, see [Top 3,4]. From Lemmas A.15 and A.16 we obtain an upper bound for the integral curvature of the cones $K_n(\nu)$. On the other hand, the Gauss-Bonnet theorem for the integral curvature of $K_n(\nu)$ can be used to obtain a lower bound for the angle sum of $K_n(\nu)$. Comparing these bounds shows that the Gauss curvature of $K_n(\nu)$ is equal to four almost everywhere and that the area of $K_n(\nu)$ is almost equal to π. Taking the limit as $n \to \infty$, we see that there exists an $(a-1)$-parameter family of surfaces $\{F\}$ that are isometric to the 2-dimensional hemisphere of curvature four with boundary γ.

It is now easy to show that the union of all the surfaces of this family is an a-dimensional surface F^a which is isometric to the a-dimensional sphere. From the previous results and $inj\ (V) \geq \frac{\pi}{2}$ it follows easily that F^a is a totally geodesic surface in M.

The proof of Theorem A.11 uses reasoning similar to the proof of Theorem A.12, but we need some topological results and, in particular, a theorem of W. Brouder.

Toponogov [Top 2-4] **conjectured** that *the manifold* $V^{2n}(-\infty, 4)$ *with extremal diameter* $diam\,(V) = \frac{\pi}{2}$ *is isometric to CROSS.*

Note that the tangent a-planes to the submanifolds $\{F(p, *)\}$ (in Theorem A.11) induce an $(a-1)$-dimensional great sphere foliation of the round sphere S_p in the tangent space T_pM. In the case $K_M > 0$ for almost each point $p \in M$ such a foliation f_p is related to the function for the sectional curvature in T_pM in the following way (compare with Definition A.6) [Top 2]:

$\bar{R}_2$) *for almost every vector* $x \in T_pM$ *there exists a unique* a*-dimensional subspace* $V \ni x$ *satisfying the condition*

$$K(x, y) = 4, \ (y \in V),$$

$\bar{R}_3$) *if an* a*-dimensional subspace* V *contains a fiber of* f_p, *then*

$$K(x, y) = 4, \ (x, \ y \in V).$$

The natural strategy to prove that *a Riemannian manifold* M *with properties* F_1) *and* F_2) *is isometric to CROSS,* (and, hence, to prove Toponogov's conjecture) is to deduce first that the induced great sphere foliations in the tangent spheres $\{S_p\}$, $(p \in M)$ are Hopf fibrations. This claim was conjectured in [Lemma 6, Top 2] for a foliation with the properties $\bar{R}_2$) and $\bar{R}_3$) and the curvature symmetries of R_1). By considering only one point of M (see the results of Section A.1) we can only show that such foliations on the tangent spheres $\{S_p\}$, $(p \in M)$, are skew-Hopf fibrations (the proof of [Lemma 6, Top 2] is wrong). In other words, the manifold in Theorem A.11, when $a = 2$, admits an almost complex structure $J : TM \to TM$ satisfying the identity

$$(\nabla_x J)x = 0, \ (x \in TM),$$

and whose holomorphic curvature is constant (the totally geodesic submanifolds $F(p, \lambda)$ are J-invariant). Note that a Hermitian manifold (M, J) satisfying the above identity is called *nearly Kählerian.* It is not known whether it is possible to prove locally that J is Hermitian in our case.

Later, Toponogov's conjecture was proven in another way [RovT], namely, by using global integral geometric methods of M. Berger and J. Kazdan as well as recent topological results for manifolds with closed geodesics.

Manifolds M with the properties F_1) and F_2) are particular cases of Blaschke manifolds. We will give a short survey of manifolds with the special behavior of geodesics.

A.2.2. Manifolds with closed geodesics

A compact Riemannian manifold M is called a C_π*-manifold* if all its geodesics are closed and of length π. This class includes (A. Allamigeon and F. Warner) *Blaschke manifolds*, for which all cut loci $Cut(p) \subset T_pM$, $(p \in M)$, are round spheres of constant radius and dimension. Examples are CROSS: a sphere or projective spaces over classical fields.

If M is a simply connected C_π-manifold, then it is homotopically equivalent to a CROSS (R. Bott and H. Samelson).

For a Blaschke manifold the exponential map $\exp_p : T_pM \to C(p)$, restricted to the sphere S_p with radius $d(p, C(p))$, defines a *great sphere foliation*; for CROSS this foliation is a Hopf fibration. Since every great sphere foliation of S^N is homeomorphic to a Hopf fibration [Sat 2], (some special cases are in [GWY] and [Ya 1,3]; some cases when foliation on a sphere is differentiable are in [Häh]), a simply connected Blaschke manifold is homeomorphic to its model CROSS.

The well-known **Blaschke conjecture**, that *any Blaschke manifold is isometric to its model CROSS*, was proved for the spherical case using the following scheme [Bes 1]:

Step 1. Using integral geometry in the space of geodesics one shows that $vol\,(M^N) \geq vol\,(S^N, can)$ with equality of volumes if and only if M is isometric to (S^N, can).

Step 2. On the other hand, one uses topological arguments to show that the *Weinstein integer* $i(M) = vol\,(M^N)/vol\,(S^N, can)$, which can be described by the cohomology of the space of oriented geodesics of the given C_π-manifold, is actually equal to one. Here (S^N, can) denote a sphere S^N with the standard metric whose geodesics are of length π.

The obvious analog of the Blaschke conjecture for C_π-manifolds is already wrong when the model CROSS is S^2, but it is unknown whether this conjecture holds in the nonspherical case.

Step 2 is related to the **weak Blaschke conjecture** by C. T. Yang, that *all Blaschke manifolds have right volumes*. This conjecture was proved in [Ya 4] for the complex projective space CP^n and in [Rez 1, 4] for C_π-manifolds homeomorphic to model CROSS.

A.2.3. Extremal theorem for a volume

In view of the facts in the previous section, the missing link for proving

Toponogov's conjecture is the following.

A.17 Theorem. [RovT] *If a Riemannian manifold* M, $(\dim M = an)$ *has the properties* F_1) *and* F_2), *then its volume is not smaller than the volume of* $KP^n(4)$, $(\dim K = a)$ *and equality holds if and only if* M *is isometric to* $KP^n(4)$.

Proof. We use the scheme of [App. D, Bes 1], but modify it in view of splitting of the Jacobi equation and the volume measure along geodesics that are tangent to the family $\{F(p, \lambda)\}$. The same proof (for a Blaschke manifold with taut geodesics) is given in [Heb 1].

Let μ be a volume measure on M; the whole measure $V(M) = \int_M d\mu$ of the manifold is called its *volume*. The volumes of the projective spaces $KP^n(4)$, $(K = \mathbb{C}, \mathbb{H}, \mathbb{C}a)$ with standard metrics and diameter $\frac{\pi}{2}$, and the volume of the sphere S^{an-1} of radius one, will be denoted by $V(KP^n)$ and $V(S^{an-1})$, respectively. Their numerical values are known. The total space of the fibration $\pi : UM \to M$ of unit spheres tangent to M is endowed with the canonical metric and measure $\mu_1 = \sigma \otimes \mu$, where σ is the standard volume measure on S^{an-1}. Thus $V(UM) = V(S^{an-1})V(M)$. The unit *geodesic* vector field Z on UM is defined: the projections of the integral curves of Z are geodesics in M and the induced dynamical system ξ on UM is called a *geodesic flow*. Note that the measure μ_1 is invariant under the geodesic flow ξ on the manifold UM [Bes 1].

Let γ_u, $(u = (p, \lambda) \in UM)$, be a unit speed geodesic with initial values $\gamma_u(0) = p$, $\gamma_u'(0) = \lambda$ and let ξ^t be an evolution map of the geodesic flow, i.e., $\pi(\xi^t(u)) = \gamma_u'(t)$.

Let $f(u, t)$, where $u \in UM$ and $t \in R_+$ denotes a volume form on M in polar coordinates, i.e., at the point $\exp(t\lambda)$ we have $d\mu = f(u, t)d\sigma \otimes dt$. This function $f(u, t)$ may be calculated with the help of Jacobi fields in the following way. Let $\{\lambda_i\}$, $(2 \le i \le an)$, be an orthonormal basis of the subspace $\lambda^\perp$ (the orthogonal complement to λ) and let $\{Y_i\}$, $(2 \le i \le an)$, be the Jacobi fields along γ_u with initial values $Y_i(0) = 0$, $Y_i'(0) = \lambda_i$. Then for all t we have $f(u, t) = |\det Y_2(t) \wedge \cdots \wedge Y_{an}(t)|$ [Bes 1]. In our case the vectors $\lambda_2, \ldots, \lambda_a$ are chosen tangent to $d(\lambda)$ and the vectors $\lambda_{a+1}, \ldots, \lambda_{an}$ are orthogonal to $d(\lambda)$. Along the geodesic γ_u the curvature transformation $R(*, \gamma_u')$, γ_u' has two invariant subspaces: the tangent space to the totally geodesic constant curvature submanifold $F(p, \lambda)$ containing γ_u, on which it is a multiplication of vectors by four, and the space orthogonal to this.

Since the Jacobi fields $Y_2(t), \ldots, Y_a(t)$ are tangent to the totally geodesic submanifold $F(p, \lambda)$ with constant curvature four, they are given by the well-known formula

$$Y_i(t) = (\frac{1}{2}\sin 2t)\bar{\lambda}_i, \quad (2 \le i \le a),$$

where $\bar{\lambda}_i \ni \lambda_i$ is a parallel vector field along γ_u. In view of F_2) the vector fields $Y_{a+1}(t), \ldots, Y_{an}(t)$ are nonzero for $0 < t < \pi$; they span the normal bundle to $F(p, \lambda)$ along the geodesic γ_u. Thus,

$$f(u,t) = f_1(u,t)|\frac{1}{2}\sin 2t|^{a-1},$$

where the function $f_1(u,t) = |\det Y_{a+1}(t) \wedge \cdots \wedge Y_{an}(t)|$ is positive for $0 < t < \pi$. Let

$$\varphi(u,t) = f_1(u,t)^{\frac{1}{an-a}},$$

i.e.,

$$f(u,t) = \varphi(u,t)^{an-a}|\frac{1}{2}\sin 2t|^{a-1}.$$

In particular, for the model CROSS $KP^n(4)$ we have $\varphi(u,t) = \sin t$.

A.18 Lemma. $\pi V^2(M) = \int_{UM}(\int_0^{\pi}(\int_0^{\pi-x} f(\xi^x(u),t)\,dt)\,dx)\,d\mu_1$.

Proof of Lemma A.18. Since the diameter and the injectivity radius of M are $\frac{\pi}{2}$, we have for all $p \in M$ and the ball $B(p, \frac{\pi}{2})$ that

$$V(M) = V(B(p, \frac{\pi}{2})) = \int_{U_pM}(\int_0^{\pi/2} f(u,t)\,dt)\,d\sigma.$$

With the help of the equality $f(-u,t) = f(u,\pi - t)$, $(\forall u \in UM)$ [App. D, Bes 1]), we obtain

$$\begin{aligned}\int_{U_pM}(\int_0^{\pi/2} f(-u,t)\,dt)\,d\sigma &= \int_{U_pM}(\int_0^{\pi/2} f(u,\pi - t)\,dt)\,d\sigma \\ &= \int_{U_pM}(\int_{\pi/2}^{\pi} f(u,t)\,dt)\,d\sigma.\end{aligned} \tag{A.9}$$

Thus the integration over the interval $[0, \pi]$ gives us the double volume

$$2V(M) = \int_{U_pM}(\int_0^{\pi} f(u,t)\,dt)\,d\sigma. \tag{A.10}$$

We integrate (A.10) over M:

$$2V^2(M) = \int_M (\int_{U_pM} (\int_0^\pi f(u,t)\,dt)\,d\sigma)\,dp = \int_{UM} (\int_0^\pi f(u,t)\,dt)\,d\mu_1$$

and then we integrate the last equality over the interval $[0,\pi]$ using the invariance of the measure μ_1 under the geodesic flow

$$2\pi V^2(M) = \int_{UM} (\int_0^\pi (\int_0^\pi f(\xi^x(u),t)\,dt)\,dx)\,d\mu_1. \tag{A.11}$$

Since $\xi^x(-u) = -\xi^{\pi-x}(u)$, $(u \in UM,\ 0 \le x \le \pi)$ and the map $u \to -u$ is a diffeomorphism of the manifold UM which preserves the measure μ_1, in view of $f(-u,t) = f(u,\pi-t)$ we have

$$\begin{aligned}
&\int_{UM} (\int_0^\pi (\int_0^{\pi-x} f(\xi^{\pi-x}(u),\pi-t)\,dt)\,dx)\,d\mu_1\\
&= \int_{UM} (\int_0^\pi (\int_{\pi-x}^\pi f(\xi^x(u),t)\,dt)\,dx)\,d\mu_1\\
&= \int_{UM} (\int_0^\pi (\int_0^{\pi-x} f(\xi^\pi(u),\pi-t)\,dt)\,dx)d\mu_1\\
&= \int_{UM} (\int_0^\pi (\int_{\pi-x}^\pi f(\xi^x(u),t)\,dt)\,dx)\,d\mu_1.
\end{aligned}$$

Thus the integral (A.11) may be broken into two equal parts. □

A.19 Lemma. *For any $u \in UM$ we have*

$$J(u) = \int_0^\pi (\int_0^{\pi-x} f(\xi^x(u),t)\,dt)\,dx \ge \pi \frac{V(KP^n)}{V(S^{an-1})}$$

with the equality only in the case of $K(\gamma_u'(x),y) \equiv 1,\ (y \perp d(\gamma_u'(x)),\ 0 \le x \le \pi)$.

Proof of Lemma A.19. We will use the Holder inequality of order $p = an - a$:

$$\int g_1 g_2 \le (\int g_1^p)^{\frac{1}{p}} (\int g_2^{\frac{p}{p-1}})^{\frac{p-1}{p}},$$

where the equality holds if and only if $g_1^p = \rho g_2^{\frac{p}{p-1}}$, $(\rho = const)$. In our case the functions are

$$g_1(t) = \varphi(t)|\frac{1}{2}\sin 2t|^{\frac{a-1}{an-a}},\quad g_2(t) = (\sin t)^{an-a-1}|\frac{1}{2}\sin 2t|^{\frac{(an-a-1)(a-1)}{an-a}}$$

and the condition $g_1 = \rho g_2^{\frac{p}{p-1}}$ takes the form $\varphi = \rho \sin$. Thus

$$J(u) = \int_0^\pi (\int_x^\pi \varphi^{an-a}(\xi^x(u), y-x)|\frac{1}{2}\sin 2(y-x)|^{a-1}\,dy)\,dx \geq$$

$$\frac{(\int_0^\pi \int_x^\pi \varphi(\xi^x(u), y-x)\sin^{an-a-1}(y-x)\,|\frac{1}{2}\sin 2(y-x)|^{a-1}\,dy\,dx)^{an-a}}{(\int_0^\pi \int_x^\pi \sin(y-x)^{an-a}\,|\frac{1}{2}\sin 2(y-x)|^{a-1}\,dy\,dx)^{an-a-1}}. \tag{A.12}$$

Since the submanifolds $\{F(p,\lambda)\}$ are totally geodesic, the function $\varphi(u,x)$ satisfies the following inequality (the proof is the same as in [App. D, Bes 1]):

$$\varphi(\xi^x(u), z) \geq \varphi(u,x)\varphi(u,x+z)\int_x^{x+z}\frac{dt}{\varphi^2(u,t)}.$$

With the help of Kazdan's Inequality [App. E, Bes 1] with weight function

$$\rho(y-x) = \sin(y-x)^{an-a-1}|\frac{1}{2}\sin 2(y-x)|^{a-1}$$

we estimate the numerator in (A.12):

$$\int_0^\pi (\int_x^\pi \varphi(\xi^x(u), y-x)\sin(y-x)^{an-a-1}\,|\frac{1}{2}\sin 2(y-x)|^{a-1}dy)dx$$

$$\geq \int_0^\pi \int_x^\pi \int_x^y \frac{\varphi(u,x)\varphi(u,y)}{\varphi^2(u,t)}\sin(y-x)^{an-a-1}|\frac{1}{2}\sin 2(y-x)\,|^{a-1}\,dt\,dy\,dx$$

$$\geq \int_0^\pi (\int_x^\pi \sin(y-x)^{an-a}\,|\frac{1}{2}\sin 2(y-x)|^{a-1}\,dy)\,dx = \beta(a,n).$$

Thus

$$J(u) \geq \beta(a,n) = \int_0^\pi (\int_x^\pi \sin(y-x)^{an-a}\,|\frac{1}{2}\sin 2(y-x)|^{a-1}\,dy)\,dx.$$

The equality holds only in the case in which (see [App. D, Bes 1])

$$K(\gamma_u'(x), y) = 1, \ (0 \leq x \leq \pi, \ y \perp d(\gamma_u'(x))).$$

We will show below that $\beta(a,n) = \pi\frac{V(KP^n)}{V(S^{an-1})}$. □

Now we continue the proof of Theorem A.17. From Lemma A.18 and Lemma A.19 it follows that

$$\pi V^2(M) \geq \int_{UM}\beta(a,n)\,d\mu_1 = V(UM)\beta(a,n).$$

Since $V(UM) = V(M)V(S^{an-1})$, we have $\pi V^2(M) \geq \beta(a,n)V(M)V(S^{an-1})$, i.e.,

$$V(M) \geq \frac{1}{\pi}\beta(a,n)V(S^{an-1}).$$

The equality holds only in the case of M of constant sectional curvature

$$K(\lambda, y) = 1, \ (\lambda \in U_pM, \ y \perp d(\lambda), \ p \in M).$$

From above it follows that our manifold M has positive $\frac{1}{4}$-pinched sectional curvature. By Theorem A.10, M is isometric to $KP^n(4)$. If we repeat the above for the model CROSS $KP^n(4)$, then we obtain the numeric value of $\beta(a,n)$:

$$V(KP^n) = \frac{1}{\pi}\beta(a,n)V(S^{an-1}) \Rightarrow \beta(a,n) = \pi\frac{V(KP^n)}{V(S^{an-1})}. \quad \square$$

From the above statement and facts about Blaschke manifolds, we get the following theorem.

A.20 Theorem. [RovT] *A Riemannian manifold M, $(\dim M = an)$ having the properties $F_1)$ and $F_2)$ is isometric to CROSS.*

Theorem A.20 and Theorem A.11 confirm Toponogov's conjecture:

A.21 Theorem. [RovT] *A Riemannian manifold $V^m(-\infty, 4)$ with extremal diameter $\frac{\pi}{2}$ is isometric to CROSS.*

A.22 Corollary. *(diameter rigidity) A complete, connected, simply connected Riemannian manifold M^{2n} with sectional curvature $0 < K_M \leq 4$ and diameter $diam\,(M) = \frac{\pi}{2}$, is isometric to CROSS.*

Note that Theorem A.21 and Corollary A.22 generalize Theorem A.10 by Berger.

Theorem A.21 has many corollaries. Below we give one of them.

The projective planes $KP^2(4)$, $(K = \mathbb{R}, \mathbb{C}, \mathbb{H}, \mathbb{C}a)$ with standard metrics are at the same time Riemannian manifolds of dimension $2a = 2, 4, 8, 16$, respectively. The totally geodesic spheres $\{KP^1(4) = S^a(4)\}$ (with dimension a and constant curvature four) play the role of *straight lines* in $KP^2(4)$ and fulfill the **axioms of projective geometry**:

P_1: *For every two different points there exists exactly one straight line which connects them.*

P_2: Every two different straight lines intersect in exactly one point.

Thus we obtain the following differential geometric test for projective planes over $\mathbb{C}$, $\mathbb{H}$, $\mathbb{C}a$.

A.23 Corollary. *Assume that M^{2a}, $(a > 1)$, is a complete Riemannian manifold satisfying the conditions P_1 and P_2. Also suppose that its straight lines are totally geodesic submanifolds isometric to the Euclidean sphere $S^a(4)$. Then $a = 2$, 4 or 8 and M^{2a} is isometric to the projective plane $KP^2(4)$, $(K = \mathbb{C}$, $\mathbb{H}$ or $\mathbb{C}a)$, with the standard metric.*

Appendix B

SUBMERSIONS OF RIEMANNIAN MANIFOLDS WITH COMPACT LEAVES

We consider a submersion $\pi : M \to \bar{M}$ of smooth manifolds M and $\bar{M}$ whose compact connected leaves $\{L_b = \pi^{-1}(b)\}_{b\in\bar{M}}$ form a fiber bundle, see Chapter I. Let M be equipped with a Riemannian metric and let $\dim \bar{M} > 1$. Recall that a horizontal vector field $Y = \pi_*^{-1}(\bar{Y})$ along a fiber L_b is called *basic*, where $\bar{Y} \in T_b\bar{M}$. Unless otherwise specified, x and x_i below are always vertical vectors, Y, Z, U, V and W are horizontal vectors or basic fields, s and s_i are horizontal vector fields, $\top$ is an orthoprojector from TM to TL and $\perp$ is an additional orthoprojector.

Riemannian submersions [O'N 3], which are characterized by the condition $h_2 = 0$, have equidistant fibers, (i.e., basic vector fields are of constant length); there exists a metric $(\ ,\ )^*$ on the base such that $\pi_*(m) : T_m L^\perp \to T_{\pi(M)}\bar{M}$ are isometries (for some generalizations, see [Ste 6]). Hopf fibrations are examples of Riemannian submersions. But already for odd-dimensional spheres there are other interesting classes of submersions, for example, skew-Hopf fibrations (see Appendix A) which are useful in studying Blaschke manifolds.

Along a leaf L_b the horizontal vector fields with summable second power of length form the vector space $\tilde{M}_b$ with the scalar product

$$< s_1, s_2 >_b = \int_{L_b} (s_1, s_2) d\,\mathrm{vol}. \qquad (B.1)$$

The basic vector fields along L_b form a $(\dim \bar{M})$-dimensional Euclidean subspace $T_b\bar{M} \subset \tilde{M}_b$ and thus $\bar{M}$ with the metric $<,>$ is a Riemannian manifold. In particular, when M is the unit tangent bundle of a C_l-manifold and $\bar{M}$ is a certain *manifold of geodesics*, the metric $<,>$ on the base was introduced in [Bes 1]. In contrast to O'Neill's method [O'N 3] for a submersion with equidistant fibers, the metric $<,>$ is defined on $\bar{M}$ for an arbitrary metric on M. In the case of equidistant fibers the metric $<,>$ is conformal to O'Neill's metric

$(\, , \,)^*$, i.e., is obtained at the point $m \in \bar{M}$ by multiplication on the volume of a fiber over this point.

The map $\pi : M \to \bar{M}$ is a smooth fibration and the spaces $\{\tilde{M}_b\}$ are the fibers of a vector bundle $p : \tilde{M} \to \bar{M}$; moreover $T\bar{M}$ (its vectors are basic fields) is a subbundle of $\tilde{M}$. Consider also the sub-bundle $T\bar{M}^{\perp} \subset \tilde{M}$, the orthogonal complement of $T\bar{M}$, and let $P^L : \tilde{M} \to T\bar{M}$ and $P^{L,\perp} : \tilde{M} \to T\bar{M}^{\perp}$ be orthoprojectors. Denote by $\Gamma(\bar{M})$ and $\Gamma(\tilde{M})$ the vector spaces of the smooth sections $T\bar{M}$ and $\tilde{M}$, respectively, and define a metric connection $\nabla^{\pi} : \Gamma(\bar{M}) \times \Gamma(\tilde{M}) \to \Gamma(\tilde{M})$ in the fibration $\tilde{M}$ by $\tilde{\nabla}$-differentiation (in points) of the horizontal vector fields:

$$(\nabla^{\pi}_{\bar{Y}} s)_m = \tilde{\nabla}_{\bar{Y}(m)} s, \quad (m \in L_b). \tag{B.2}$$

The horizontal connection $\tilde{\nabla}$ and horizontal curvature operator $\tilde{R}$ were defined in Section 2.4.2. The curvature tensor $R^{\pi} : T\bar{M} \times T\bar{M} \times \tilde{M} \to \tilde{M}$ of the connection ∇^{π} is calculated by a well-known formula and the following equality holds:

$$< R^{\pi}(\bar{Y}, \bar{Z}) s_1, s_2 > \; = \int_{L_b} (\tilde{R}(Y, Z) s_1, s_2) d\,\mathrm{vol}. \tag{B.3}$$

The connection and the curvature tensor of a metric $< \, , \, >$ on $T\bar{M}$ are denoted by $\bar{\nabla}$ and $\bar{R}$, respectively. The second fundamental form $h^L : T\bar{M} \times T\bar{M} \to T\bar{M}^{\perp}$ of the submersion π (sub-bundle $T\bar{M} \subset \tilde{M}$) is defined by the formula

$$h^L(\bar{Y}, \bar{Z}) = P^{L,\perp}(\tilde{\nabla}_Y Z). \tag{B.4}$$

B.1 Proposition. *$h^L(\bar{Y}, \bar{Z}) = h^L(\bar{Z}, \bar{Y})$ (symmetry of h^L).*

Proof. Let $Y, Z \subset TM$ be horizontal lifts of the vector fields $\bar{Y}, \bar{Z} \subset T\bar{M}$. The vector field $[Y, Z]$ is equal to the sum of a basic field (the horizontal lift of $[\bar{Y}, \bar{Z}]$) and the vertical vector field $[Y, Z]^{\top}$. Thus we have

$$h^L(\bar{Y}, \bar{Z}) - h^L(\bar{Z}, \bar{Y}) = P^{L,\perp}(\tilde{\nabla}_Y Z) - P^{L,\perp}(\tilde{\nabla}_Z Y) = P^{L,\perp}([Y, Z]^{\perp}) = 0. \tag{B.5}$$

□

By Lemma B.9 (see below) we have

$$\nabla^{\pi}_{\bar{Y}} \bar{Z} = \bar{\nabla}_{\bar{Y}} \bar{Z} + h^L(\bar{Y}, \bar{Z}). \tag{B.6}$$

By analogy with the theory of Riemannian submanifolds with degenerate second fundamental form (see Chapter V) we define the main class of submersions in Appendix B.

B.2 Definition. When $h^L = 0$ a submersion π is said to be *totally geodesic*; in other words, in view of (B.5), if Y and Z are basic fields, then $\tilde{\nabla}_Y Z$ is a basic field corresponding to the vector $\bar{\nabla}_{\bar{Y}} \bar{Z}$. (From this it follows that if Y, Z and W are basic fields, then $\tilde{R}(Y, Z)W$ is also a basic field corresponding to the vector $\bar{R}(\bar{Y}, \bar{Z})\bar{W}$).

B.3 Example. Let $\pi : M \to \bar{M}$ be a submersion with compact leaves and let $f : M \to R_+$ be a smooth function on a Riemannian manifold satisfying the following conditions:

$$(\text{grad } f)^{\perp} = 0, \ \ f * (Y, Z) \text{ is constant along the leaves for basic fields } Y, \ Z. \tag{B.7}$$

Then $\tilde{\nabla}_Y Z$ is a basic field for the basic fields Y and Z [Gr 2]. So for $\bar{M}$ equipped with the metric $< \, , \, >$, we have $h^L = 0$, i.e., π is a totally geodesic submersion. (The case of $f \equiv 1$ corresponds to a Riemannian submersion). Such a submersion appears under the natural projection of the domain

$$G_{a,b} = \{x \in \mathbb{R}^{2n+2} \approx \mathbb{C}^{n+1} : 0 < a \leq |x| \leq b\}$$

onto CP^n and analogously for HP^n.

Strongly parabolic curvature invariant submanifolds endowed with a totally geodesic foliation were studied by many authors, see Chapter V. Thus it is natural to introduce the following class which includes the class of totally geodesic submersions.

B.4 Definition. A submersion π is said to be *curvature invariant* if the following property is satisfied:

$$P^{L,\perp}(R^{\pi}(\bar{Y}, \bar{Z})\bar{W}) = 0. \tag{B.8}$$

In other words, if the fields Y, Z and W are basic, then $\tilde{R}(Y, Z)W$ is also basic and corresponds to the vector $\bar{R}(\bar{Y}, \bar{Z})\bar{W}$.

Foliations, modelled by such classes of submersions, are also interesting, but little is known about them. We will obtain local characterizations of the above submersions (Corollaries B.13 and B.15), and prove their existence with the help of twisted products (defined in Section 2.3.2).

B.5 Definition. The *relative nullity-space* at a point $b \in \bar{M}$ is defined by the rule

$$N_b = \{\bar{Y} \in T_b\bar{M} : h^L(\bar{Y}, \bar{Z}) = 0, \ \forall \, \bar{Z} \in T_b\bar{M}\}.$$

A submersion π is said to be *strongly parabolic* if $N_b \neq 0$ at each point b. The number $\mu(h^L) = \min\{\dim N_b\}$ is called the *relative index of nullity* of h^L. Let $\bar{G}(h^L) \subset \bar{M}$ denote the regularity domain, which consists of points with $\mu(h^L)$–dimensional relative nullity-spaces.

The following class of submersions is closely connected to the submersions above.

B.6 Definition. A submersion π is said to be of *codimension* ≤ 1 at the point $b \in \bar{M}$, if $\dim N_1(b) \leq 1$, where $N_1(b) \subset \tilde{M}_b$ (*the first normal space*) is the subspace generated by the vectors $h^L(\bar{Y}, \bar{Z})$.

Totally geodesic submersions are of codimension ≤ 1.

B.7 Problem. *To classify totally geodesic submersions of a round sphere S^{2n+1}. Are there skew-Hopf fibrations among them?*

For a C_l-manifold M consider the submersion $\pi : T_1M \to CM$ with fibers $L = S^1$, where T_1M denotes the unit tangent bundle and CM denotes the manifold of closed geodesics. For $M = S^n$ and RP^n a submersion π is Riemannian [Bes 1].

B.8 Problem. *What are the properties of the second fundamental form h^L of the submersion $\pi : T_1M \to CM$ for the C_l-manifolds $M = CP^n$, HP^n or CaP^2?*

The map P^L allows us to characterize geodesics in $(\bar{M}, <\ ,\ >)$ in the following manner.

B.9 Lemma. *The curve $\bar{\Gamma}(t) \subset \bar{M}$ with $\bar{Y}(t) = \frac{d}{dt}\bar{\Gamma}(t)$ is geodesic iff*

$$P^L(\tilde{\nabla}_{Y(t)}Y(t)) = 0$$

along every leaf $L_t = \pi^{-1}(\bar{\Gamma}(t))$ in M.

The proofs of Lemma B.9, Lemma B.10 and Lemma B.11 below are analogous to the corresponding ones in [Bes 1] and we omit them.

The following lemma explains what the Riemannian connection $\bar{\nabla}$ of a metric $<\ ,\ >$ on $\bar{M}$ is.

B.10 Lemma. *Let $\bar{Z} \subset T\bar{M}$ be a vector field on a neighborhood of $b \in \bar{M}$ and let $\bar{Y}$, $\bar{W} \in T_b\bar{M}$. Then*

$$< \bar{\nabla}_{\bar{Y}}\bar{Z}, \bar{W} > \ = \int_{L_b} (\tilde{\nabla}_Y Z, W) d\,vol. \qquad (B.9)$$

The equation analogous to Gauss's equation for embedded submanifolds holds for the subbundle $T\bar{M} \subset \tilde{M}$.

B.11 Lemma. *Let $\bar{\sigma} \subset T_b\bar{M}$ be a 2-dimensional tangent plane with orthonormal basis $\bar{Y}, \bar{Z}$. Then*

$$\bar{K}(\bar{\sigma}) = \int_{L_b} (\tilde{R}(Y, Z)Y, Z) d\, vol$$

$$+ < h^L(\bar{Y}, \bar{Y}), h^L(\bar{Z}, \bar{Z}) > - < h^L(\bar{Y}, \bar{Z}), h^L(\bar{Z}, \bar{Y}) > . \qquad (B.10)$$

From Gauss's equation (B.10) follows the simple formula with curvature of M and $\bar{M}$ in the case of a totally geodesic submersion:

$$\bar{K}(\bar{Y}, \bar{Z}) = \int_{L_b} (\tilde{R}(Y, Z)Y, Z) d\, \text{vol}, \ (\bar{Y}, \bar{Z} \in T_b\bar{M}), \qquad (B.10')$$

generalizes a formula by [O'N 3], because $\tilde{R}$ may be written in terms of R and O with the help of (2.34a). For an arbitrary submersion, the Codazzi and Ricci equations also hold; deducing them is not difficult, but we do not need them below.

Submersions with the skew-symmetric tensor O, defined by (2.11), have the following property: *if Y, Z are basic fields, then $\tilde{\nabla}_Y Z$ is also a basic field.* Note that for a basic vector field y, equation (3.19') is equivalent to

$$(\nabla_x y)^{\perp} = O_y x, \quad (x \in TL). \qquad (B.11)$$

Thus, it is natural to assume that the horizontal component of $(\nabla_x(\tilde{\nabla}_Y Z) - O_{\tilde{\nabla}_Y Z} x$ depends only on the tensor T and the second fundamental form h_2.

B.12 Lemma. *The following equality is true:*

$$(\nabla_x(\tilde{\nabla}_Y Z) - O_{\tilde{\nabla}_Y Z} x, W) = ((\nabla_W h_2)(Z, Y) - (\nabla_Y h_2)(Z, W)$$

$$-(\nabla_Z h_2)(Y, W), x) + (h_2(Y, Z), T_x W) - (h_2(Y, W), T_x Z) - (h_2(Z, W), T_x Y). \qquad (B.12)$$

Proof. We write the formula for the curvature tensor

$$(\nabla_{[x,Y]} Z - \nabla_x \nabla_Y Z + \nabla_Y \nabla_x Z, W) = (R(x, Y)Z, W),$$

where Y and Z are basic fields, $x \subset TL$ is a vertical vector field and $[x, Y]^{\perp} = 0$. We calculate each term with the help of (B.11) and Theorem 2.24:

$$(\nabla_{[x,Y]^\top} Z, W) = (O_Z([x, Y]^\top), W) = -(O_Z W, T_x Y) + (O_Z W, (\nabla_Y x)^\top),$$

$$(\nabla_x \nabla_Y Z, W) = (\nabla_x(\tilde{\nabla}_Y Z), W) - (O_Y Z, T_x W),$$

$$(\nabla_Y \nabla_x Z, W) = (\nabla_Y(O_Z x), W) - (O_Y W, T_x Z)$$

$$= ((\nabla_Y O)_Z x + O_{\tilde{\nabla}_Y Z} x, W) - (O_Z W, (\nabla_Y x)^\top) - (O_Y W, T_x Z). \quad (B.13)$$

$((\nabla_Y O)_Z x, W) = -((\nabla_Y O)_Z W, x)$, which follows from (2.11) and (2.12), we obtain

$$(\nabla_x \tilde{\nabla}_Y Z - O_{\tilde{\nabla}_Y Z} x, W) = (R(Z, W)Y, x)$$

$$-((\nabla_Y O)_Z W, x) + (O_Y Z, T_x W) - (O_Y W, T_x Z) - (O_Z W, T_x Y). \quad (B.14)$$

By (2.34b) the right hand side of (B.14) is equal to

$$((\nabla_W O)_Z Y - (\nabla_Z O)_W Y - (\nabla_Y O)_Z W, x) +$$

$$(O_Y Z, T_x W) - (O_Y W, T_x Z) - (O_W Z, T_x Y).$$

In view of (2.35) and the above, the right hand side of (B.14) is equal to

$$((\nabla_W h_2)(Z, Y) - (\nabla_Y h_2)(Z, W) - (\nabla_Z h_2)(W, Y), x)$$

$$+(O_Y Z, T_x W) - (O_Y W, T_x Z) - (O_W Z, T_x Y).$$

The field $\tilde{\nabla}_Z Y - \tilde{\nabla}_Y Z = [Y, Z]^\perp$ is basic, so in view of (B.11), the right hand side of (B.14) is symmetric in Y and Z, i.e.,

$$(O_Y Z, T_x W) - (O_Y W, T_x Z) - (O_W Z, T_x Y)$$

$$= (O_Z Y, T_x W) - (O_Z W, T_x Y) - (O_W Y, T_x Z).$$

From this and the identity

$$2(h_2(Y, W), T_x Z) + 2(h_2(Z, W), T_x Y)$$

$$= [(O_Y W, T_x Z) + (O_W Z, T_x Y)] + [(O_Z W, T_x Y) + (O_W Y, T_x Z)]$$

it follows that

$$(O_Y Z, T_x W) - (O_Y W, T_x Z) - (O_W Z, T_x Y)$$

$$= (h_2(Y, Z), T_x W) - (h_2(Y, W), T_x Z) - (h_2(Z, W), T_x Y). \qquad \square$$

From Lemma B.12 follows a local test of totally geodesic submersions in terms of the second fundamental form h_2 and the tensor T on M.

B.13 Corollary. *The condition $h^L = 0$ is equivalent to the equation*

$$((\nabla_Z h_2)(Y,Y) - 2(\nabla_Y h_2)(Z,Y), x) + (h_2(Y,Y), T_x Z) - 2(h_2(Z,Y), T_x Y) = 0. \qquad (B.15)$$

For a totally geodesic $\{L\}$, equation (B.15) is true, for example, when the second fundamental form of the distribution $TL^\perp$ is parallel

$$(\nabla_W h_2)(Y, Z) = 0. \qquad (B.16)$$

Submanifolds with parallel second fundamental form (called *symmetric submanifolds*) are classified in some cases (see the survey [Lum 2]). Corollary B.13 allows us to construct examples of totally geodesic submersions.

B.14 Example. Let $B \times_f P$ be a twisted product (see Section 2.3.2). It is not difficult to see that (B.16) is true for π_2, when the submanifolds $\{\pi_1^{-1}(b)\}$ have parallel second fundamental form (i.e., their mean curvature vector H is parallel). The last condition is equivalent to the identity $f(b,p) = \lambda(b)\mu(p)$. In this case π_2 is a totally geodesic submersion, which for $\lambda(b) \neq const$ is not a Riemannian submersion. In particular, for warped product with nonconstant function f the projection π_2 corresponds to a totally geodesic submersion.

From Lemma B.12 follows a local test for submersions with codimension ≤ 1.

B.15 Corollary. *A submersion π has codimension ≤ 1 in the case*

$$(\nabla_{x_2} F(x_1, Y, W)) F(x_1, Z, W) = (\nabla_{x_2} F(x_1, Z, W)) F(x_1, Y, W), \qquad (B.17)$$

where

$F(x, Y, W) =$

$((\nabla_W h_2)(Y,Y) - 2(\nabla_Y h_2)(Y,W), x) + (h_2(Y,Y), T_x W) - 2(h_2(Y,W), T_x Y)$.

Proof. In view of

$$2h^L(\bar Y, \bar Z) = h^L(\bar Y + \bar Z, \bar Y + \bar Z) - h^L(\bar Y, \bar Y) - h^L(\bar Z, \bar Z),$$

it suffices to analyze the problem when the vectors $\{h^L(\bar Y, \bar Y)\}$, $(\bar Y \in T_b \bar M)$, form a 1-dimensional subspace. Since the vectors $h^L(\bar Y, \bar Y)$ and $h^L(\bar Z, \bar Z)$ are parallel, there exists a function λ, which is constant on L_b such that the field $\tilde\nabla_Y Y - \lambda(\tilde\nabla_Z Z)$ is basic. For this field it suffices to apply (B.12) and to write the condition $\nabla_{x_2}\lambda = 0$.

To deduce a local test of curvature invariant submersions, we need the following lemma.

B.16 Lemma. *For a submersion (and also for a foliation) we have*

$$-(\nabla_x(\tilde{R}(Y,Z)W)-O_{\tilde{R}(Y,Z)W}x,U) = ((\nabla_Y\nabla_U h_2)(Z,W)-(\nabla_Y\nabla_Z h_2)(W,U)$$
$$-(\nabla_Y\nabla_W h_2)(Z,U)-(\nabla_Z\nabla_U h_2)(Y,W)+(\nabla_Z\nabla_Y h_2)(W,U)+(\nabla_Z\nabla_W h_2)(Y,U),x)$$
$$+((\nabla_Y h_2)(Z,W)-(\nabla_Z h_2)(Y,W),T_xU)-((\nabla_Y h_2)(Z,U)-(\nabla_Z h_2)(Y,U),T_xW)$$
$$-(\nabla_Y h_2)(W,U),T_xZ)+(\nabla_Z h_2)(W,U),T_xY)-(h_2(Z,U),(\nabla_Y T)_xW)$$
$$+(h_2(Z,W),\nabla_Y T)_xU)-(h_2(W,U),(\nabla_Y T)_xZ-(\nabla_Z T)_xY)$$
$$-(h_2(Y,W),(\nabla_Z T)_xU)+(h_2(Y,U),(\nabla_Z T)_xW). \tag{B.18}$$

Proof. Let us write the formula for the horizontal curvature tensor:

$$\tilde{R}(Y,Z)W = \tilde{\nabla}_{[Y,Z]^\perp}W - \tilde{\nabla}_Y\tilde{\nabla}_Z W + \tilde{\nabla}_Z\tilde{\nabla}_Y W.$$

We differentiate each term with respect to x at the point $p \in M$ and consider a scalar product by a vector $U \in T_pL^\perp$:

$$(\nabla_x\tilde{\nabla}_Y\tilde{\nabla}_Z W,U) = (\nabla_x\nabla_Y(\tilde{\nabla}_Z W) - \nabla_x(O_Y(\tilde{\nabla}_Z W)),U)$$
$$= (-R(x,Y)(\tilde{\nabla}_Z W) + \nabla_Y\nabla_x(\tilde{\nabla}_Z W) + \nabla_{[x,Y]^\top}(\tilde{\nabla}_Z W) - T(O_Y(\tilde{\nabla}_Z W)),U).$$

Transposing with the help of (B.12) we get:

$$(\nabla_x(\tilde{\nabla}_Z W),U) = (O_{\tilde{\nabla}_Z W}x,U) + \delta_1.$$

Finally we have

$$(\nabla_x\tilde{\nabla}_Y\tilde{\nabla}_Z W,U)$$
$$= (-R(x,Y)(\tilde{\nabla}_Z W) + \nabla_Y(O_{\tilde{\nabla}_Z W}x)\nabla_{[x,Y]^\top}(\tilde{\nabla}_Z W) - T(O_Y(\tilde{\nabla}_Z W)),U)$$
$$+ \nabla_Y\delta_1.$$

Analogously we have

$$(\nabla_x\tilde{\nabla}_Z\tilde{\nabla}_Y W,U)$$
$$= (-R(x,Z)(\tilde{\nabla}_Y W) + \nabla_Z(O_{\tilde{\nabla}_Y W}x) + \nabla_{[x,Z]^\top}(\tilde{\nabla}_Y W) - T(O_Z\tilde{(}\nabla_Y W)),U)$$
$$+ \nabla_Z\delta_2.$$

The following equality is also true:

$$(\nabla_x \tilde{\nabla}_{[Y,Z]^\perp} W, U) = (O_{\tilde{\nabla}_{[Y,Z]^\perp} W} x), U) + \delta_3.$$

Note that

$$(\nabla_Y (O_{\tilde{\nabla}_Z W} x), U) = ((\nabla_Y O)_{\tilde{\nabla}_Z W} x + O_{\tilde{\nabla}_Y \tilde{\nabla}_Z W} x + O_{\tilde{\nabla}_Z W} (\nabla_Y x)^\top, U),$$

$$(\nabla_Z (O_{\tilde{\nabla}_Y W} x), U) = ((\nabla_Z O_{\tilde{\nabla}_Y W} x + O_{\tilde{\nabla}_Z \tilde{\nabla}_Y W} x + O_{\tilde{\nabla}_Y W} (\nabla_Z x)^\top, U).$$

Suppose that we have the following equalities at the point p:

$$\begin{aligned} &\nabla_Y U^\perp = \nabla_Z U^\perp = \nabla_Z x^\top = \nabla_Y x^\top = \\ &\nabla_Y Z^\perp = \nabla_Y W^\perp = \nabla_Z Y^\perp = \nabla_Z W^\perp = 0. \end{aligned}$$

Then we obtain $\delta_3 = 0$ and

$$(\nabla_x (\tilde{R}(Y,Z)W) - O_{\tilde{R}(Y,Z)W} x, U) = \nabla_Z \delta_2 - \nabla_Y \delta_1,$$

i.e., equality (B.18) is true.

B.17 Corollary. *A submersion π is curvature invariant iff*

$$\begin{aligned} &((\nabla_Y \nabla_U h_2)(Z,W) - (\nabla_Y \nabla_Z h_2)(W,U) - (\nabla_Y \nabla_W h_2)(Z,U) - (\nabla_Z \nabla_U h_2)(Y,W) \\ &\quad + (\nabla_Z \nabla_Y h_2)(W,U) + (\nabla_Z \nabla_W h_2)(Y,U), x) \\ &\quad + ((\nabla_Y h_2)(Z,W) - (\nabla_Z h_2)(Y,W), T_x U) \\ &\quad - ((\nabla_Y h_2)(Z,U) - (\nabla_Z h_2)(Y,U), T_x W) - ((\nabla_Y h_2)(W,U), T_x Z) \\ &\quad + ((\nabla_Z h_2)(W,U), T_x Y) - (h_2(Z,U), (\nabla_Y T)_x W) + (h_2(Z,W), (\nabla_Y T)_x U) \\ &\quad - (h_2(W,U), (\nabla_Y T)_x Z - (\nabla_Z T)_x Y) - (h_2(Y,W), (\nabla_Z T)_x U) \\ &\quad + (h_2(Y,U), (\nabla_Z T)_x W) = 0. \qquad (B.19) \end{aligned}$$

For totally geodesic fibers $\{L\}$, i.e., $T = 0$, the equality (B.19) is true, for example, when

$$(\nabla_Y \nabla_Z h_2)(V, W) = 0. \qquad (B.19')$$

Hypersurfaces in $\mathbb{R}^n$ which satisfy the condition $\nabla\nabla h = 0$ are completely classified: they include symmetric submanifolds and also cylinders over a plane curve – *clothoid* (*Cornu spiral*) $k(s) = as + b$, see [Lum 2].

B.18 Example. We consider again the twisted product $B \times_f P$ and let $\{e_i\}$, $(1 \leq i \leq \nu = \dim B)$ be a local basis of the vertical distribution. Then the mean curvature vector H of the leaves $\pi_1^{-1}(b)$ is calculated by the formula $H = \sum_i (e_i \log f) e_i$. For a horizontal vector Y we obtain $(\nabla_Y H)^\top = -\sum Y(e_i \log f) e_i$. Therefore, for any horizontal vector Z we have

$$(\nabla_Z((\nabla_Y H)^\top))^\top = -\sum_i Z(Y(e_i \log f)) e_i. \tag{B.20}$$

So $(\nabla_Z((\nabla_Y H)^\top))^\top = 0$ (see (B.19)) is equivalent to the equation $Z(Y(e_i \log f)) = 0$, which the following functions satisfy:

$$f(b, p) = \mu_0(p)\lambda(b)\exp(b_1\mu_1(p) + \cdots + b_\nu\mu_\nu(p)), \tag{B.21}$$

where b_i are local coordinates on B. The functions $\mu_i(p)$, for example, have support in the chart. From (B.21) follows condition (B.19'). If at least one function $\mu_i(p)$, $(1 \leq i \leq \nu)$ is nonconstant, then for $n > 1$ the submersion π_2 is curvature invariant.

The notion of the relative index of nullity is used in the theorems about totally geodesic submersions below.

B.19 Theorem. *Let $\pi : M \to \bar{M}$ be a curvature invariant submersion of a compact Riemannian manifold M with compact connected leaves $\{L\}$ which satisfies the following conditions: $\mu(h^L) > \frac{1}{2}\dim \bar{M}$ and*

$$\tilde{K}(Y, Z) > 0, \ (Y, Z \in TL^\perp). \tag{B.22}$$

Then $h^L = 0$ and π is a totally geodesic submersion.

Proof. The sub-bundle $T\bar{M} \subset \tilde{M}$ is curvature invariant. Hence a distribution N on the domain $\bar{G}(h^L) \subset \bar{M}$ is tangent to a totally geodesic foliation with complete leaves in a metric $<\ ,\ >$, see Section 2.3.1. In view of (B.10) and (B.22) the mixed sectional curvature is positive. The leaves are compact, because their sectional curvature is positive (see (B.10) and (B.22)) and $\bar{M}$ is compact. From Corollary 3.31 with $q = 1$ (or [Tan 2] and [Fr 1]) we have either $\mu(h^L) < \frac{1}{2}\dim \bar{M}$ or $\mu(h^L) = \dim \bar{M}$. $\square$

B.20 Definition. A vector $\bar{Y} \in T\bar{M}$ is called *h^L-asymptotical*, if $h^L(\bar{Y}, \bar{Y}) = 0$, or equivalently if for all $Z \in TL^\perp$

$$((\nabla_Z h_2)(Y, Y) - 2(\nabla_Y h_2)(Z, Y), x) + (h_2(Y, Y), T_x Z) - 2(h_2(Y, Z), T_x Y) = 0. \tag{B.23}$$

The condition $\mu(h^L) > \frac{1}{2} \dim \bar{M}$ holds, for example, if π is a submersion of codimension ≤ 1 and at each point $b \in \bar{M}$ there exists a subspace of h^L–asymptotical vectors whose dimension is at least $\frac{3}{4} \dim \bar{M}$.

B.21 Definition. Vectors $\bar{Y} \in T_b\bar{M}$, satisfying the condition $h_2(Y, Z) = 0$ for all Z, form an h_2-null subspace $\bar{N}_2(b)$; $\bar{G}_2 \subset \bar{M}$ denotes the regularity domain, where such subspaces have minimal dimension $\mu(h_2)$, called the *h_2-null-index* of the submersion π.

If the inequality $0 < \mu(h_2) < \dim \bar{M}$ holds, then the regular h_2-null distribution $\bar{N}_2$ is given on $\bar{G}_2$; its horizontal lift into M is denoted by N_2. The equality $\mu(h_2) = \dim \bar{M}$ means that π is a Riemannian submersion.

B.22 Lemma. *If $\bar{Y} \in \bar{N}_2(b)$ and $\bar{Z} \in T_b\bar{M}$, then $(Y, Z) = const$ along L_b.*

Proof. $\nabla_x(Y, Z) = (O_Y x, Z) + (O_Z x, Y) = -2(h_2(Z, Y), x) = 0$.

B.23 Theorem. *Let $\pi : M \to \bar{M}$ be a totally geodesic submersion with compact connected leaves $\{L\}$ and $\mu(h_2) \in (0, \dim \bar{M})$. Then the regularity domain $\bar{G}_2 \subset \bar{M}$ is foliated by $\mu(h_2)$-dimensional totally geodesic submanifolds in a metric $< , >$, and the distribution N_2 is geodesic in M. Moreover, if M is complete, then the leaves in $\bar{G}_2$ are also complete in the metric $<, >$.*

Proof. Consider the vector fields $\bar{Y}, \bar{W} \subset \bar{N}_2$ and $\bar{Z} \perp \bar{N}_2$. From (B.12) it follows that $(h_2(Z[Y, W]^\perp, x) = 0$. Thus $[Y, W]^\perp \subset \bar{N}_2$, i.e., the distribution $\bar{N}_2$ is integrable. Next consider the vector fields $\bar{Y},\ \bar{Z} \subset \bar{N}_2$ and $\bar{W} \perp \bar{N}_2$. By (B.12) we have $(h_2(\nabla_Y Z + \nabla_Z Y), x) = 0$, i.e., $\tilde{\nabla}_Y Z + \tilde{\nabla}_Z Y \subset \bar{N}_2$ (the vector fields $\tilde{\nabla}_Y Z$ and $\tilde{\nabla}_Z Y$ are basic because $h^L = 0$). Thus integral submanifolds of the distribution $\bar{N}_2$ are totally geodesic.

Let $\bar{c}_t \subset \bar{M}$ be a geodesic tangent to $\bar{N}_2(b)$ and $c_t \subset M$ its horizontal lift at some point. Then c_t is a geodesic in M. Actually, for any $\bar{Y}_t = \bar{c}_t'$ the vector field Y_t has unit length by Lemma B.22. In this case we have $\bar{\nabla}_{\bar{Y}_t} \bar{Y}_t = 0$, $h^L(\bar{Y}_t, \bar{Y}_t) = 0$ and $h_2(Y_t Y_t) = 0$, i.e., $\nabla_{Y_t} Y_t = 0$. Consequently, an integral curve c_t of a vector field Y_t is a geodesic in M and so N_2 is a geodesic distribution on M, possibly integrable.

We will prove that the leaves of the distribution $\bar{N}_2$ are complete when M is complete. Let $\bar{W} \subset \bar{N}_2$ and $\bar{Y} = \bar{Z} \perp \bar{N}_2$ be vector fields along $\bar{c}_t$, and let W, Y and Z be their horizontal lifts along c_t. Then from (B.12), in view of $\bar{\nabla}_Y W = 0$, it follows that

$$(\nabla_W(h_2(Y, Y)) - h_2(Z, (\tilde{\nabla}_W Y))$$

$$-h_2(Y, (\tilde{\nabla}_W Z)) + 2h_2(Y, (\tilde{\nabla}_Y W)), x) + (h_2(Y, Y), T_x W) = 0.$$

With the help of $\tilde{\nabla}_W Y = \tilde{\nabla}_Y W$, we then deduce the equation

$$(\nabla_W(h_2(Y, Y)), x) + (h_2(Y, Y), T_x W) = 0. \tag{B.24}$$

Since (B.24) is a linear differential equation along the curve c_t that is homogeneous for $h_2(Y, Y)$ and has nonzero initial value, then we deduce that $h_2(Y, Y) \neq 0$ along a curve c_t. Thus, the dimension of the subspaces of $\bar{N}_2$ is a nonincreasing function along $\bar{c}_t$. By the properties of continuity and the completeness of $\bar{G}_2$, the dimension of $\bar{N}_2$ is also a nondecreasing function. Thus we may extend $\bar{c}_t$ up to infinity into $\bar{G}_2$; hence integral submanifolds for $\bar{N}_2$ are complete. □

B.24 Corollary. *Let $\pi : M \to \bar{M}$ be a totally geodesic submersion of a compact Riemannian manifold M with compact connected leaves $\{L\}$ and suppose that $(B.22)$ holds. Then from $\mu(h_2) \geq \frac{1}{2} \dim \bar{M}$ it follows that $\mu(h_2) = \dim \bar{M}$ and π is a Riemannian submersion.*

Proof. The proof of this corollary is analogous to the proof of Theorem B.19. The regularity domain $\bar{G}_2 \subset \bar{M}$ (in view of Theorem B.19) is foliated by totally geodesic leaves, which are complete in the metric $<,>$. In view of (B.10') the mixed sectional curvature is positive. The leaves are compact because their sectional curvature is positive and M is compact. Finally, from Corollary 3.31 with $q = 1$ (or [Tan 2] and [Fr 1]) it follows that either $\mu(h_2) < \frac{1}{2} \dim \bar{M}$ or $\mu(h_2) = \dim \bar{M}$. □

B.25 Remark. In view of Lemma B.22 and Corollary 3.31, the Corollary B.24 holds with (B.22) replaced by positiveness of transversal $q-$Ricci curvature and with the inequality $\mu(h_2) \geq \frac{1}{2}(\dim \bar{M} + q - 1)$.

Appendix C

FOLIATIONS BY CLOSED GEODESICS WITH POSITIVE MIXED SECTIONAL CURVATURE

Hopf fibrations provide the example of the foliations of *odd-dimensional* Riemannian manifolds on closed geodesics $\{L = S^1\}$ with constant positive mixed sectional curvature. The problem of the existence of *even-dimensional* Riemannian manifold foliated on closed geodesics with $(\delta \approx 1)-$pinched positive mixed sectional curvature was formulated in Section 1.4. This problem is important in view of result by Ferus (see Section 1.4), which, in particular, means the nonexistence of such a foliation for $\delta = 1$, and in relations with the results in Chapters 4,6, and 7, where the foliations with positive mixed curvature are investigated under some additional assumptions.

Theorem C1 below follows from Lemmas C3 and C4 and gives the positive answer to the above problem for any $\delta \in (0, 1)$.

Theorem C1. *For any* $\delta \in (0, 1)$ *there exists Riemannian manifold* M^{2n+2}, *where* $n \geq \frac{\sqrt{\delta}}{1-\sqrt{\delta}}$, *with the fibration on closed geodesics* $\{L = S^1\}$ *and with positive* $\delta-$*pinched mixed sectional curvature.*

Remark C2. Our hypothesis is that Theorem C1 holds when $\delta \to 1$ and the dimension n is fixed. In the proof of Theorem C1, we construct on M^{2n+2} the metric with mixed curvature $(\frac{n}{n+1})^2 \leq K_{mix} \leq 1$, and with the length of geodesics $l(L) = 2\pi(n + 1)$, that runs to infinity when $\delta \to 1$. Can anyone obtain for any $\delta \in (0, 1)$ the example with the length $l(L) \leq c$ (for instance, with $c = 2\pi$) under standard inequalities $\delta \leq K_{mix} \leq 1$? Theorem C1 is *local,* meaning directions transversal to some leaf $L = S^1$, and the problem of the existence of analogue foliation on *compact* Riemannian manifold M^{2n+2} is also open.

Lemma C3. *Let the symmetric matrix $R(t)$ and the nondegenerate matrix $Y(t)$ have the order $n \times n$ and are T-periodic, and the Jacobi equation is satisfied*

$$\ddot{Y}(t) + R(t) \cdot Y(t) = 0, \qquad (0 \leq t \leq T).$$

Then there exists the Riemannian metric on the product $M^{n+1} = S^1 \times B^n(r)$ of the circle S^1 and n-dimensional ball $B^n(r)$ of radius r with the following properties:
a) the closed curves $\{\gamma_z(t) = (t, Y(t)z)\}_{z \in B^n(r)}$ are the geodesics,
b) the components of the mixed sectional curvature $R(*, \dot{\gamma}_o)\dot{\gamma}_o$ along $\gamma_o(t)$ is expressed by the formula $R(z, \dot{\gamma}_o)\dot{\gamma}_o = R(t)z, \quad (z \perp \dot{\gamma}_o)$,
c) the Jacobi tensor of the foliation $\{\gamma_z\}$ has the form $Y(t)$ for some parallel orthonormal base along γ_o.

Proof. Together with the coordinate system $t = z_0, z = (z_1, \ldots, z_n)$ of the *direct product*, we consider on M^{n+1} the *adapted coordinates* $t = y_0, y = (y_1, \ldots, y_n)$, with the following relation $z = Y(t)\, y$. For the adapted coordinates, the curves $\{\gamma_z\}$ are the coordinate lines $\{y = const\}$. The property a) means that the following Christoffel symbols vanish

$$\Gamma^i_{oo}(t, y) \equiv 0, \qquad (0 \leq i \leq n). \tag{C.1}$$

In view of formulas which express the Christoffel symbols through the coefficients of the metric $\{g_{ij}(t, y)\}$ and their derivatives [KN]

$$2 \sum_s g_{sk} \Gamma^s_{ij} = g_{jk,i} + g_{ki,j} - g_{ij,k}\,, \tag{C.2}$$

we obtain from (C.1) the equalities

$$g_{oo,i}(t, y) \equiv 2 g_{oi,o}(t, y) \qquad (i \geq 0). \tag{C.3}$$

In view of (C.3) for $i = 0$, the coefficient $g_{oo}(t, y)$ does not depend on t. Also we need the following "initial" conditions along γ_o:

$$g_{oi}(t, 0) \equiv \delta_{oi}, \qquad (i \geq 0), \tag{C.4}$$

$$g_{ij}(t, 0) \equiv \left(Y(t)^{\top} \cdot Y(t)\right)_{ij} \qquad (i, j \geq 1). \tag{C.5}$$

The conditions (C.4)–(C.5) mean that the coordinate vector fields along γ_o in the product coordinates (t, z) are orthonormal. Since the matrix $B(t) := \dot{Y}(t) \cdot Y^{-1}(t)$ must correspond to the structural tensor of foliation $\{\gamma_z\}$ along γ_o

in the product coordinate system (t, z), we need the following properties for the matrix $\Gamma_o(t, 0) := \{\Gamma^k_{oi}(t, 0)\}$ in the adapted coordinates (t, y):

$$\Gamma_o(t, 0) = Y^{-1}(t) \cdot B(t) \cdot Y(t) = Y^{-1}(t) \cdot \dot{Y}(t). \qquad (C.6)$$

We write the formulas (C.2), which relate the metric with the above Christoffel symbols, in the appropriate form for our purposes

$$2\sum_s g_{sj}(t, 0) \cdot \Gamma^s_{oi}(t, 0) - g_{ij,o}(t, 0) = g_{jo,i}(t, 0) - g_{io,j}(t, 0). \qquad (C.7)$$

We will show that the RHS of (C.7) does not depend on t.
To prove this, we first show that the matrix $C := \dot{Y}^\top(t) \cdot Y(t) - Y^\top(t) \cdot \dot{Y}(t)$ is constant, since its derivative with respect to t vanishes:

$$\begin{aligned} \dot{C} = & \ddot{Y}^\top(t) \cdot Y(t) + \dot{Y}^\top(t) \cdot \dot{Y}(t) - \dot{Y}^\top(t) \cdot \dot{Y}(t) - Y^\top(t) \cdot \ddot{Y}(t) = \\ & (-R(t) \cdot Y(t))^\top \cdot Y(t) - Y(t)^\top \cdot (-R(t) \cdot Y(t)) = \\ & - Y(t))^\top \cdot R(t) \cdot Y(t) + Y(t)^\top \cdot R(t) \cdot Y(t) = 0. \end{aligned}$$

Hence we put $C := \dot{Y}^\top(0) \cdot Y(0) - Y^\top(0) \cdot \dot{Y}(0)$. In view of (C.5) the derivatives of the coefficients of the metric $g_{ij,o}(t, 0)$ form the following matrix

$$(Y^\top(t) \cdot Y(t))' = \dot{Y}^\top(t) \cdot Y(t) + Y^\top(t) \cdot \dot{Y}(t) = 2\dot{Y}^\top(t) \cdot \dot{Y}(t) + C,$$

and, in view of (C.5) and (C.6), the functions $2\sum_s g_{sj}(t, 0) \cdot \Gamma^s_{oi}(t, 0)$ are the elements of the matrix

$$2(Y^\top(t) \cdot Y(t)) \cdot (Y^{-1}(t) \cdot \dot{Y}(t)) = 2Y^\top(t) \cdot \dot{Y}(t).$$

Hence the RHS of (C.7) represents the coefficients of the constant matrix $-C$, and does not depend on t. Moreover, the matrix C is skew-symmetric. Since (C.7) was reduced to the equalities

$$g_{jo,i}(t, 0) - g_{io,j}(t, 0) = -C_{ij}, \qquad (C.8)$$

we can assume $g_{io,j}(t, y) := \begin{cases} 0, & j \le i \\ C_{ij}, & j > i \end{cases}$. We integrate these expressions and, in view of (C.4), have

$$g_{io}(t, y) = \sum_{j>i} C_{ij} y_j, \qquad (i \ge 1). \qquad (C.9)$$

Let us substitute (C.9) into (C.3) so that in view of (C.4) we obtain that the functions $g_{oo}(t, y)$ are constant and equal to 1. The condition (C.4) obviously holds. Finally, in view of (C.5), we can define other coefficients by the formula

$$g_{ij}(t, y) := \left(Y(t)^{\top} \cdot Y(t)\right)_{ij}, \qquad (i, j \geq 1). \tag{C.10}$$

For sufficiently small $r > 0$ the metric $\{g_{ij}(t, y)\}$ on M^{n+1} is positive. Note that the property b) follows from the formulas (C.6) and the property c) follows from the formulas (C.5). □

Lemma C4. *For any $n \in \mathbb{N}$ there exists the matrix $Z_{n,n+1}(t, s)$ of the order $(2n+1) \times (2n+1)$ with the elements*

$$z_{jk} = \begin{cases} a_{jk}\cos(\frac{t}{n}) + b_{jk}\sin(\frac{t}{n}), & j \leq n, \\ a_{jk}\cos(\frac{s}{n+1}) + b_{jk}\sin(\frac{s}{n+1}), & n+1 \leq j \leq 2n+1, \end{cases}$$

with the property $\det Z_{n,n+1}(t, s) = \cos(t - s)$. *In particular, the determinant of the matrix $Y_{2n+1}(t) := Z_{n,n+1}(t, t)$ is identically* 1.

Proof. 1. The structure of the matrix $Z_{n,n+1}(t, s)$ is given for n odd and n even on Table 2. Consider the case of odd n. Note that $z_{1,1} = \cos(\frac{t}{n})$, $z_{1,n+1} = -\sin(\frac{t}{n})$ and in both cases of n the quadratic matrices A_1 and A_2 have even order. Let the matrix $\tilde{A}_1$ is obtained from A_1 (with the size $n-1$) by deleting the last column and completing with the column $\{z_{j,1}\}_{2 \leq j \leq n}$ inside of its first column. Analogously, let the matrix $\tilde{A}_2$ be obtained from A_2 (with the size $n+1$) by replacing of its first column by the column $\{z_{j,n}\}_{j>n}$. For short notations let

$$c_n := \cos\left(\tfrac{t}{n}\right), \quad s_n := \sin\left(\tfrac{t}{n}\right), \qquad c_{n+1} := \cos\left(\tfrac{s}{n+1}\right), \quad s_{n+1} := \sin\left(\tfrac{s}{n+1}\right).$$

In this way the elements of initial n rows of the matrix $Z_{n,n+1}(t, s)$ are linear combinations of c_n, s_n, and the elements in the other (lower) $n+1$ rows are linear combinations of c_{n+1}, s_{n+1}. Obviously, we have, see Table 2,

$$\det Z_{n,n+1}(t, s) = c_n \det A_1 \cdot \det A_2 + s_n \det \tilde{A}_1 \cdot \det \tilde{A}_2.$$

To proof the equality $\det Z_{n,n+1}(t, s) = \cos(t) \cdot \cos(s) + \sin(t) \cdot \sin(s)$ we will provide (in steps 2–4) the following equalities

$$\det A_1 = \cos(t)/c_n, \qquad \det A_2 = \cos(s),$$

$$\det \tilde{A}_1 = \sin(t)/s_n, \qquad \det \tilde{A}_2 = \sin(s).$$

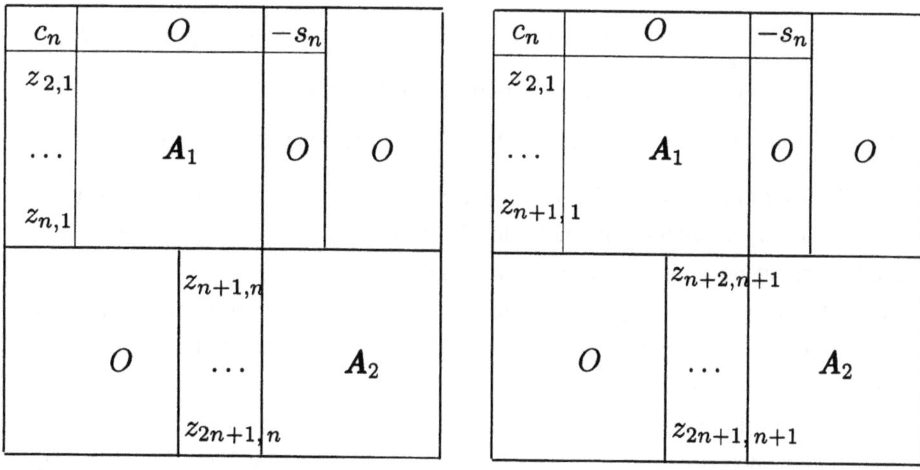

for n odd   for n even

Table 2.

2. We separate the real and imaginary parts in Moivre's formula

$$\cos(mx) + \boldsymbol{i}\sin(mx) = (\cos x + \boldsymbol{i}\sin x)^m = \sum_{k=0}^{m} \boldsymbol{i}^k C_m^k \cos^{m-k}(x) \cdot \sin^k(x)$$

and find the polynomials $p_m(a, b)$, $q_m(a, b)$ of degree m

$$p_m(a, b) = a^m - C_m^2 a^{m-2} b^2 + C_m^4 a^{m-4} b^4 - C_m^6 a^{m-6} b^6 + \dots,$$
$$q_m(a, b) = C_m^1 a^{m-1} b - C_m^3 a^{m-3} b^3 + C_m^5 a^{m-5} b^5 + \dots,$$

with the following properties

$$\cos(t) = p_n(c_n, s_n), \qquad \sin(t) = q_n(c_n, s_n),$$
$$\cos(s) = p_{n+1}(c_{n+1}, s_{n+1}), \quad \sin(s) = q_{n+1}(c_{n+1}, s_{n+1}).$$

It is easy to check, that the roots of the polynomial $p_n(\lambda, 1)$ are the following $\lambda_i = \cot(\frac{\pi(2i+1)}{2n})$, where $i = 1..n$, and the roots of the polynomial $q_n(\lambda, 1)$ are $\lambda_i = \cot(\frac{\pi i}{n})$, where $i = 1..n$.

Proposition C5. *For any polynomial $P_m(\lambda)$ of degree m there exists the constant matrix D_m of order $m \times m$ whose characteristic polynomial* $\det(D_m - \lambda \cdot I_m)$ *is $P_m(\lambda)$.*

Proof of Proposition. Consider the factoring of $P_m(\lambda)$ over the field $\mathbb{R}$ onto (linear and quadratic) multipliers, assuming for simplicity the absence of multiple roots, $P_m(\lambda) = \prod_\alpha(\mu_\alpha - \lambda) \cdot \prod_\beta((\lambda_\beta - \lambda)\cdot(\bar{\lambda}_\beta - t))$, where $\lambda_\beta = u_\beta + i v_\beta$, and let $D_\beta = \begin{pmatrix} u_\beta & -v_\beta \\ v_\beta & u_\beta \end{pmatrix}$ be the matrices of the order 2×2. Then the block-diagonal matrix $D_n = [\dots \mu_\alpha, \dots D_\beta, \dots]$ has the desired property. □

In view of Proposition C5, we choose matrices $A_1,\ A_2$ satisfying $c_n \det A_1 = \cos(t),\ \det A_2 = \cos(s)$. Namely, let $A_1 = s_n \cdot D_{n-1} - c_n \cdot I_{n-1}$, where the characteristic polynomial of the matrix D_{n-1} is $p_n(\lambda, 1)/\lambda$, and let $A_2 = s_{n+1} \cdot D_{n+1} - c_{n+1} \cdot I_{n+1}$, where the characteristic polynomial of the matrix D_{n+1} is $p_{n+1}(\lambda, 1)$. Note that $A_1,\ A_2$ are the block-diagonal matrices, all their blocks have the size 2×2.

3. We will obtain the equalities $s_n \det \tilde{A}_1 = \sin(t),\ \det \tilde{A}_2 = \sin(s)$. Note that the determinants of the matrices $A_1,\ A_2$ keep their values, if we replace some zero elements in the right side of blocks with arbitrary real numbers.

Obviously, the determinant of the matrix $\tilde{A}_2$ is a linear function of variables $\{a_{j,n},\ b_{j,n}\}_{j=n+1\dots 2n+1}$. We collect the coefficients of the monomials $(c_{n+1})^k \cdot (s_{n+1})^{n+1-k}$, where $k = 0 \dots n+1$, in the equation $\det \tilde{A}_2 = q_{n+1}(c_{n+1}, s_{n+1})$ and obtain the system of $n+1$ equations with $2n-2$ variables $\{a_{j,n},\ b_{j,n}\}$. Although we will not prove it here, it is true that by replacing some of the elements of A_2 in a way that preserves its block triangle form, this linear system would be compatible. Its arbitrary solution $\{a_{j,n},\ b_{j,n}\}$ (which was calculated using the package $MAPLE$ in the examples with $n = 3$, 4 and 5 below) defines the desired matrix $\tilde{A}_2$.

Analogously, in the case $\tilde{A}_1$ we obtain the compatible system of $n+2$ equations with $2n+2$ variables $\{a_{j,1},\ b_{j,1}\}_{j=2\dots n}$. Its arbitrary solution (which was calculated using the package $MAPLE$ in the examples with $n = 3$, 4 and 5 below) defines the desired matrix $\tilde{A}_1$.

So, for odd n the matrix $Z_{n,n+1}(t,s)$ with the determinant $\cos(t-s)$ is constructed. The case of even n is analogous. Note that the matrix $Y_{2n+1}(t)$ satisfies the Jacobi differential equation

$$\ddot{Y}_{2n+1}(t) + R_{2n+1} \cdot Y_{2n+1}(t) = 0,$$

where R_{2n+1} is diagonal matrix with elements $\frac{1}{n^2}$ and $\frac{1}{(n+1)^2}$ on its diagonal. Hence $\delta = (\frac{n}{n+1})^2 \to 1$ when $n \to \infty$. □

For $n = 1$ the matrix $Z_{1,2}(t,s) = \begin{pmatrix} \cos(t) & 2\sin(t) & 0 \\ 0 & \sin(\frac{s}{2}) & \cos(\frac{s}{2}) \\ \sin(\frac{s}{2}) & -\cos(\frac{s}{2}) & -\sin(\frac{s}{2}) \end{pmatrix}$, and the matrices $Y_3(t),\ B_3(t) = Y_3(t)' \cdot (Y_3(t))^{-1}$, given with some modification in Section 3.1, satisfy the example with $\delta = 0.25$.

Consider some more examples, where for simplicity we denote by $C_n := \cos(\frac{t}{n})$, $S_n := \sin(\frac{t}{n})$. The following matrix $Y_5(t) := Z_{2,3}(t,t)$ was constructed using the package $DERIVE$ on the different way, than in the proof of Lemma C4, it satisfies the case $\delta = 4/9 \approx 0.(44)$

$$Y_5(t) = \begin{pmatrix} 0 & S_3 & -C_3 & 0 & 0 \\ S_3 & C_3 & 0 & 3C_3 & 0 \\ C_3 & 3S_3 & 0 & S_3 & 0 \\ 0 & 0 & -2S_2 & C_2 & S_2 \\ 0 & 0 & 0 & S_2 & C_2 \end{pmatrix}.$$

The following matrices were constructed using the package $MAPLE$ by the scheme in the proof of Lemma C4.

The matrix $Y_7(t) := Z_{3,4}(t,t)$ realizes the case $\delta = \frac{9}{16} \approx 0.5625$, where

$$Y_7(t) = \begin{vmatrix} C_3 & 0 & 0 & -S_3 & 0 & 0 & 0 \\ S_3 & C_3 & 3S_3 & 0 & 0 & 0 & 0 \\ 3C_3 & S_3 & C_3 & 0 & 0 & 0 & 0 \\ 0 & 0 & 0 & C_4+S_4 & -2S_4 & 0 & 0 \\ 0 & 0 & 0 & C_4 & C_4-S_4 & 0 & 2C_4 \\ 0 & 0 & 0 & 0 & 0 & C_4+S_4 & 2S_4 \\ 0 & 0 & C_4-S_4 & 0 & 0 & C_4 & C_4-S_4 \end{vmatrix}$$

The matrix $Y_9(t) := Z_{4,5}(t,t)$ realizes the case $\delta = \frac{16}{25} \approx 0.64$, where

$$Y_9(t) =$$

$$\begin{vmatrix} C_5 & 0 & 0 & 0 & 0 & -S_5 & 0 & 0 & 0 \\ -\frac{1}{5}S_5 & C_5 & 0 & 0 & S_5 & 0 & 0 & 0 & 0 \\ 0 & 0 & C_5 & S_5 & 0 & 0 & 0 & 0 & 0 \\ 15S_5 & C_5 & 5S_5 & C_5 & 5S_5 & 0 & 0 & 0 & 0 \\ -5C_5 & S_5 & 0 & S_5 & C_5 & 0 & 0 & 0 & 0 \\ 0 & 0 & 0 & 0 & 0 & C_4{+}S_4 & -2S_4 & 0 & 0 \\ 0 & 0 & 0 & 0 & 0 & C_4 & C_4{-}S_4 & 0 & 2C_4 \\ 0 & 0 & 0 & 0 & 0 & 0 & 0 & C_4{+}S_4 & 2S_4 \\ 0 & 0 & 0 & 0 & S_4-C_4 & 0 & 0 & C_4 & C_4{-}S_4 \end{vmatrix}$$

The matrix $Y_{11}(t) := Z_{5,6}(t,t)$ realizes the case $\delta = \frac{25}{36} \approx 0.7$, where

$$Y_{11}(t) = \begin{vmatrix}
C_5 & 0 & 0 & 0 & 0 & -S_5 & 0 & 0 & 0 & 0 & 0 \\
\frac{1}{5}S_5 & C_5 & 0 & 0 & S_5 & 0 & 0 & 0 & 0 & 0 & 0 \\
0 & 0 & C_5 & S_5 & 0 & 0 & 0 & 0 & 0 & 0 & 0 \\
15S_5 & C_5 & 5S_5 & C_5 & 5S_5 & 0 & 0 & 0 & 0 & 0 & 0 \\
-5C_5 & S_5 & 0 & S_5 & C_5 & 0 & 0 & 0 & 0 & 0 & 0 \\
0 & 0 & 0 & 0 & 0 & C_6 & S_6 & 0 & 0 & 0 & 0 \\
0 & 0 & 0 & 0 & -6C_6 & S_6 & C_6 & 4C_6 & 0 & 0 & 0 \\
0 & 0 & 0 & 0 & \frac{256}{33}C_6 & 0 & 0 & C_6 & S_6 - 4C_6 & 0 & 0 \\
0 & 0 & 0 & 0 & 0 & 0 & 0 & S_6 & -C_6 & 4C_6 & 0 \\
0 & 0 & 0 & 0 & \frac{256}{33}C_6 & 0 & 0 & 0 & 0 & C_6 & 4C_6 + S_6 \\
0 & 0 & 0 & 0 & -\frac{68}{33}C_6 & 0 & 0 & 0 & 0 & S_6 & -C_6
\end{vmatrix}$$

In view of the small size of the page we can not place here the matrix $Y_{13}(t) := Z_{6,7}(t,t)$ of the order 13×13, which realize the case $\delta = \frac{36}{49} \approx 0.73$, etc. We also do not give here the matrices $B_{2n+1}(t) = Y_{2n+1}(t)' \cdot (Y_{2n+1}(t))^{-1}$, where $n = 2, 3, 4 \dots$.

REFERENCES

[Ab 1] Abe K., *Characterization of totally geodesic submanifolds of S^N and CP^N by an inequality*, Tohoku Math. J., **23** (1971), 219–244.

[Ab 2] ———, *A complex analogue of Hartman-Nirenberg cylinder theorem*, J. Diff. Geom., **7** (1972), 453–460.

[Ab 3] ———, *Application of a Riccati type differential equation to Riemannian manifolds with totally geodesic distributions*, Tohoku. Math. J., **25** (1973), 425–444.

[Ab 4] ———, *Some remarks on a class of submanifolds in space of non-negative curvature*, Math. Ann., **247** (1980), 275–278.

[Ab 5] ———, *A characterization of totally geodesic hypersurfaces of S^n and CP^n*, Proc. A.M.S., **81** (1981), 603–606.

[AB] Abe K. and Blair D., *Ruled hypersufaces of Euclidean space*, Rocky Mountain J. of Math., **17** (1987), 697–708.

[AM 1] Abe K. and Magid M., *Relative nullity foliations and indefinite isometric immersions*, Pacific J. Math., **124** (1986), 1–20.

[AM 2] ———, *Indefinite Kähler submanifolds with positive index of relative nullity*, Ann. Global Anal. Geom., **6** (1988), 231–258.

[AH] Abe K. and Haas A., *Isometric Immersions of H^n into H^{n+1}*, A.M.S., Proc. of Symp. in Pure Math., "Diff. Geometry: Riemannian Geometry" **54** (1993), 23–30.

[AHo] Anderson L. and Howard R., *Comparison and Rigidity theorems in Semi-Riemannian Geometry*, Comm. Anal. Geom., dg-ga/9707020 **54** (1993), 1–46 (to appear).

[AGr] Abresch U. and Gromoll D., *On complete manifolds with nonnegative Ricci curvature*, J. of A.M.S., **3** (1990), 355–374.

[AM] Abresch U. and Meyer W., *Injectivity radius estimates and sphere theorems* (1995), 1–38 (to appear).

[Ada] Adachi M., *Embeddings and immersions (chapter 5: Foliations on open manifolds)*, Transl. of Math. Monographs. A.M.S.), vol. 124, 1984, pp. 183.

[Adm] Adams J., *Vector fields on spheres*, Ann. Math., **75** (1962), 603–632.

[Ad] Adams S., *Rank rigidity for foliations by manifolds of nonpositive curvature*, Diff. Geom. and its Appl., **3** (1993), 47–70.

[AF] Adams S. and Freire A., *Nonnegatively curved leaves in foliations*, J. Diff. Geom., **34** (1991), 681–700.

[AS] Adams S. and Stuck G., *Splitting of nonnegatively curved leaves in minimal sets of foliations*, Duke Math. J., **71** (1993), 71–92.

[ALS] Ahn S., Lee S. and Suh Y., *On ruled real hypersurfaces in a complex space form*, Tsukuba J. Math., **17(2)** (1993), 311–322.

[Ak] Akivis M., *On multi-dimensional strongly parabolic submanifolds*, Soviet Mathematics (Izv. VUZ), **31** (1987), no. 5, 1–11.

[AkG] Akivis M. and Goldberg V., *Projective Differential Geometry of Submanifolds*, North-Holland, 1993, pp. 335.

[Ale] Alexander S., *Cylindricity of Euclidean immersions*, Notice A.M.S., **16** (1969), 682–685.

[AMa] Alexander S. and Maltz R., *Isometric immersions of Riemannian products in Euclidean space*, J. of Diff. Geom., **11(1)** (1976), 47–57.

[AM] Al-Medhadi N. and Marzouk M., *An integral formula and applications*, Bull. Calcutta Math. Soc., **84** (1992), 137–144.

[AP] Alexander S. and Portnoy E., *Cylindricy of isometric immersions between hyperbolic spaces*, Trans. A. M. S., **237** (1978), 311–329.

[Al 1] Alvarez-Lopez J. A., *On Riemannian foliations with minimal leaves*, Ann. Inst. Fourier, (Grenoble), **40** (1990), 163–176.

[Al 2] ———, *The basic component of the mean curvature of Riemannian foliations*, Ann. Global Anal. Geom., **10** (1992), 179–194.

[Am] Aminov Yu., *Geometry of vector field*, Moscow: Nauka, 1990, pp. 208. (Russian)

[An] Anderson M., *Hausdorf perturbations of Ricci-flat manifolds and splitting theorem*, Duke Math. J., **68** (1992), 67–82.

[AGl] Asimov D. and Gluck H., *Morse-Smale fields of geodesics*, Global Theory of Dynamical Systems, Springer Lect. Notes in Math., vol. 819, 1980, pp. 1–17.

[AY] Aso K. and Yorozu S., *A generalization of Clairaut's theorem and umbilic foliations*, Nihouka Math. J., **2** (1991), 139–153.

[AsD] Asperti A. and Dajczer M., *N-dimensional submanifolds of R^{N+1} and S^{N+1}*, Illinois J. of Math., **28** (1984), 621–645.

[Au 1] Aumann G., *Zur Theorie der $(k+1)$-Gratregelflächen*, Sitzungsber. Österr. Akad. Wiss. Wien **189** (1980), 101–127.

[Au 2] ———, *Untersuchungen über verallgemeinerte regelflächen im Euclidischen raum E_n*, RAD Jugosl. Akad. znan. i umjetn. Mat. znzn., **5** (1986), 1–7.

[Au 3] ———, *Die Minimalhyperregelflächen*, Manuscripta Math., **34** (1981), 293–304.

[Bai] Baikoussis C., *Ruled submanifolds with finite type Gauss map*, J. of Geom., **49 (1–2)** (1994), 42–45.

[BaK] Baikoussis C. and Koufogiorgos T., *On a type of contact manifolds*, J. Geom., **46** (1993), 1–9.

[BBK] Baikoussis C., Blair D. and Koufogiorgos T., *A decomposition of the curvature tensor of a contact manifolds with ξ in $N(k)$*, Math. Technical Report, (1992), Univ. of Ioannina (Greece).

[BaW1] Baird P. and Wood J., *Harmonic morphisms and conformal foliations by geodesics of three-dimensional space forms*, J. Austral. Math. Soc. Ser. A, **51** (1991), 118–153.

[BaW 2] ———, *Harmonic morphisms, conformal foliations and shear-free ray congruences*, dg-ga/9603005, (1996), 1–40.

[BG] Brunella M. and Ghys E., *Umbilical foliations and transversely holomorphic flows*, J. Diff. Geom., **41(1)** (1995), 1–20.

[BGrS] Ballmann W., Gromov M. and Schroder V., *Manifolds of nonpositive curvature*, Birkhäuser, 1985.

[BM] Banaszczyk M. and Majchrzak R., *An integal formula for a Riemannian manifold with three orthogonal distributions*, Acta Sci. Math., **54** (1990), 201–207.

[Ban 1] Bangert V., *The existence of gaps in minimal foliations*, Aequationes Math., **34** (1987), 153–166.

[Ban 2] ———, *Minimal foliations and laminations*, Proc. of Int. Congr. of Math., (1994, Zürich), **1** (1995), Birkhäuser Verlag, 453–464.

[BD] Barbosa L. and Delgado J., *Ruled submanifolds in space forms with mean curvature of nonzero constant length*, Amer. J. of Math., **106** (1984), 763–780.

[BDJ] Barbosa L., Dajczer M. and Jorge L., *Minimal ruled submanifolds in spaces of constant curvature*, Indiana Univ. Math. J., **33** (1984), 531–547.

[BDT] Barbosa L., Dajczer M. and Tojeiro R., *Isometric immersions of Riemannian products revisited*, Comment. Math. Helv., **69** (1984), 281–290.

[BKO] Barbosa L., Kenmotsu K. and Oshikiri G., *Foliations by hypersurfaces with constant mean curvature*, Math. Z., **207** (1991), 97–107.

[BGoS] Barbosa J., Gomes J., and Silveira A., *Foliation of 3-dimensional space forms by surfaces with constant mean curvature*, Bol. Soc. Brasil. Mat. **18** (1987), 1–12.

[BaB] Baikoussis C. and Blair D., *On the Gauss map of ruled surfaces*, Glasgow Math. J., **34 (3)** (1992), 355–359.

[BCV] Baikoussis C., Chen B. and Verstraelen L., *Ruled surfaces and tubes with finite Gauss map*, Tokyo Math. J., **16 (2)** (1993), 341–349.

[Bas] Basmajian A., *The geometry of totally geodesic hypersurfaces in hyperbolic manifolds*, Diff. geometry: Riemannian geometry (Los Angeles, CA, 1990), Proc. Sympos. Pure Math., vol. 54, A. M. S., Providence, RI, 1993, pp. 81–87.

[BW] Basmajian A. and Walschap G., *Metric flows of space forms of nonpositive curvature*, Proc. A.M.S., **123(10)** (1995), 3177–3181.

[BE] Beem D. and Ehrlich P., *Global Lorentzian Geometry*, Pure and Applied Math.: A Series of Monographs and Textbooks, 2-d edition, vol. 202, Dekker, New York, 1996, pp. 664.

[Ber] Berestovskii V., *"Submetries" of space forms of nonnegative curvature*, Sibirsk. Mat. Zh., **28(4)** (1987), 552–562 (44–56).

[Be 1] Berger M., *Les varietes riemanniennes $\frac{1}{4}$-pincees*, Ann. Scuola Norm. Sup. Pisa, **14** (1960), 161–170.

[Be 2] ———, *On the diameter of some Riemannian manifolds*, Technical report, (1962), Univ. of California.

[Be 3] ———, *Sur les varietes riemanniennes pincees juste au-dessous de $\frac{1}{4}$*, Ann. Inst. Fourier, **33** (1983), 135–150.

[Be 4] ———, *Geometrie*, Cedic/Fernand Nathan, Paris, 1977/1978.

[BB] Bergery B. and Bourguignon J., *Laplacians and Riemannian submersions with totally geodesic fibers*, Illinois J. of Math., **26** (1982), 181–200.

[BVa] Berndt J. and Vanhecke L., *Aspects of the geometry of the Jacobi operator*, Riv. Mat. Univ. Parma, **5(3)** (1994), 91–108.

[Bes 1] Besse A., *Manifolds all of whose Geodesics are Closed*, Springer-Verlag, 1978.

[Bes 2] ———, *Einstein manifolds*, Ergebisse Grenzgeb. Math., vol. 1,2, Springer-Verlag, 1987.

[Bi] Binding P., *Hermitian forms and the fibrations of spheres*, Proc. A.M.S., **94** (1985), 581–584.

[BiN] Bishop R. and O'Neill B., *Manifolds of negative curvature*, Trans. A.M.S., **149** (1969), 1–49.

[Bi 1] Bishop R., *Clairaut submersions*, Diff. Geom. in honor of Yano, Tokyo, 1972, pp. 21–31.

[Bi 2] ———, *An infinitesimal cylindricity theorem*, Tensor (N.S.), **38** (1982), 147–153.

[Biv] Bivens I., *Orthogonal geodesic and minimal distributions*, Trans. A.M.S., **275** (1983), 397–408.

[BK] Blair D. and Koufogiorgos T., *Ruled surfaces with vanishing second Gaussian curvature*, Monatsh. Math., **113 (3)** (1992), 177–181.

[BV] Blair D. and Vanstone J., *A generalization of the helicoid*, Minimal Submanifolds and Geodesics, Kaigai publ., Tokyo, 1978, pp. 13–16.

[Bl 1] Blumenthal R., *Riemannian foliations with parallel curvature*, Nagoya Math. J., **90** (1983), 145–153.

[Bl 2] ———, *Transverse curvature of foliated manifolds*, Asterisque, **116** (1984), 25–30.

[Bl 3] ———, *Foliations with locally reductive normal bundle*, Illinois J. Math., **28** (1984), 691–702.

[Bl 4] ———, *Stability theorems for conformal foliations*, Proc. A.M.S., **91** (1984), 485–491.

[Bl 5] ———, *Sprays, fibre spaces and product decompositions*, Diff. Geom. Res. Notes in Math., vol. 131, Boston, 1985, pp. 156–161.

[BlH 1] Blumenthal R. and Hebda J., *De Rham decomposition theorems for foliated manifolds*, Ann. Inst. Fourier, **33** (1983), 183–198.

[BlH 2] ———, *Complementary distributions which preserve the leaf geometry and applications to totally geodesic foliations*, Quarterly J. Math. Oxford, **35** (1984), 383–392.

[BlH 3] ———, *Ehresmann connections for foliations*, Indiana Univ. Math. J., **33** (1984), 597–611.

[BlH 4] ———, *Un analogue de nappe d'holonomie pour une variété feuilletée*, C. R. Acad. Sci. Paris, **303** (1986), 931–934.

[BlH 5] ———, *A sufficient condition for the leaves of a totally umbilic foliation to be conformally complete*, Ann. Global Anal. Geom., **6** (1988), 165–175.

[BlH 6] ———, *An analogue of the holonomy bundle for a foliated manifold*, Tohoku Math. J., **40** (1988), 189–197.

[Boe1] Boeckx E., *Foliated semi-symmetric spaces*, Doctoral Thesis, (1994), Leuven.

[Boe 2] ———, *Asymptotically Foliated Semi-Simmetric Spaces*, Bolletino U. M. I., **9-B(7)** (1995), 975–1019.

[Bö 1] Böning R., *Completeness of curvature surfaces of submanifolds in complex space forms*, Results in Math., **27** (1995), 27–34.

[Bö 2] ———, *Curvature surfaces of submanifolds of pseudoriemannian manifolds with applications to complex space forms* (to appear).

[Bor 1] Borisenko A., *Complete l-dimensional submanifolds of nonpositive extrinsic curvature in Riemannian space*, Math. Sbornic, **104** (1977), 559–577. (Russian)

[Bor 2] ———, *About external geometrical properties of parabolic submanifolds and topological properties of saddle submanifolds in symmetric spaces of rank one*, Math. Sbornic (N.S.), **116(158)** (1981), no. 3(11), 440–457. (Russian)

[Bor 3] ———, *Submanifolds of nonpositive extrinsic curvature in spaces of constant curvature*, Math. Sbornic (N.S.), **114(156)** (1981), 339–354. (Russian)

[Bor 4] ———, *On complete parabolic submanifolds*, Plenum, **48** (1990), no. 1, January, 6–14.

[Bor 5] ———, *On extremal properties of compact parabolic submanifolds in Riemannian space*, Math. Sbornic, **133(175)** (1987), no. 1(5), 112–126. (Russian)

[Bor 6] ———, *Normal curvature and topological structure of multidimensional surfaces in a spherical space*, Plenum, **54** (1991), no. 1, March, 675–678.

[Bor 7] ———, *On cylindrical multi-dimensional submanifolds in Lobachevsky space*, Plenum **53** (1991), no. 5, February, 474–480.

[Bor 8] ———, *The foliations of extrinsic negative curvature in a Riemannian space*, Conf. on Diff. Geometry and Applications, Abstracts, Brno, Czech Republic, 1995, pp. 5–6.

[Bor 9] ———, *On structure of l-dimensional submanifolds with degenerate second fundamental form*, Ukrain. Geom. Sbornic, **13** (1972), 18–27. (Russian)

[Bor 10] ———, *On isometric embeddings of pseudo-Riemannian spaces of constant curvature*, Ukrain. Geom. Sbornic, **19** (1976), 11–18.

[BN] Borisenko A. and Nikolaevskii Yu., *Grassmannian manifolds and Grassmann's image of submanifolds*, Uspehi Math. Nauk, **46(2)** (1991), Moscow, 41–83.

[BU 1] Borisenko A. and Ushakov V., *One characteristic property of multi-dimensional cylindrical surfaces*, J. Sov. Math., **54 (1)** (1991), 678-683.

[BU 2] ———, *An isometric embedding of strong parabolic metrics in class of strong parabolic submanifolds*, Ukrain. Geom. Sbornic, **33** (1990), 27–36. (Russian)

[BY 1] Borisenko A. and Yampolskii A., *The Sasaki metric of the normal bundle of a submanifold in a Riemannian space*, Math. Sbornic (N.S.), **134(176)** (1987), 158–176. (Russian)

[BY 2] ———, *Riemannian geometry of bundles*, Russian Math. Surveys **46** (1991), 55–106.

[BS] Borowskii Y. and Shefel S., *On the theorem by Chern and Kuiper*, Siberian Math. J., **19(6)** (1978), 1386–1387. (Russian)

[Bot] Bott R., *Lectures on characteristic classes and foliations*, Springer Lect. Notes in Math., vol. 279, 1972, pp. 1–94.

[Bou 1] Boualem H., *Theoreme de decomposition d'un feuilletage riemannien singulier transversalement integrable*, C. R. Acad. Sci. Paris, **316** (1993), no. 1, 59–62.

[Bou 2] ———, *Feuilletages riemanniens singuliers transversalement integrables*, Compositio Math., **95** (1995), 101–125.

[Br 1] Brito F., *Une obstruction geometrique a l'existence de feuilletages de codimension 1 totalement geodesiques*, J. Diff. Geom., **16** (1981), 675–684.

[Br 2] ———, *A remark on minimal foliations of codimension two*, Tohoku Math. J. (2), **36** (1984), 341–350.

[BLR 1] Brito F., Langevin R. and Rosenberg H., *Integrales de courbure sur une variete feuilletees*, C. R. Acad. Sci. Paris, **285** (1977), 533–536.

[BLR 2] ———, *Integrales de courbure sur une variete feuilletees*, J. Diff. Geom., **16** (1981), 19–50.

[BrW] Brito F. and Walczak P., *Totally geodesic foliations with integrable normal bundle*, Bol. Soc. Bras. Mat., **17** (1986), 41–44.

[Broo] Brooks R., *Convexite et courbure de Ricci*, C. R. Acad. Sci. Paris, **319** (1994), 73–75.

[Bro] Browne H., *Codimension 1 totally geodesic foliations of H^n*, Tohoku Math. J., **36** (1984), 315–340.

[BuZ] Burago Yu. and Zalgaller V., *Introduction to Riemannian geometry*, Russian Acad. Sci., S.-Petersburg, 1994, pp. 318. (Russian)

[BZ] Borzellino J. and Zhu S.-H., *The splitting theorem for orbifolds*, Illinois J. of Math., **38** (1994), 679–691.

[CGL] Cai M., Galloway G. and Liu Z., *Local splitting theorems for Riemannian manifolds*, Proc. A.M.S., **120 (4)** (1994), 1231–1239.

[Ca 1] Cairns G., *Geometrie globale des feuilleteges totalement geodesiques*, C. R. Acad. Sci. Paris, **297** (1983), 525–527.

[Ca 2] ———, *Feuilletages totalement geodesiques de dimension 1, 2 ou 3*, C. R. Acad. Sci. Paris, **298** (1984), 341–344.

[Ca 3] ———, *Aspects cohomologoques des feuilletages totalement geodesiques*, C. R. Acad. Sci. Paris, **299** (1984), 1017–1019.

[Ca 4] ———, *A general description of totally geodesic foliations*, Tohoku Math. J., **38** (1986), 37–55.

[Ca 5] ———, *Some properties of a cohomology group associated to a totally geodesic foliations*, Math. Z., **192** (1986), 391–403.

[Ca 6] ———, *Totally umbilic Riemannian foliations*, Mich. Math. J., **37** (1990), 145–159.

[Ca 7] ———, *Compact 4-manifolds that admit totally umbilic metric foliations*, Differential geometry and its applications (Brno, 1989), World Sci. Publ., 1990, pp. 9–16.

[Ca 8] ———, *Feuilletages geodesibles*, These, Acad. de Montpellier, Univ. des Sci. et Techniques du Languedoc, 1987, pp. 105.

[CG] Cairns G. and Ghys E., *Totally geodesic foliations on 4-manifolds*, J. Diff. Geom., **23**

(1986), 241–254.

[CN] Camacho C. and Neto A., *Geometric Theory of Foliations*, Progress in Math., Birkhäuser, Boston, 1985, pp. 205.

[CaM] Cao H. and Mok N., *Holomorphic immersions between compact hyperbolic space forms*, Invent. Math., **100** (1990), 49–61.

[CDo] Cordero L. and Dodson C., *Differential geometry of frame bundles*, 1989.

[CD'A] Carfagna D'Andrea A., *Helical and 2-symmetric submanifolds* (to appear).

[CMR] Carfagna D'Andrea A., Marzzocco R. and Romani G., *Some characterizations of 2-symmetric submanifolds in spaces of constant curvature*, Czechosl. Math. J., **44(119)** (1994), 691–711.

[CDM] Carmo H., Dajczer M. and Mercuri F., *Compact conformally flat hypersurfaces*, Trans. A.M.S., **288** (1985), 189–203.

[CDa 1] Carmo H. and Dajczer M., *A rigidity theorem for higher codimension*, Math. Ann., **274** (1986), 577–583.

[CDa 2] ———, *Conformal rigidity*, Amer. J. Math., **109** (1987), 963–985.

[CDa 3] ———, *Local isometric immersion of R^2 into R^4*, J. reine angev Math., **442** (1993), 205–219.

[CDa 4] ———, *Rotation hypersurfaces in spaces of constant curvature*, Trans. A.M.S., **277** (1983), 685–709.

[doC] do Carmo M., *Riemannian geometry*, 2 ed., 1992.

[CG] Carriere Y. and Ghys E., *Feuilletages totalement geodesiques*, Anais da Acad. Bras. de Ciencias, **53(2)** (1981), 427–432.

[Car] Carriere Y., *Flots riemannies*, Asterisque, **116** (1984), 31–52.

[Cas] Cass Daniel M., *Minimal leaves in foliations*, Trans. A.M.S., **287** (1985), 201–213.

[Car 1] Cartan E., *La déformation des hypersurfaces dans l'espace conforme réel a n dimensions*, Bull. Soc. Math. France, **44** (1916), 65–99.

[Car 2] ———, *Sur les varietes de courbure constante d'un space euclidien on non-euclidien*, Bull. Soc. Math. France, **47** (1919), 125–160.

[Car 3] ———, *Sur les varietes de courbure constante d'un space euclidien on non-euclidien*, Bull. Soc. Math. France, **48** (1920), 132–208.

[ChC 1] Cheeger J. and Colding H., *Almost rigidity of warped products and the structure of spaces with Ricci curvature bounded below*, C. R. Acad Sci. Paris, **320** (1995), 353–357.

[ChC 2] ———, *Lower bounds on Ricci curvature and the almost rigidity of warped products* **144** (1996), 189–237.

[ChG 1] Cheeger J. and Gromoll D., *The splitting theorem for manifolds of nonnegative Ricci curvature*, J. Diff. Geom., **6** (1971), 119–128.

[ChG 2] ———, *On the structure of complete manifolds of nonnegative curvature*, Ann. of Math., **96** (1972), 413–443.

[CE] Cheeger J. and Ebin D., *Comparison theorems in Riemannian geometry*, vol. 9, North. Holl. Publ. Co., Amsterdam; Oxford, 1975, pp. 179.

[Ch 1] Chen B.Y., *Geometry of submanifolds*, Marcel Dekker, New York, 1973, pp. 298.

[Ch 2] ———, *Geometry of submanifolds and its applications*, Science Univ. of Tokyo, 1981, pp. 96.

[Ch 3] ———, *Total mean curvature and submanifolds of finite type*, World Scientific, Syngapore, 1984.

[Ch 4] ———, *Some open problems and conjectures on submanifolds of finite type*, Soochow Math. J., **17** (1991), 169–188.

[Ch 5] ———, *A report on submanifolds of finite type*, Soochow J. of Math., **22(2)** (1996), 117–337.

[CDVV] Chen B., Dillen F., Verstraelen L. and Vrancken L., *Ruled surfaces of finite type*, Bull. Austral. Math. Soc. **42(3)** (1990), 447–453.

[CK] Chern S. and Kuiper N., *Some theorems on the isometric imbedding of compact Riemannian manifolds in Euclidean Space*, Ann. Math., **56** (1952), 422–430.

[CL 1] Chern S. and Lashov R., *On the total curvature of immersed manifolds*, Amer. J. Math., **79** (1957), 306–318.

[CL 2] ———, *On the total curvature of immersed manifolds, II*, Michigan Math. J., **5** (1958), 5–12.

[CE] Chicone C. and Ehrlich P., *Gradient-like vector fields on R^2*, Manuscripta Math., **49** (1984), 141–164.

[Chi] Chinea D., *Almost contact metric submersions*, Rend. del Circolo Mat. di Palermo, ser. II, **34** (1985), 89–104.

[Cho] Choi S., *On the Gauss map of ruled surfaces in a 3–dimensional Minkowski space*, Tsukuba J. Math., **19(2)** (1995), 285–304.

[Chr] Chrusciel P., *Sur les feuilletages "conformement minimax" des varietes Riemanniennes de dimension trois*, C. R. Acad. Sci. Paris, **301** (1985), 609–612.

[Chu] Chu I., *Riemannian fibrations of Euclidean spaces*, Thesis, SUNY at Stony Brook (1981).

[CM] Clifton Y. and Maltz R., *K-nullity spaces of the curvature operator*, Mich. Math. J., **17** (1970), 85–89.

[Cŏm] Cŏmić I., *Curvature theory of vector bundles and subbundles*, Filomat, **7** (1993), 55–66.

[Con] Connes A., *Sur la theorie noncommutative de l'integration*, Algebres d'Operateurs, Lecture Notes in Math., vol. 725, Springer, 1979, pp. 19–143.

[Co] Coulton P., *An invariant formula for orthogonal distributions*, Ann. Global Anal. Geom., **6** (1988), 47–54.

[Cr] Croke B., *Volumes of balls in manifolds without conjugate points*, Int. J. Math., **3** (1992), 455–467.

[CrK] Croke C. and Keliner B., *A warped product splitting theorem*, Duke Math. J., **67(3)** (1992), 571–574.

[Cu] Curras-Bosch C., *The geometry of totally geodesic foliations admitting Killing field*, Tohoku Math. J., **40** (1988), 535–548.

[DD] Dajczer M. and Dombrowski P., *Examples of 1-dimensional non totally geodesic isometric immersions of Pseudo-Riemannian space forms with the same positive constant curvature and the same space-like rank*, Global Diff. Geom. and Global Abalysis, Lecture Notes Math., vol. 1156, Berlin, 1985, pp. 59-73.

[DG 1] Dajczer M. and Gromoll D., *On spherical submanifolds with nullity*, Proc. A.M.S., **93(1)** (1985), 99–100.

[DG 2] ———, *Gauss parametrization and rigidity aspects of submanifolds*, J. Diff. Geom., **22** (1985), 1–12.

[DG 3] ———, *Rigidity of complete Euclidean hypersurfaces*, J. Diff. Geom., **31** (1990), 401–416.

[DR 1] Dajczer M. and Rodrigues L., *Substantial codimension of submanifolds*, Bull. London Math. Soc., **19** (1987), 467–473.

[DR 2] ———, *On isometric immersions into complex space forms*, preprint, (1992).

[DR 3] ———, *Complete real Kähler minimal submanifolds*, J. Reine Angew. Math., **419** (1991), 1–8.

[DH] Dazord P., Hector G., *Integration symplectique des variétés de Poisson totalement asphériques*, Symplectic Geometry, Groupoids and Integrable Systems, Seminare Sud Rhodanien á Berkeley (1989), Springer Math. SRIP, Springer-Verlag, vol. 21, 1991, pp. 38–72.

[Di] Dillen F., *Ruled submanifolds of finite type*, Proc. A.M.S., **114(3)** (1992), 795–798.

[DWV] Dillen F., Van de Woestyne I. and Verstraelen L., *Ruled surfaces of finite type in a 3−dimensional Minkowski space*, Results Math., **27** (1995), 250–255.

[Dol] Dold A., *Lectures on Algebraic Topology*, Springer-Verlag, 1972.

[Dom] Dombrowski P., *Jacobi fields, totally geodesic foliations and geodesic differential forms*, Result. Math., **1** (1979), 156–194.

[Dm] Dominguez D., *Finiteness and tenseness theorems for Riemannian foliations*, Abstr. A.M.S., **15** (1994), 583.

[DK] Duchamp T. and Kalka M., *Complex foliations*, Recent developments in geometry (Los Angeles, CA, 1987), Contemp. Math., vol. 101, A.M.S., Providence, RI, 1989, pp. 323–338.

[Eb] Eberlein P., *Manifolds of nonnegative curvature*, A.M.S., Proc. of Symp. in Pure Math. "Diff. Geometry: Riemannian Geometry", Part 3, vol. 54, 1993, pp. 179–227.

[Ed] Edge W., *The theory of ruled surfaces*, Cambridge, 1931.

[EMS] Edwards R., Millett K. and Sullivan D., *Foliations with all leaves compact*, Topology **16** (1977), 13–32.

[EG] Efremovič V. and Gorelik E., *Metric fibrations of Lobachevsky–Bolyai space*, Colloq. Math. Soc. Janos Bolyai, Diff. Geom., Eger (Hungary), vol. 56, 1989, pp. 223–230.

[EVGL] Efremovič V., Vainstein A., Gorelik E. and Loginov E., *Metric fibrations of Riemannian manifolds*, Colloq. Math. Soc. J. Bolyai (1979), 195–198.

[ER] Ehresmann C. and Reeb G., *Sur les champs d'elements complement integrables*, C. R. Acad. Sci. Paris, **218** (1944), 955–957.

[Ehr 1] Ehresmann C., *Les connexions infinitesimales dans un expace fibre differentiable*, Colloque de Topologie, Bruxelles, (1950), 29–55.

[Ehr 2] ———, *Structures feuilletees*, Proc. 5th Canad. Math. Congress, Montreal, 1961, pp. 109–172.

[EhK] Ehrlich P. and Kim S., *From the Riccati Inequality to the Raychaudhuri Equation*, Contemporary Math., **170** (1994), 65–78.

[EL] Eells J. and Lemaire L., *Another report on harmonic maps*, Bull. London. Math. Soc., **20** (1988), 385–524.

[Enn] Enneper A., *Die cyklischen Flächen*, J. Math. Phys., **14** (1869), 393–421.

[Eps] Epstein C., *Orthogonally integrable line fields in* $\mathbb{H}^3$, Comm. Pure Appl. Math., **38** (1985), 599–608.

[Epd] Epstein D., *Foliations with all leaves compact*, Ann. Inst. Fourier (Grenoble), **26** (1976), 265–282.

[EMT] Epstein D., Millett K. and Tischler D., *Leaves without holonomy*, J. London. Math. Soc. (2), **16** (1977), 548–552.

[Er] Erguet M., *The drall and the scalar normal curvature of* $(r+1)$*-dimensional generalised ruled surfaces*, Commun. Fac. Sci. **38 no. 1/2** (1989), Univ. Ankara, Ser. A1, 115–125.

[EH 1] Eschenburg J. and Heintze E., *An elementary proof of the Cheeger-Gromoll splitting theorem*, Ann. Glob. Anal. and Geom., **2** (1984), 141–151.

[EH 2] ———, *Comparison theory for Riccati equation*, Manuscripta Math., **68** (1990), 209–214.

[ES] Eschenburg J. and Sullivan J., *Jacobi tensors and Ricci curvature*, Math. Ann., **252** (1980), 1–26.

[Es 1] Escobales R., *Submersions from spheres*, Bull. A.M.S., **79** (1973), 71–74.

[Es 2] ———, *Riemannian submersions with totally geodesic fibers*, J. Diff. Geom., **10** (1975), 253–276.

[Es 3] ———, *Riemannian submersion from complex projective space*, J. Diff. Geom., **13** (1978), 93–107.

[Es 4] ———, *Sufficient conditions for a bundle-like foliation to admit a Riemannian submersion onto its leaf space*, Pros. A.M.S., **84** (1982), 280–284.

[Es 5] ———, *The integrability tensor for bundle-like foliations*, Trans. A.M.S., **270** (1982), 333–339.

[Es 6] ———, *Bundle-like foliations with Kählerian leaves*, Trans. A.M.S., **276** (1983), 853–859.

[Es 7] ———, *Riemannian foliations of the rank one symmetric spaces*, Proc. A.M.S., **95** (1985), 495–498.

[Es 8] ———, *The mean curvature cohomology class for foliations and the infinitesimal geometry of the leaves*, Diff. Geom. Appl., **2** (1992), 167–178.

[EP] Escobales R. and Parker P., *Geometric consequences of the normal curvature cohomology class in umbilic foliations*, Indiana Univ. Math. J., **37** (1988), 389–408.

[FVV] Falbel E., Veloso J. and Verderesi J., *Constant curvature models in sub-Riemannian geometry*, Mathem. Contemporanea, Soc. Bras. Mat, **4** (1993), 119–125.

[Fa] Farafonova N., *Riemannian metrics in fibred bundles, I*, Russian Mathematics (Izv. VUZ), **398(7)** (1995), 65–77.

[Fe 1] Ferus D., *On the type number of Hypersurfaces in spaces of constant curvature*, Math. Ann., **187** (1970), 310–316.

[Fe 2] ———, *Totally geodesic foliations*, Math. Ann., **188** (1970), 313–316.

[Fe 3] ———, *On the completeness of nullity foliations*, Mich. Math. J., **18** (1971), 61–64.

[Fe 4] ———, *On isometric immersions between hyperbolic spaces*, Math. Ann., **205** (1973), 193–200.

[Fe 5] ———, *Isometric Immersions of Constant Curvature Manifolds*, Math. Ann., **217** (1975), 155–156.

[Fe 6] ———, *Symmetric submanifolds of Euclidean spaces*, Math. Ann., **247** (1980), 81–93.

[FGL] Ferrandez A., Garay O. and Lucas P., *Finite type ruled manifolds shaped on spherical submanifolds*, Arch. Math. (Basel), **57** (1991), 97–104.

[Fi] Fischer A., *Riemannian maps between Riemannian manifolds*, Math. aspects Class. Field Theory: Proc. A.M.S. – IMS – SIAM Jt. Summer Res. Conf., Seattle, Wash., 1991–1992, pp. 331–366.

[FW] Fisher G. and Wu H., *Developable complex analytic submanifolds*, Internat. J. of Math., **6** (1995), 229–272.

[Flo 1] Florit L., *A sharp estimate for the index of relative nullity*, 8 School on Diff. Geometry, Mat. Contemp., **298** (1993), 127–129.

[Flo 2] ———, *On submanifolds with nonpositive extrinsic curvature*, Math. Ann., **298** (1994), 187–192.

[Fuj] Fujiki A., *Remarks on the extremal Kähler metrics on ruled manifolds*, Nagoya Math. J., **126** (1992), 89–101.

[Fu] Fuks D., *Foliations*, Itogi Nauki i Techniki, ser. Algebra. Topology. Geometry, vol. 18, Moscow: VINITI, 1981, pp. 151–213. (Russian)

[Fr 1] Frankel T., *Manifolds with positive curvature*, Pacific J. Math., **11** (1961), 165–171.

[Fr 2] ———, *On the fundamental group of a compact minimal submanifold*, Pacific J. Math., **83** (1966), 66–73.

[Ga] Gage M., *A note on the skew-Hopf fibrations*, Proc. A.M.S., **93** (1985), 145–150.

[GR] Galledo E. and Reventos A., *Curvature and plane fields*, C. R. Acad. Sci. Paris, **306** (1988), 675–679.

[GHL] Gallo S., Hulin D. and Lafontaine J., *Riemannian geometry*, Springer, 1987, pp. 248.

[Gal] Gallowy G., *A generalization of the Cheeger-Gromoll splitting theorem*, Arch. Math., **47** (1986), 372–375.

[Gar] Gardner R., *New viewpoints in the geometry of submanifolds of R*, Bull. A.M.S., **83** (1977), 1–35.

[Gau 1] Gauchman H., *The nullity spaces of the conformal curvature tensor*, Tensor (N.S.), **35** (1981), 91–98.

[Gau 2] ———, *On a decomposition of Riemannian manifolds*, Houston J. Math., **7** (1981), 365–372..

[Gę 1] Gębarowski A., *On Einstein warped products*, Tensor, N.S., **52(3)** (1983), 204–207.

[Gę 2] ———, *On conformally flat double warped products*, Soochow Math. J., **21(1)** (1995), 125–129.

[Gh 1] Ghys E., *Classification des feuilletages totalement geodesiques de codimension un*, Comment. Math. Helv., **58** (1983), 543–572.

[Gh 2] ———, *Feuilletages riemmanian sur les varietes simplement connexes*, Ann. Inst. Fourier (Grenoble), **34(4)** (1984), 203–223.

[Gil] Gil-Medrano O., *Geometric properties of some classes of Riemannian almost-product manifolds*, Rend. Circ. Mat. Palermo **32(3)** (1983), 315–329.

[GN 1] Gil-Medrano O. and Naveira A., *The Gauss-Bonnet integrand for a class of Riemannian manifolds admitting two orthogonal complementary foliations*, Canad. Math. Bull., **26(3)** (1983), 358–364.

[GN 2] ———, *Some remarks about the Riemannian curvature operator of Riemannian almost-product manifold*, Rev. Roumanine Math. Pures Appl., **30** (1985), 647–658.

[Gla] Glazyrin V., *Topological and metric properties of k-saddle submanifolds*, Docladi Acad. Sci SSSR, **233** (1977), 1028–1030.

[Gl 1] Gluck H., *Can space be filled by geodesics, and if so, how?*, open letter, (1979).

[Gl 2] ———, *Dynamical behavior of geodesic fields*, Global theory of Dynamical Systems, Springer Lect. Notes in Math., vol. 819, 1980, pp. 190–215.

[GW] Gluck H. and Warner F., *Great circle fibrations of the three sphere*, Duke Math. J., **50** (1983), 107–132.

[GWY] Gluck H., Warner F. and Yang C., *Division algebras, fibrations of spheres by great circles and the topological determination of space by the cross behavior of its geodesics*, Duke Math. J., **50** (1983), 1041–1076.

[GWZ 1] Gluck H., Warner F. and Ziller W., *The geometry of the Hopf fibrations*, Enseign. Math., **32** (1986), 173–198.

[GWZ 2] ———, *Fibrations of spheres by parallel great spheres and Berger's ridigity theorem*, Ann. of Global Analysis and Geom., **5** (1987), 53–92.

[GZ] Gluck H. and Ziller W., *On the volume of a unit vector field on the three-sphere*, Comment. Math. Helv., **61** (1986), 177–192.

[God] Godbillon C., *Feulletages: etudes geometriquess*, Public. de l'Inst. Kuherche Math. Avancee, Strasbourg, 1985–86, Progress in Math., vol. 98, Birkhäuser, 1991, pp. 474.

[GK] Goldberg S. and Kobayashi S., *On holomorphic bisectional curvature*, J. Diff. Geom., **1** (1967), 225–233.

[GV] Gonzalez-Davila J., Vanhecke L., *Geometry of tubes and isometric flows*, Math. J. Toyama Univ., **18** (1995), 9–35.

[GGV 1] Gonzalez-Davila J., Gonzalez-Davila M., and Vanhecke L., *Reflections and isometric flows*, Kyungpook Math. J., **35** (1995), 113–144.

[GGV 2] ———, *Classification of Killing-transversally symmetric spaces*, Tsukuba J. Math., **20** (1996), 321–347.

[GGV 3] ———, *The Gelfand theorem in flow geometry*, C. R. Math. Rep. Acad. Sci. Canada, **15** (1993), 281–285.

[GGV 4] ———, *Geodesic spheres and isometric flows*, Colloq. Math., **67** (1994), 223–240.

[GGV 5] ———, *Normal flow space forms and their classification*, Publ. Math. Debrecen, **48/1–2** (1996), 151–173.

[Gor] Gorelik E., *Metric fibrations of Lobachevsky space*, Docl. Acad. Nauk SSSR, **301(6)** (1988), 1301–1304.

[Got 1] Gotoh T., *Harmonic foliations on a complex projective space*, Tsukuba J. Math., **14** (1990), 99–106.

[Got 2] ———, *A remark on foliations on a complex projective space with complex leaves*, Tsukuba J. Math., **16** (1992), 169–172.

[Gr 1] Gray A., *Spaces of constancy of curvature operators*, Proc. A.M.S., **17** (1966), 897–902.

[Gr 2] ———, *Pseudo-Riemannian almost product manifolds and submersions*, J. of Math. and Mech., **16** (1967), 715–737.

[Gr 3] ———, *Isometric immersions in symmetric Spaces*, J. Diff. Geom., **3** (1969), 237–244.

[Gr 4] ———, *Some relations between curvature and characteristic classes*, Math. Ann., **184** (1970), 257–267.

[Gr 5] ———, *Tubes*, Addison-Wesley, Reading, 1990.

[GHe] Gray A. and Hervella L., *The sixteen classes of almost product-hermitian manifolds and their linear invariants*, Ann. di Math. Pura ed Appl., **73** (1980), 35–58.

[Gree] Greene R., *Some recent development in Riemannian geometry*, Contemporary Math., **101** (1989), 1–30.

[Gre] Green L., *A Theorem of E. Hopf*, Mich. Math. J. **5** (1958), 31–34.

[Grl] Gromoll D., *Spaces of nonnegative curvature*, A.M.S., Proc. of Symp. in Pure Math. "Diff. Geometry: Riemannian Geometry", vol. 54, 1993, pp. 337–356.

[GG 1] Gromoll D. and Grove K., *One-dimensional metric foliations in constant curvature spaces*, Diff. Geom. and Complex Analysis, H. Rauch Memorial Vol., 1985, pp. 165–168.

[GG 2] ———, *The low-dimensional metric foliations on Euclidean spheres*, J. Diff. Geom., **28** (1988), 143–156.

[GG 3] ———, *A generalization of Berger's rigidity theorem for positively curved manifolds*, Ann. Sci. Ec. norm. super., **20** (1987), 227–239.

[GKM] Gromoll D., Klingenberg W. and Meyer W., *Riemannsce Geometrie in Grosen*, Lect. Notes in Math., vol. 55, Springer-Verlag, 1968.

[Grv 1] Gromov M., *Partial Differential Relations*, Berlin, Heidelberg, New York: Springer, 1986.

[Grv 2] ———, *Stability and pinching*, Geom. Seminars. Sessions on Topology and Geometry of Manifolds (Bologna, 1990), Univ. Stud. Bologna, Bologna, 1992, pp. 55–97.

[GrH] Grundhöfer T. and Hähl H., *Fibrations of spheres by great spheres over division algebras and their differentiability*, J. Diff. Geom., **31** (1990), 357–363.

[Gu] Guimaraes F., *The integral of the scalar curvature of complete manifolds without conjugate points*, J. Diff. Geom., **36** (1992), 651–662.

[Hae 1] Haefliger A., *Structures feuilletees et cohomologie a valeurs dans un faisceau de groupoides*, Comment. Math. Helv., **32** (1958), 249–329.

[Hae 2] ———, *Varietes feuilletees*, Ann. Scuola Norm. Sup. Pisa, **16** (1962), 367–397.

[Hae 3] ———, *Some remarks on foliations with minimal leaves*, J. Diff. Geom., **15** (1980), 269–284.

[Häh] Hähl H., *Differentiable fibrations of the* $(2n-1)$*-sphere by great* $(n-1)$*-spheres and their coordinatization over quasifields*, Result. Math., **12** (1987), 99–118.

[Háj] Hájková V., *Foliated semi-symmetric spaces in dimension* 3, Doctoral Thesis, (1995), Praque (Czech).

[Han 1] Hangan T., *On totally geodesic distributions of planes*, Topics in diff. geometry, Colloquium: Diff. geometry; Debrecen, 1984, Colloquia Math. Soc. Janos Bolyai, vol. 1, Amsterdam-New York, 1988, pp. 519–530.

[Han 2] ———, *Sur la courbure totale d'integrabilite d'un espace riemannien non holonome compact*, Rend. Sem. Fac. Sci. Univ. Cagliari **58** (1988), 79–94.

[HL] Hangan T. and Lutz R., *Champs d'hyperplans totalement geodesiques sur les spheres*, Third Schnepfenried geom. conf. Asterisque, vol. 1, Soc. Math. France, Paris, 1983, pp. 189–200.

[Har] Hartman P., *On isometric immersion in Euclidean space of manifolds with non-negative sectional curvatures, II*, Trans. A.M.S., **147** (1970), 529–540.

[HarE] Har'El Z., *Harmonic mappings and distortion theorems*, Tensor, **30(1)** (1976), 47–54.

[HV] Hasanis T. and Vlaskos T., *A classification of ruled surfaces of finite type in* S^3, J. Diff. Geom., **50 no. 1–2** (1994), 84–94.

[Has] Hasserblatt B., *Horospheric foliations and relative pinching*, J. Diff. Geom., **39** (1994), 1087–1091.

[Heb 1] Hebda J., *Blaschke manifolds with taught geodesics*, Duke Math. J., **48** (1981), 85–91.

[Heb 2] ———, *Curvature and focal points in Riemannian foliations*, Indiana Univ. Math. J., **35** (1986), 321–331.

[Heb 3] ———, *An example relevant to curvature pinching theorem for Riemannian foliations*, Proc. A.M.S., **114** (1992), 195–199.

[Hec] Hector G., *Une nouvelle obstruction a l'integrabilité des variétés de Poisson régulieres*, Hokkaido Math. J., **21** (1991), 159–185.

[HH] Hector G. and Hirsch U., *Introduction to the Geometry of Foliations*, Vieweg, Braunschweig, vol. A, B, 1981, 1983.

[Hel] Helfer A., *Conjugate points on spacelike geodesics, or pseudo-self-adjoint Morse-Sturm-Liouville systems*, Pacific Math. J., **164(2)** (1994), 321–350.

[Hen 1] Henke W., *Isometrische Immersionen der Kodimension* 2 *von Raumformem*, Manuscripta Math., **19** (1976), 165–188.

[Hen 2] ———, *Über die isometrischer Fortsetzbarkeit isometrische Immersionen der Standard* m*-Sphäre* S^m $(\subset \mathbf{R}^{m+1})$ *in* $\mathbf{R}^{m+2}$), Math. Ann., **219** (1976), 261–276.

[Her 1] Hermann R., *A sufficient condition that a mapping of Riemannian manifolds be a fiber bundle*, Proc. A.M.S., **11** (1960), 236–242.

[Her 2] ———, *The differential geometry of foliations, I*, Ann. Math., **72** (1960), 445–457.

[Her 3] ———, *The differential geometry of foliations, II*, J. Math. Mech., **11** (1962), 305–315.

[HR] Hiepko S. and Reckziegel H., *Über sphärische Blätterungen und die Vollständigkeit ihrer Blätter*, Manuscripta Math., **31** (1980), 269–283.

[Hi] Hirsch M., *Differential topology*, Graduate texts in Math., Springer-Verlag, 1976.

[Hop 1] Hopf H., *Über die Abbildungen der dreidimeusionalen Sphäre auf die Kugelfläche*, Math. Ann., **104** (1931), 637–665.

[Hop 2] ———, *Über die Abbildungen von Sphären auf Sphären niedriqerer Dimension*, Fund. Math., **25** (1935), 427–440.

[Hos] Hoschek J., *Globale geometrie der Regelflächen*, Contributions to Geometry, Proc. Geom. Symp. held in Siegen, 1978, pp. 362–370.

[Hs] Hsiang W., *Global differential geometry of submanifolds*, Algebra, Analysis and Geometry (Taipei, 1988), World Sci. Publ., Teaneck, NJ, 1989, pp. 59–75.

[Hus] Husmoller D., *Fibre bundles*, NY: McGraw-Hill, 1966.

[Hw] Hwang C., *On cerfain hypersurfaces immersed in conformally flat manifold*, J. Nanjing Univ. Math. Biguarterly, **1** (1984), 1–10.

[Iga] Igarachi M., *On quaternion Kählerian Blaschke manifolds*, Math. J. Okayama Univ., **32** (1990), 223–226.

[INR] Ikawa T., Nakagawa H. and Romero A., *Product complex submanifold of indefinite complex space form*, Rocky Mountain J. Math., **18** (1988), 601–615.

[In 1] Innami N., *Splitting theorems of Riemannian manifolds*, Compos. Math., **47** (1982), 237–247.

[In 2] ———, *Totally flat foliations and peakless functions*, Arch. Math., **41** (1983), 464–471.

[In 3] ———, *A proof of the existence of the stable Jacobi field*, Proc. Japan Acad. Sci., ser. A, **63** (1987), 39–41.

[In 4] ———, *Application of the differential equation of Riccati to Riemannian geometry*, The study of dynamical systems (Kyoto, 1989), World Sci. Adv. Ser. Dyn. Syst., vol. 7, World Sci. Publ., Teaneck, NJ, 1989, pp. 105–112.

[In 5] ———, *Manifolds without conjugate points and with integral curvature zero*, J. Math. Soc. Japan, **41** (1989), 251–261.

[In 6] ———, *Jacobi vector fields along geodesic flows*, Dynamical systems and related topics (Nagoya, 1990),, ser. Adv. Ser. Dyn. Syst., vol. 9, World Sci. Publ., River Edge, NJ, 1991, pp. 166–174.

[In 7] ———, *The class of second order equations which Riemannian geometry can be applied to*, J. Math. Soc. Japan, **45** (1993), 89–103.

[Is] Ishihara T., *Radii of immersed manifolds and nonexistence of immersions*, Proc. of A.M.S., **78(2)** (1980), 276–279.

[Jac] Jacobowitz M., *Isometric embedding of a compact Riemannian manifold into a Euclidean space*, Proc. A.M.S., **40(1)** (1973), 245–246.

[Jag] Jagne N., *A note on the sectional curvature of Legendre foliations*, Yokogama Math. J., **41** (1994), 153–161.

[Jagy] Jagy W., *Minimal hypersurfaces foliated by spheres*, Michigan Math. J., **38** (1991), 255–270.

[Jan] Janenko N., *Some questions of embedding theory of Riemannian metrics into Euclidean spaces*, Uspehi Math. Nauk, **8(1)** (1953), Moscow, 21–100.

[Jo 1] Johnson D., *Kähler submersions and holomorphic connections*, J. Diff. Geom., **15** (1980), 71–79.

[Jo 2] ———, *Deformations of totally geodesic foliations*, Geom. and Topol., Lect. Notes in Pure Appl. Math., vol. 105, New-York, 1987, pp. 167–178.

[JW 1] Johnson D. and Whitt L., *Totally geodesic foliations*, J. Diff. Geom., **15** (1980), 225–235.

[JW 2] ———, *Totally geodesic foliations on 3-manifolds*, Proc. A.M.S., **76** (1979), 355–357.

[JN] Johnson D. and Naveira A., *A topological obstruction to the geodesibility of foliation of odd dimension*, Geom. Dedicata, **11** (1981), 347–357.

[Kam] Kamada H., *Foliations on manifolds with positive constant curvature*, Tokyo J. Math., **16** (1993), 49–60.

[KT 1] Kamber F. and Tondeur Ph., *Foliated bundles and characteristic classes*, Springer Lect. Notes in Math., vol. 494, 1975.

[KT 2] ———, *Harmonic foliations*, Proc. NSF Conf. on Harmonic Maps, Tulane, 1980, Lect. Notes in Math., vol. 949, Springer, 1982, pp. 87–121.

[KT 3] ———, *Curvature properties of harmonic foliations*, Illinois J. Math., **28** (1984), 458–471.

[KT 4] ———, *Foliations and metrics*, Proc. of a year in Diff. Geometry, Univ. of Maryland, Progress in Math., vol. 32, Birkhäuser, 1983, pp. 103–152.

[KiT] Kim H. and Tondeur P., *Riemannian foliations on manifolds with non-negative curvature*, Manuscr. Math., **74** (1992), 39–45.

[KTT] Kamber F., Tondeur Ph. and Toth G., *Transversal Jacobi Fields for harmonic foliations*, Michigan Math. J., **34** (1987), 261–266.

[Kar] Karcher H., *Riemannian comparison constructions*, M. A. A. Studies in Mathematics (S. S. Chern, eds.), Global Diff. Geometry, vol. 27, 1989, pp. 170–222.

[Kas] Kashiwada T., *On the integrability of the nullity distribution of the Bochner curvature tensor field*, Tensor (N.S.), **52** (1993), 184–188.

[KP] Kauko V. and Pekonen O., *Hyperpallo*, Arkhimedes, **46** (1994), 24–38.

[KiB] Kim B., *On Escobales-Parker's theorem*, Proc. Japan. Acad., ser. A, **65** (1989), 151–153.

[KX 1] Kenmotsu K. and Xia C., *Hadamard-Frankel type theorems for manifolds with partially positive curvature*, Pacific J. Math., **27** (1994).

[KX 2] ———, *Intersections of minimal submanifolds in manifolds of partially positive curvature*, Kodai Math. J., **18(2)** (1995), 242–249.

[KK] Ki U. and Kim N., *Ruled real hypersurfaces of a complex space form*, Acta Math. Sin., New Ser., **10** (1994), no. 4, 401–409.

[KW] Kim H. and Walschap G., *Riemannian foliations on compact hyperbolic manifolds*, Indiana Univ. Math. J., **41** (1992), 37–42.

[Kimu 1] Kimura M., *Sectional curvatures of a holomorphic plane in $P_n(C)$*, Math. Ann., **276** (1987), 487–497.

[Kimu 2] ———, *Minimal hypersurfaces foliated by geodesics of n-dimensional space forms*, Tokyo J. Math., **16** (1993), 241–260.

[Kit 1] Kitahara H., *The completeness of a Clairaut foliations*, Ann. Sci. Kanazawa Univ., **11** (1974), 37–40.

[Kit 2] ———, *Differential geometry of Riemannian foliations*, Lect. Notes, Kyungpook National Univ., 1986.

[KK] Ki U. and Kim N., *Ruled real hypersurfaces of a complex space form*, Acta Math. Sin., New Ser., **10** (1994), no. 4, 401–409.

[KN] Kobayashi S. and Nomizu K., *Foundations of differential geometry*, vol. 1,2, Springer, 1963, 1969.

[Kli] Klingenberg W., *Riemannian geometry*, 2nd ed., Berlin, New York, Walter de Gruyter, 1995.

[KL] Kocak H. and Laidlaw D., *Computer Graphics and geometry of S^3*, The Math. Intellegencer, **9** (1987), 8–10.

[Koc] Koch R., *Regelflächen in euklidischen Räumen*, Giering O. (ed.) et al., Geometrie und ihre Anwendungen. München: Carl Hanser Verlag, 71–106 pp., 1994.

[Ko 1] Koike N., *Foliations on Riemannian manifold and Ehresmann connections*, Indiana Univ. Math. J., **40** (1991), 277–292.

[Ko 2] ———, *Totally umbilic foliations and decomposition theorems*, Saitama Math. J., **8** (1990), 1–19.

[Ko 3] ———, *Totally umbilic orthogonal nets and decomposition theorems*, Saitama Math J., **10** (1992), 1–19.

[Ko 4] ———, *Ehresmann connections for a foliation on a manifold with boundary*, SUT J. of Math., **30** (1994), 147–158.

[Ko 5] ———, *On tubes with nonconstant radius*, Tsukuba J. Math., **17** (1993), 286–267.

[Kou] Koufogiorgos T., *Contact metric manifolds*, Ann. of Global Anal. Geom., **11** (1993), 25–34.

[KoK] Kowalski O. and Külish I., *Generalized symmetric submanifolds of Euclidean spaces*, Math. Ann., **277** (1987), 67–78.

[KS] Kowalski O. and Sekizawa M., *Riemannian 3-manifolds with c-conullity two*, Bolletino U.M.I. (1996).

[Kü] Kühnel W., *Ruled W-surfaces*, Ann. Math., **62** (1994), 475–480.

[Kup] Kupeli D., *On existence and comparison of conjugate points in Riemannian and Lorentzian geometry*, Math. Ann., **276** (1986), 67–79.

[Lan 1] Langevin R., *Tight foliations*, Lect. Notes Math., vol. 858, 1981, pp. 181–186.

[Lan 2] ———, *A list of questions about foliations*, Diff. topology, foliations, group actions: Workshop on Topology Rio de Janeiro, Brasil, Contemporary Math., A.M.S., vol. 161, 1992, pp. 59–80.

[Lan 3] ———, *Courbures, feuilletages ets surfaces (measures et distributions de Gauss)*, These de Doctorat D'Etat Es Sciences Math., vol. 57, Univ. Paris-Sud, Centre D'Orsay, reimpreession Dijon, 1994, 1980, pp. 145.

[LL 1] Langevin R. and Levitt G., *Courbure totale das feuilletages des surfaces*, Comment. Math. Helv., **57** (1982), 175–195.

[LL 2] ———, *Sur la courbure totale des feuilletages des surfaces a bord*, Bol. Soc. Brasil. Mat., **16** (1985), 1–13.

[LP] Langevin R. and Possani C., *Total curvature of foliatons*, Illinois Math. J., **37** (1993), 508–524.

[LR] Langevin R. and Rosenberg H., *Qnand deux sous-varietes sont forcess par leur geometrie a se rencontrer*, Bull. Soc. Math. France, **120** (1992), 227–235.

[Las] Lasso de la Vega C., *Groupoide fondamental et groupoid d'holonomie de certains feuilletages réguliers*, Publ. Mat. UAB, **33(3)** (1989), 431–443.

[Law 1] Lawson H., *Foliations*, Bull. A.M.S., **80** (1974), 369–418.

[Law 2] ———, *Complete minimal surfaces in S^3*, Annl. of Math., **92** (1970), 335–374.

[LMM] de Leon M., Marin-Slano J. and Marrero J., *Ehresmann connections in classical field theories*, Diff. Geom. and its Appl., Zn. Fis. Monogr. 2, CIEMAT, Madrid, 1995, pp. 73–89.

[Lim] Lima E., *Commuting vector fields on S^3*, Ann. of Math., **81** (1965), 70–81.

[Li] Li Zhi-bo, *Garmonic foliations on sphere*, Tsukuba J. Math., **15** (1991), 397–407.

[LS] Li Z. and Sun Z., *Complex foliations on a complex projective spaces*, Acta Math. Sin., **37** (1994), no. 6, 762–766. (Chinese)

[LiV] van Lindt D. and Verstraelen L., *A survey on axioms of submanifolds in Riemannian and Kählerian geometry*, Colloq. Math., **54(2)** (1987), 193–213.

[Lis] Lisitsa V., *Embedding of n-dimensional cylindrical metrics as cylindrical submanifolds in Lobachevsky space*, Ukrain. Geometr. Sbornic, **34** (1991), 18–27.

[LV] Loo B. and Verjovsky A., *The Hopf fibrations over S^8 admits no S^1-subfibrations*, Topology, **31** (1992), 239–254.

[Lu] Lu D., *Homogeneous foliations of spheres*, Trans. A.M.S., **340** (1993), 95–102.

[Lum 1] Lumiste Yu., *Multi-dimensional ruled submanifolds in Euclidean space*, Math. Sbornic, **55(4)** (1961), 411–420. (Russian)

[Lum 2] ———, *Semi-symmetric submanifolds*, Itogi Nauki i Techniki, ser. geometry, vol. 23, Moscow: VINITI, 1991, pp. 3–28. (Russian)

[Lum 3] ———, *Die n-dimensionalen Minimalflächen mit einer $(n-1)$-dimensionalen asymptotischen Richtung im jedem Punkte*, Tartu Riikl. Ül. Toimetised, **62** (1958), 117–141. (Russian)

[Lun] Lundell A., *A short proof of the Frobenius theorem*, Proc. A.M.S., **116** (1992), 1131–1133.

[Mac] Machigashira Y., *Complete open manifolds of non-negative radial curvature*, Pacific J. Math., **165(1)** (1994), 153–160.

[Mae] Maeda S., *Geometry of submanifolds which are neither Kähler nor totally real in a complex projective space*, Bull. Nagoya Inst. Technol., **45** (1993), 1–50.

[MO] Maeda S. and Ogiue K., *Geometry of submanifolds in terms of behavior of geodesics*, Tokyo J. Math., **17** (1994), 347–354.

[Mak] Mackenzie K., *Lie groupoids and Lie algebroids in differential geometry*, vol. 124, London Math. Soc. Lect. Note Series, Cambridge-New York, 1987, pp. xvi+327.

[Mal 1] Maltz R., *The nullity spaces of the curvature-operator*, Cahiers de topologie et geometrie differentielle, **8** (1966), 1–20.

[Mal 2] ———, *The nullity spaces of curvature-like tensors*, J. Diff. Geometry., **7** (1972), 519–525.

[Mal 3] ———, *Cylindricity of isometric inmersions into Euclidean space*, Proc. A.M.S., **53** (1975), 428–432.

[MaU] Maeda S. and Udagawa S., *Real hypersurfaces of a complex projective space in terms of holomorphic distribution*, Tsukuba J. Math., **14** (1990), no. 1, 39–52.

[MT] Marenich V. and Toponogov V., *Open manifolds of nonnegative curvature*, Itogi Nauki i Techniki, ser. Problems of geometry, vol. 21, Moscow: VINITI, 1989, pp. 67–91. (Russian)

[Mar] Martinez A., *Ruled real hypersurfaces in quaternion projective space*, An. Stiint. Univ., **34** (1988), 73–78.

[Mas] Masa X., *Duality and minimality in Riemannian foliations*, Comm. Math. Helv., **67** (1992), 17–27.

[MV-SC] Masias-Virgos E. and Sanmartin-Carbon E., *Minimal foliations on the Lie groups*, Indagat Math., **3** (1992), 41–46.

[MR 1] Mazzocco R. and Romani G., *Multihelicoids in standard spaces of constant curvature*, Riv. Mat. Univ. Parma, **1(5)** (1992), 163–174.

[MR 2] ———, *Symmetry and minimality properties for generalized ruled submanifolds*, Rend. Semin. Mat. Univ. Padova, **92** (1994), 1–8.

[MR 3] ———, *Symmetry properties for submanifolds and totally geodesic maps*, 1–9 (to appear).

[Miq] Miquel V., *Some examples of Riemannian almost-product manifolds*, Pacific J. of Math., **III** (1984), 163–178.

[Mil] Millett K., *Compact foliations*, Differential topology and geometry, (Proc. Colloq., Dijon, 1974) Lect. Notes in Math., vol. 484, Springer, Berlin, 1975, pp. 277–287.

[Mln] Milnor J., *Characteristic classes*, Annals of Math. Studies, vol. 76, Princeton, New Jersey, 1974.

[MRT] Min-oo M., Ruh E. and Tondeur P., *Transversal curvature and tautness for Riemannian foliation*, J. Reine Angew. Math., **415** (1991), 167–174.

[Mrz] Mirzoyan V., *Normal nullity of a submanifold in a Riemannian manifold*, Acta et Comment. Univ. Tartuensis **734** (1986), 63–79.

[Mok] Mok N., *Foliation techniques and vanishing theorems*, Contemp. Math., **49** (1986), 79–118.

[Mo] Mok K., *On the differential geometry of frame bundles of a Riemannian manifold*, J. of reine angew., **302** (1978), 16–31.

[Mol] Molino P., *Riemannian foliations*, Progr. in Math., vol. 73, Birkhäuser, 1988, pp. 339.

[Mol 2] ———, *Orbit-like foliations*, Proc. "Geometric Study of Foliations", Tokyo, 1993, World Sci., Singapore, 1994, pp. 97–119.

[Mon 1] Montesinos A., *On certain of almost product structures*, Mich. Math. J., **30** (1983), 31–36.

[Mon 2] ———, *Conformal curvature for the normal bundle of a conformal foliation*, Ann. Inst. Fourier, **32** (1982), 261–274.

[Moe] Moore D., *Conformally flat submanifolds of Euclidean space*, Math. Ann., **225** (1977), 89–97.

[Moo 1] Moore J., *Isometric immersions of Riemannian products*, J. Diff. Geom., **5** (1971), 159–168.

[Moo 2] ———, *Submanifolds of constant positive curvature, I*, Duke Math. J., **44** (1977), 449–484.

[Moo 3] ———, *An application of second variation to submanifold theory*, Duke Math. J., **42** (1975), 191–193.

[MS] Moore C. and Schochet C., *Global analysis on foliated spaces*, Math Sci. Res. Inst. Publ., vol. 9, New York e.a.: Springer, 1988, pp. 337.

[Mor] Morgan A., *Holonomy and metric properties of foliations in higher codimension*, Proc. A.M.S., **58** (1976), 255–261.

[MZ] Morvan J. and Zafindratafa G., *Conformally flat submanifolds*, Ann. Fac. des Sci de Toulouse, **8** (1986–1987), 331–348.

[Mrv] Morvan J., *Distance of two submanifolds of a manifold with positive curvature*, Rend. mat. e appl., **3** (1983), 357–366.

[Mos] Moser J., *Topics in Calculus of Variations, Lecture Notes in Math.,*, vol. 1365, Springer Verlag, Berlin and New York, 1988, pp. 62–99.

[Nag] Nagamine S., *A remark on minimal foliations of Lie groups*, Tsukuba J. Math., **9** (1985), 317–320.

[NT] Nakagawa H. and Takagi R., *Harmonic foliations on a compact Riemannian manifold of non-negative constant curvature*, Tohoku Math. J., **40** (1988), 465–471.

[Nar] Narita F., *Riemannian submersion with isometric reflections with respect to the fibers*, Kodai Math. J., **16** (1993), 416–427.

[Nav] Naveira A., *A classification of Riemannian almost product manifolds*, Rend. Math., **3** (1983), 577–592.

[NR] Naveira A. and Rocamora A., *A geometric obstruction to the existence of two totally umbilic complementary foliations in compact manifolds*, Springer Lect. Notes in Math., vol. 1139, 1985, pp. 263–270.

[Nom 1] Nomizu K., *Characteristic roots and vectors of a differentiable family of symmetric matrices*, Linear and multilinear algebra, **1** (1973), 159–162.

[Nom 2] ———, *Isometric immersions of the hyperbolic plane into the hyperbolic space*, Math. Ann., **205** (1973), 181–192.

[No] Nore T., *Second fundamental form of a map*, Ann. Mat. pura ed. appl, **1436** (1987), 281–310.

[Nor 1] Noronha H., *A splitting theorem for complete submanifolds with nonnegative curvature*, Proc. A.M.S., **105** (1989), 979–985.

[Nor 2] ———, *Nonnegatively curved submanifolds in codimension two*, Trans. A.M.S., **332** (1992), 351–364.

[Okr 1] Okrut S., *Riemannian Gilbert submanifolds of nonpositive extrinsic curvature*, Plenum, **53** (1991), no. 5, February, 526–532.

[Okr 2] ———, *Structure of curvature of Riemannian manifold with axiom of hyperplanes*, Math. Physics, Analysis, Geometry, **1(2)** (1994), 227–231.

[Okr 3] ———, *Warped products in Kähler geometry*, Math. Physics, Analysis, Geometry, **3(3)** (1996).

[O'N 1] O'Neill B., *Isometric immersion of flat Riemannian manifolds in Euclidean space*, Mich. Math. J., **9** (1962), 199–205.

[O'N 2] ———, *Umbilics of constant curvature immersions*, Duke Math. J., **32** (1965), 149–159.

[O'N 3] ———, *The fundamental equations of a submersion*, Michigan Math. J., **13** (1966), 1–20.

[O'N 4] ———, *Submersions and geodesics*, Duke Math. J., **34** (1967), 363–373.

[O'N 5] ———, *Semi-Riemannian Geometry*, Academic Press, New York, 1983.

[O'N 6] ———, *Immersion of manifolds of non-positive curvature*, Proc. A.M.S., **11** (1960), 132–134.

[O'NS] O'Neill B. and Stiel E., *Isometric immersions of constant curvature manifolds*, Mich. Math. J., **10** (1963), 335–339.

[OR] Ornea L. and Romani G., *The fundamental equations of conformal submersions*, Beitrage zur Algebra Geom., **34** (1993), 233–243.

[Os 1] Oshikiri G., *A remark on minimal foliations*, Tohoku Math. J., **33** (1981), 133–137.

[Os 2] ———, *Jacobi fields and the stability of leaves of codimension-one minimal foliations*, Tohoku Math. J., **34** (1982), 417–424.

[Os 3] ———, *Totally geodesic foliations and Killing fields*, Tohoku Math. J., **35** (1983), 387–392.

[Os 4] ———, *Totally geodesic foliations and Killing fields*, 2, Tohoku Math. J., **38** (1986), 351–356.

[Os 5] ———, *Some remarks on minimal foliations*, Tohoku Math. J., **39** (1987), 223–229.

[Os 6] ———, *Mean curvature functions of codimension-one foliations*, Comment. Math. Helv., **65** (1990), 79–84.

[Os 7] ———, *Mean curvature functions of codimension-one foliations, 2*, Comment. Math. Helv., **66** (1991), 512–520.

[Os 8] ———, *On codimension one foliations of constant mean curvature*, Math. Z., **203** (1990), 105–113.

[Ots] Otsuki T., *On the existence of solutions of a system of quadratic equations and its geometrical application*, Proc. Japan Acad., **29** (1953), 99–100.

[Pak] Pak H., *On one-dimensional metric foliations in Einstein spaces*, Illinois J. Math., **36** (1992), 594–599.

[Pan] Pantilie R., *A simple proof of the de Rham decomposition theorem*, Bull. Math. Sci. Math. Roumanie (N.S.), **36(84)** (1992), no. 3-4, 341–343.

[Pap] Papantonion B., *Contact Riemannian manifolds satisfying $R(\xi, x) * R = 0$ and $\xi \in (k, \mu)$-nullity distribution*, The Yokohama Math. J., **40** (1993), 149–161.

[Per] Perelman G., *Proof of the soul conjecture of Cheeger and Gromoll*, J. Diff. Geom., **40** (1994), 209–212.

[Pe] Perez J.-D., *Some conditions on real hypersurfaces of quaternionic projective space*, An. Stiint. Univ. Al. I. Cuza Iasi, Mat., **38** (1992), no. 1, 103–110.

[Pet] Petro J., *Great sphere fibrations of manifolds*, Rocky Mountain J. Math., **17** (1987), 865–885.

[Ph 1] Phillips A., *Foliations of open manifolds, I*, Comment. Math. Helv., **43** (1968), 204–211.

[Ph 2] ———, *Foliations of open manifolds, II*, Comment. Math. Helv., **44** (1969), 367–370.

[PR] Ponge R. and Reckziegel H., *Twisted products in pseudo-Riemannian geometry*, Geom. Dedicata, **48** (1993), 15–25.

[Pos] Possani C., *A theorem of the total curvature of two orthogonal foliations of a hyperbolic surface*, An. Acad. Brasil. Cienc., **60** (1988), 253–254. (Portuguese)

[Pyo] Pyo Y.-S., *On ruled real hypersurfaces in a complex space form*, Math. J. Toyama Univ., **17** (1994), 55–71.

[Ra 1] Ranjan A., *On a remark of O'Neill*, Duke Math. J., **53** (1986), 113–115.

[Ra 2] ———, *Structural equations and an integral formula for foliated manifolds*, Geom. Dedicata, **20** (1986), 85–91.

[Ra 3] ———, *Substantial Riemannian Submersions of S^{15} with 7−dimensional fibres*, MRR 22-95, dg-ga/9510009, (1995), 1–10.

[Ra 4] ———, *Almost diameter rigidity for the Cayley Plane*, MRR 07-97, dg-ga/9707003, (1997), 1–8.

[Ras] Ras-Sabido A., *Killing fields preserving totally geodesic codimension-one foliations*, Kodai Math. J., **14** (1991), 477–484.

[RW] Ratner M. and Witte D., *Rigidity of horospherical foliations*, Ergodic theory and dynamical systems **9** (1989), 191–205.

[Rec 1] Reckziegel H., *On the eigenvalues of the shape operator of an isometric immersion into a space of constant curvature*, Math. Ann., **243** (1979), 71–82.

[Rec 2] ———, *Completness of curvature surfaces of an isometric immersions*, J. Diff. Geom., **14** (1979), 7–20.

[Rec 3] ———, *A fiber bundle theorem*, Manuscripta Math., **76** (1992), 105–110.

[Rec 4] ———, *On the decomposition theorem of Blumenthal and Hebda*, Geom. and topology of submanifolds (Leuven, 1991), vol. IV, World Sci. Publ., 1992, pp. 174–182.

[Rem] Remsing C.-C., *Introducere in teoria geometrica a foliatiilor*, Monogr. Matematice, Univ. din Timisoara, Fac. a de Stiinte ale Naturii, Sectia Matematica, Timisoara, vol. 23, 1984, pp. xii+154. (Romanian)

[Ree 1] Reeb G., *Sur certaines proprietes topologiques des varietes feuilletages*, Actualite Sci. Indust., vol. 1183, Hermann, Paris, 1952.

[Ree 2] ———, *Sur les structures feuilletees de codimension un et sur un theoreme de M.A.Denjoy*, Ann. Inst. Fourier, **11** (1961), 185–200.

[Re] Reid W., *Riccati differential eguations*, Academic Press, 1972.

[Rei 1] Reinhart B., *Foliated manifolds with bundle-like metrics*, Ann. of Math., **69** (1959), 119–132.

[Rei 2] ———, *The second fundamental form of a plane field*, J. Diff. Geom., **12** (1977), 619–627.

[Rei 3] ———, *Differential geometry of foliations*, Ergeb. Math., vol. 99, Springer-Verlag, New York, 1983, pp. 194.

[Rez 1] Reznikov A., *The weak Blaschke conjecture for HP^n*, Docl. Acad. Nauk SSSR, **283(2)** (1985), 308–312.

[Rez 2] ———, *Totally geodesic fibrations of Lie groups*, "Diff. geometry of families of figures" **16** (1985), Kaliningrad Univ., 67–70. (Russian)

[Rez 3] ———, *Lower bound on volumes of vector fields*, Arch. Math., **58** (1992), 509–513.

[Rez 4] ———, *The weak Blaschke conjecture for CP^n*, Invent. Math., **117** (1994), 447–454.

[Rie] Riemann B., *Über die Fläche vom Kleinsten Inhalt bei gegebener Begrenzung*, Leipzig, Gesammelte Math. Werke, (1876), 283–315.

[Roc 1] Rocamora A., *Some geometric consequences of the Weitzenbock formula on Riemannian almost-product manifolds; Weak-harmonic distributions*, Illinois J. of Math., **32** (1988), 654–671.

[Roc 2] ———, *Weak-harmonic distributions with respect to the characteristic connection*, Rev. Roumane Math. Pures Appl., **35** (1990), 563–576.

[Rd] Rodriguez L., *Immersions of non-zero relative nullity in manifolds of constant positive curvature*, Arch. Math., **32** (1979), 181–184.

[Rod] Rodionov E., *On homogeneous Z-manifolds* **32** (1980), VINITI, Moscow (deposited 12.09.80 No. 5394). (Russian)

[RAT] Rozenfeld B., Atanasyan S. and Timoshenko T., *The geometry of Hopf fibrations*, Soviet Mathematics (Izv. VUZ), **31** (1987), no. 6, 65–72.

[Ros 1] Rosenthal A., *Riemannian manifolds of constant nullity*, Mich. Math. J., **14** (1967), 469–480.

[Ros 2] ———, *Kähler manifolds of constant nullity*, Mich. Math. J., **75** (1968), 433–440.

[Rov 1] Rovenskii V., *Totally geodesic foliations*, Siberian Math. J., **23(1)** (1982), 217–219. (Russian)

[Rov 2] ———, *Quasi-developing and parabolic submanifolds in a sphere*, Soviet Math. Dokl., **30** (1984), 373–375.

[Rov 3] ———, *Geodesic foliations on the three-dimensional sphere*, Theoretical and applied problems in differential equations, CMP– 90 13, Karagand. Gos. Univ., Karaganda, 1986, pp. 115–119. (Russian)

[Rov 4] ———, *Remarks on complete parabolic surfaces*, Plenum, **51** (1990), no. 5, October, 2565–2566.

[Rov 5] ———, *Metric structure of ruled and parabolic surfaces in S^m and CP^m*, Siberian Math. J., **30** (1989), 783–793.

[Rov 6] ———, *Uniform Riemannian manifolds, admitting periodic functions*, Siberian Math. J., **30(1)** (1989), 156–159.

[Rov 7] ———, *Decomposition of ruled and parabolic surfaces*, Soviet Math. Dokl., **39(2)** (1989), 366–368.

[Rov 8] ———, *Characterization of totally geodesic submanifolds S^m and CP^m*, J. Sov. Math., **48(1)** (1990), 87–94.

[Rov 9] ———, *Totally geodesic foliations with nearly-integrable orthogonal distribution*, Siberian Math. J., **32(1)** (1991), 161–165.

[Rov 10] ———, *Twisted Hopf bundles*, J. Soviet Math., **53** (1991), 532–541.

[Rov 11] ———, *Conditions for the decomposition of ruled and parabolic surfaces in S^m and CP^m*, J. Sov. Math., **59(2)** (1992), 763–770.

[Rov 12] ———, *Pseudo-Riemannian ruled and parabolic submanifolds*, Geometry of multidimensional spaces, Altaisk. Gos. Univ., Barnaul, 1991, pp. 54–66. (Russian)

[Rov 13] ———, *Application of a Riccati differential equation to Riemannian manifolds with distributions*, preprint (1991), 12.

[Rov 14] ———, *Totally geodesic foliations that are close to Riemannian*, Ukrain. Geometr. Sbornic, **35** (1992), 114–118. (Russian)

[Rov 15] ———, *Metric decomposition of foliations with nonnegative curvature*, English Transl.: J. Math. Sci., 1994, Russian Acad. Sci. Dokl., Math., **49** (1994), 202–205.

[Rov 16] ———, *Classes of submersions of Riemannian manifolds with compact leaves*, Siberian Math. J., **35(5)** (1994), 1027–1035.

[Rov 17] ———, *Theorem on totally geodesic foliations with positive mixed sectional curvature and its applications*, Geometry of multi-dimensional spaces, Altaisk. Gos. Univ., Barnaul, 1994, pp. 50–58. (Russian)

[Rov 18] ———, *A local theorem on distributions whose mixed sectional curvature is positive*, Russian Mathematics (Izv. VUZ), **38** (1994), no. 4, 27–30.

[Rov 19] ———, *Decomposition of foliations and ruled submanifolds with nonnegative mixed curvature*, Izvestija VUZov, Math. **398(7)** (1995), 38–50.

[Rov 20] ———, *The integral of mixed scalar curvature along a leaf of foliation*, Differential Geometry and Applications, Proc. Conf., Aug. 28–Sept. 1, 1995 Brno, 1996, pp. 93–101.

[Rov 21] ———, *Local curvature properties of Hopf fibrations*, Deposited in VINITI **7.08.85, N 5909** (1985), 17 pp.

[Rov 22] ———, *Submanifolds and foliations with restrictions on partial Ricci curvature*, Problems of mathematical analysis, Krasnoyarsk Gos. Technical Univ., 1996, pp. 53–62. (Russian)

[Rov 23] ———, *On the role of partial Ricci curvature in geometry of submanifolds and foliations*, Ann. Polon. Math., **68** (1998) (to appear).

[RovT] Rovenskii V. and Toponogov V., *Geometric characteristics of a complex projective space*, Geometry and topology of homogeneous spaces, Altaisk. Gos. Univ., Barnaul, 1988, pp. 98–104. (Russian)

[Ru 1] Rummler H., *Quelques notions simplesen geometrie riemannienne et leur applications aux feuilletages compacts*, Comment. Math. Helv., **54** (1979), 224–239.

[Ru 2] ———, *Differential forms, Weitzenbock formula and foliations*, Publ. Mathem., **33** (1989), 543–554.

[Ryz] Ryzhkov V., *On tangential degenerate submanifolds*, Docladi Acad. Sci SSSR, **135(1)** (1960), 20–22.

[Sab] Sabuncuoğlu A., *Ricci curvature tensor of $(k+1)$-ruled surface in E^n*, Commun. Fac. Sci. Univ. Aukara, Ser. A1, Math. Statist. **37** (1991), 17–22.

[San] Sanini A., *Minimal and totally geodesic foliations*, Rend. Sem. Math. Univ. Politec., Torino, **43** (1985), 511–517.

[Sac 1] Sacksteder R., *On hypersurfaces with no negative sectional curvature*, Amer. J. Math.,

82(3) (1960), 609–630.

[Sac 2] ———, *The rigidity of hypersurfaces*, J. Mech., **11** (1962), 929–940.

[Sar] Sarkaria K., *A finiteness theorems for foliated manifolds*, J. Math. Soc. Japan, **30** (1978), 687–696.

[Sas 1] Sasaki S., *On the differential geometry of tangent bundle of a Riemannian manifold, I,*, Tohoku Math. J., **10** (1958), 338–354.

[Sas 2] ———, *On the differential geometry of tangent bundle of a Riemannian manifold, II,*, Tohoku Math. J., **14** (1962), 145–155.

[Sat 1] Sato H., *On topological Blascke conjecture, I*, Geometry of geodesics and related topics, Adv. Stud. in Pure Math., vol. 3, Tokyo, 1984, pp. 231–238.

[Sat 2] ———, *On topological Blascke conjecture, III*, Lect. Notes Math., vol. 1201, 1986, pp. 242–253.

[Sbr] Sbrana U., *Sulle varietá ad $n-1$ dimensione deformabili nello spazio euclideo ad n dimensione*, Rend. Circ. Mat. Palermo, **27** (1909), 1–45.

[Scw] Schwartz J., *Differential Geometry and Topology*, Notes on Math. and its Appl., New York, 1968.

[Sch 1] Schweitzer P., *Some problems in foliation theory and related areas*, Springer Lect. Notes in Math., vol. 625, 1978, pp. 240–252.

[Sch 2] ———, *Existence of codimension-one foliations with minimal leaves*, Ann. Global Anal. and Geom., **9** (1991), 77–81.

[Sch 3] ——— (ed.), *Differential topology, foliations and group actions*, Rio de Janeiro, 1994, pp. 287.

[Sch 4] ———, *Codimension-one foliations without compact leaves*, Comment. Math. Helv., **70(2)** (1995), 171–209.

[Se 1] Sekizawa M., *The nullity spaces of the conformal curvature tensor*, J. Math. Soc. Japan., **25** (1973), 125–131.

[Se 2] ———, *The completness of k-nullity foliation*, J. Diff. Geom., **11** (1976), 451–455.

[ST] Sekizawa M. and Tachibana S., *On the k-nullity spaces of Riemannian curvature tensor*, Tohoku Math. J., **27** (1975), 25–30.

[Sha] Sharma R., *On the curvature of contact metric manifolds*, J. of Geometry, **53(1/2)** (1995), 179–190.

[She] Shefel S., *About two classes of k-dimensional submanifolds in n-dimensional Euclidean space*, Siberian Math. J., **10(2)** (1969), 459–467.

[Shen] Shen Z., *On complete Manifolds of Nonnegative k th–Ricci curvature*, Math. Sci. Research Inst., (1991), Berkeley, California, 1–37.

[ShW] Shen Z. and Wei G., *Volume Growth and Finite Topological Type*, MSRI, preprint, (1990), 1–12.

[Shi] Shioya T., *Splitting theorems for nonnegatively curved open manifolds with large ideal boundary*, Math. Z., **212** (1993), 223–238.

[Shu] Shurygin V., *Ehressman connection for canonical foliation on manifold over local algebra*, Math. Zametki (Russia), **59(2)** (1996), 303–310.

[Si] Singh H., *Splitting of Jacobi equation and decomposition of index form*, Indian Appl. Math., **23** (1992), 277–284.

[Sol] Solomon B., *On foliations of R^{N+1} by minimal hypersurfaces*, Comment. Math. Helv., **61** (1986), 67–83.

[So 1] Solovjev A., *The second fundamental form of a distribution*, Mat. Zametki [Akad. Nauk SSSR] **31** (1982), 139–146. (Russian)

[So 2] ———, *Curvature of a hyperdistribution and contact metric manifolds*, Mat. Zametki [Akad. Nauk SSSR] **38** (1985), 450–462.

[So 3] ———, *Riemannian foliations with constant transversal curvature*, Sibirsk. Mat. Zh., **28** (1987), 160–166.

[So 4] ———, *Pontryagin classes of completely geodesic subfoliations*, Siberian Math. J., **29** (1988), 818–823 (510–512).

[Sp] Spivak M., *A comprehensive introduction to differential geometry*, vol. I–V, Berkeley Publish of Perish, 1978.

[Ste 1] Stepanov S., *A class of Riemannian almost-product structures*, Soviet Mathematics (Izv. VUZ), **33** (1989), no. 7, 51–59.

[Ste 2] ———, *A geometrical obstruction to the existence of totally umbilical distribution on a compact manifold*, Webs and quasigroups, Kalinin. Gos. Univ., 1990, pp. 135–137. (Russian)

[Ste 3] ———, *Bochner's technique in the theory of Riemannian almost product structures*, Math. Notes, **48** (1990), 778–781.

[Ste 4] ———, *A minimal hyperdistribution on a compact manifold*, Diff. Geom. Mnogoobrazii Figur, vol. 22, Univ., Kaliningrad, 1991, pp. 101–104. (Russian)

[Ste 5] ———, *A class of Riemannian almost-product structures*, Tensor, N.S. **55** (1994), 209–214.

[Ste 6] ———, *Weyl submersions*, Russian Mathematics (Izv. VUZ), **36** (1992), no. 5, 87–89.

[Ste 7] ———, *On theory of maps of Riemannian manifolds in the large*, Russian Mathematics (Izv. VUZ), **38** (1994), no. 10, 77–84.

[Ste 8] ———, *On projective submersions and immersions in the large*, Diff. Geom. Mnogoobrazii Figur, vol. 24, Univ., Kaliningrad, 1993, pp. 97–101. (Russian)

[Ste 9] ———, *On the global theory of some classes of mappings*, Annals of Global Anal. and Geom., **13** (1995), 239–249.

[Str] Strake H., *A splitting theorem for open nonnegatively curved manifolds*, Manuscripta Math., **61** (1988), 315–325.

[StW] Strake M. and Walschap G., *Σ-flat manifolds and Riemannian submersions*, Manuscripta Math., **64** (1989), 213–226.

[St] Strübing W., *Symmetric submanifolds of Riemannian manifold*, Math. Ann., **245** (1979), 37–44.

[Suh] Suh Y., *A characterization of ruled real hypersurfaces in CP^n*, J. Korean Math. Soc., **29** (1992), no. 2, 351–359.

[Sul 1] Sullivan D., *Cycles for the dynamical study of foliated manifolds and complex manifolds*, Invent. Math., **36** (1976), 225–255.

[Sul 2] ———, *A foliation of geodesics is characterized by having no tangent homologies*, J. Pure Appl. Algebra, **13** (1978), 101–104.

[Sul 3] ———, *A homological characterization of foliations consisting of minimal surfaces*, Comment. Math. Helv., **54** (1979), 218–223.

[Sz] Szabo Z., *A short topological proof for the symmetry of 2-point homogeneous spaces*, Invent. Math., **106** (1991), 61–64.

[TY] Takagi R. and Yorozu S., *Minimal foliations on Lie groups*, Tohoku Math. J., **36** (1984), 541–554.

[Tam] Tamura I., *Topology of foliations: An introduction*, Transl. from the 1976 Japanese ed. and with an afterword by K. Hudson. With a foreword by T. Tsuboi, Transl. of Math. Monographs, vol. 97, A.M.S., Providence, RI, 1992, pp. xii+193.

[Tan 1] Tanno S., *A theorem on totally geodesic foliations and its applications*, Tensor, **24** (1972), 116–122.

[Tan 2] ———, *Totally geodesic foliations with compact leaves*, Hokkaido Math. J., **1** (1972), 7–11.

[Te] Terng C., *Recent Progress in Submanifold Geometry*, "Diff. Geometry: Partial Differential equations on manifolds", A.M.S., Proc. of Symp. in Pure Math., part 1, vol. **54**, 1991, pp. 439–484.

[Th 1] Thas G., *Minimal monosystems*, Yokohama Math. J., **26** (1978), 157–167.

[Th 2] ———, *Minimal n-submanifolds of* $R^{n+1}(k)$ *foliated on* $(n-1)$*-totally geodesic submanifolds* $R^{n+1}(k)$, Kodai Math. J., **2** (1979), 287–293.

[Th 3] ———, *A theory of non-developable generalised ruled surfaces in the elliptic space* E^m, Czechosl. Math. J., **34(109)** (1984), 609–618.

[Th 4] ———, *A geometric structure in a generic generating space of a non-developable generalised ruled surface in the hyperbolic space* H^m, J. of Geometry, **25** (1985), 131–146.

[Tho] Thom R., *Generalization de la theorie de Morse aux varietes feuilletees*, Ann. Inst. Fourier (Grenoble), **14** (1964), 173–190.

[Ton 1] Tondeur P., *Foliations on Riemannian manifolds*, Universitext, New York: Springer, 1988, pp. 247.

[Ton 2] ———, *Riemannian Foliations and Tautness*, "Diff. Geometry: Riemannian Geometry", Part 3, A.M.S., Proc. of Symp. in Pure Math., vol. 54, 1993, pp. 667–672.

[Ton 3] ———, *Geometry of Riemannian Foliations*, Keio Univ., 1994, pp. ii+110.

[TV 1] Tondeur P. and Vanhecke L., *Transversally symmetric Riemannian foliations*, Tohoku Math. J., **42** (1990), 307–317.

[TV 2] ———, *A characterization of Riemannian foliations and totally umbilic submanifolds*, Bull. Austral. Math. Soc., **48** (1993), 101–108.

[TV 3] ———, *Jacobi fields, Riccati equation and Riemannian foliations*, Illinois J. Math. **40(2)** (1996), 211–225.

[Top 1] Toponogov V., *Metric structure of Riemannian spaces of nonnegative curvature, which contain straight lines*, Siberian Math. J., **5(6)** (1964), 1358–1369. (Russian)

[Top 2] ———, *A characteristic property of a four-dimensional symmetric space of rank* 1, Siberian Math. J., **13(4)** (1973 (1972)), 616–628.

[Top 3] ———, *Extremal theorems for Riemannian spaces with curvature bounded above*, Siberian Math. J., **15(6)** (1975 (1974)), 954–971.

[Top 4] ———, *Riemannian spaces of diameter* π, Siberian Math. J., **16(1)** (1975), 99–105.

[Top 5] ———, *Extremal theorems for Riemannian spaces with curvature bounded from above*, Soviet Math. Docl., **10(1)** (1969), 88–90.

[Ts] Tsukada K., *Totally geodesic submanifolds of Riemannian manifolds and curvature-invariant subspaces*, Kodai Math J., **19** (1996), 395–437.

[Uc] Ucci J., *On the nonexistence of a Riemannian submersions from* CP^7 *and* HP^3, Proc. A.M.S., **88** (1983), 698–700.

[Us] Ushakov V., *The base of a strongly parabolic Riemannian manifold*, Ukrain. Geom. Sbornic, **34** (1991), 112–121. (Russian)

[Vai] Vainstein A., *The uniquiness of metric fibration on* 3*-sphere*, Uspehi Math. Nauk, **40(3)** (1985), Moscow, 201–202. (Russian)

[VGE] Vainstein A. Gorelik E. and Efremovič V., *On metric fibrations*, Proc. of Moscow Univ. Sem. of Vector and Tensor Analysis, vol. 20, 1981, pp. 55–65. (Russian)

[Va 1] Vaisman I., *Varietes riemanniennes feuilletes*, Czech. Math. Jour., **21** (1971), 46–75.

[Va 2] ———, *Conformal foliations*, Kodai Math. J., **2** (1979), 26–37.

[Van] Vanhecke L., *Geometry in normal and tubular neighbourhood*, Lecture Notes, Rendiconti Sem. Fac. Sci. Univ. Cagliari, Suplemento al 26 (1988), 74–176.

[Ver] Verhoczki L., *Special isoparametric orbits in Riemannian Symmetric spaces*, Geom. Dedicata, **55** (1995), 305–317.

[VE] Ver-Eecke P., *Introduction a la theorie des varietes feuilletees*, Esquisses Math., Univ. d'Amiens, U.E.R. de Math., Amiens, vol. 31, 1982, pp. 206.

[Vil] Vilms J., *Totally geodesic maps*, J. Diff. Geom., **4** (1970), 73–79.

[Vin] Viner E., *A characteristic property of four dimensional rank one symmetric space*, Siberian Math. J., **16(2)** (1975), 165–173.

[Ve] Verstraelen L., *Foliations of quasi-umbilical submanifolds*, Simon Stevin, **51** (1977/78), 65–69.

[Vid] Vidal E. and Vidal-Costa E., *Special connections and almost foliated metrics*, J. Diff. Geom., **8** (1973), 297–304.

[Vo] Vogt E., *Existence of foliations of Euclidean spaces with all leaves compact*, Math. Ann., **296(1)** (1993), 159–178.

[Wal 1] Walczak P., *On minimal Riemannian foliations*, Inst. Math. Pol. Acad. Sci., Warsaw, **253** (1981).

[Wal 2] ———, *On foliations with leaves satisfying some geometrical condition*, Dissert. Math., (Rozprawy Mat.), **226** (1983), 1–47.

[Wal 3] ———, *Mean curvature functions for codimension one foliations with all leaves compact*, Czech. Math. J., **34** (1984), 146–155.

[Wal 4] ———, *Dynamics of the geodesic flow of a foliation*, Ergodic Theory Dynamical Systems, vol. 8, 1988, pp. 637–650.

[Wal 5] ———, *On the geodesic flow of a foliation of a compact manifold of negative constant curvature*, Proc. of the Winter School on Geometry and Physics, (Srni, 1988) Rend. Circ. Mat. Palermo, 2 Suppl. **21** (1989), 349–354.

[Wal 6] ———, *An integral formula for a Riemannian manifold with two orthogonal complementary distributions*, Collog. Math., **LVIII** (1990), 243–252.

[Wal 7] ———, *A finiteness theorem for Riemannian submersions*, Ann. Pol. Math., **57** (1992), 283–290.

[Wal 8] ———, *Foliations invariant under the mean curvature flow*, Illinois J. of Math., **37** (1993), 609–623.

[Wal 9] ———, *Jacobi operator for leaf geodesics*, Colloq. Math., **65** (1993), 213–226.

[Wal 10] ———, *Existence of smooth invariant measures for geodesic flows of foliations of Riemannian manifolds*, Proc. A.M.S., **120** (1994), 903–906.

[Wal 11] ———, *An estimate for the second fundamental tensor of a foliation.*, Proc. Sixth Int. Colloq. on Diff. Geom. (Santiago de Compostela, 1988), vol. 61, Univ. Santiago de Compostela, 1989, pp. 247–252.

[Wls 1] Walschap G., *Metric foliations and curvature*, J. Geom. Anal., **2** (1992), 373–381.

[Wls 2] ———, *Foliations of symmetric spaces*, Amer. J. Math., **115** (1993), 1189–1195.

[Wls 3] ———, *Some rigidity aspects of Riemannian fibrations*, "Diff. Geometry: Riemannian Geometry", Part 3, A.M.S., Proc. of Symp. in Pure Math., vol. 54, 1993, pp. 667–672.

[Wls 4] ———, *Soul-preserving submersions*, Mich. Math. J., **41(3)** (1994), 609–617.

[Wan] Wang P., *Decomposition theorems of Riemannian manifolds*, Trans. A.M.S., **184** (1973), 291–311.

[Wat 1] Watson B., *Almost hermittian submersions*, J. Diff. Geom., **11** (1976), 147–165.

[Wat 2] ———, *The differential geometry of two types of almost contact metric submersions*, The Math. heritage of Gauss, Collect. Pap. Mem. Gauss, 1991, pp. 827–861.

[Wil] Wilhelm F., *The radius rigidity theorem for manifolds of positive curvature*, J. Diff. Geom., **44** (1996), 634–665.

[Wilk] Wilker J., *Inversive geometry and the Hopf fibration*, Studia Sciet. Math. Hungarica, **21** (1986), 91–101.

[Win 1] Winkelnkemper H., *The graph of a foliation*, Ann. Global Anal. Geom., **1** (1983), 51–75.

[Win 2] ———, *The number of ends of the universal leaf of a Riemannian foliation*, Diff. Geometry Proc., Special Year, Maryland 1981-82 (R. Brooks, eds.), Birkhäuser, 1983, pp. 89–101.

[Wol 1] Wolak R., *Foliations admitting transverse systems of differential equations*, Compositio Math., **67** (1988), 89–101.

[Wol 2] ———, *Le graphe d'un feuilletage admettant un systeme d'equations differentielles*, Math. Z., **201** (1989), 177–182.

[Wol 3] ———, *Compact leaves in minimal Riemannian foliations*, Indag. Math. (N.S.), **5(3)** (1994), 375–379.

[Wol 4] ———, *Pierrot's theorem for singular Riemannian foliations*, Publ. Mat., **38** (1994), 433–439.

[Wol 5] ———, *The graph of a totally geodesic foliation*, Ann. Polonici Matem., **LX 3** (1995), 241–247.

[Wol 6] ———, *Geometric Structures on Foliated Manifolds*, Univ. de Santiago de Compostela (1989).

[Wol 7] ———, *Foliated G-structures and Riemannian foliations*, Manuscripta Math., **66** (1989), 45–59.

[Wol 8] ———, *Le graphe d'un feuilletage admettant un systeme transverse d'equations differentielles*, Math. Z., **201** (1989), 177–182.

[Wol 9] ———, *Graphs, Ehresmann connections and vanishing cycles*, Differential Geometry and Applications, Proc. of 6th Int. Conf., Aug. 28–Sept. 1, 1995, Brno, 1996, pp. 345–352.

[Wol 10] ———, *Ehresmann connections and lagrangian foliations*, J. of Geometry and Physics, **17** (1995), 310–320.

[Wu 1] Wu H., *On the de Rham decomposition theorem*, Illinois J. Math., **8** (1964), 291–311.

[Wu 2] ———, *Manifolds of partially positive curvature*, Indiana Univ. Math. J., **36** (1987), 525–548.

[Wu 3] ———, *Isoparametric submanifolds in Hyperbolic space*, Trans. A.M.S., **331(2)** (1992), 609–626.

[Yam 1] Yampolskii A., *On totally geodesic vector fields on a submanifold*, Math. Physics, Analysis and Geometry, **1** (1994), no. 3/4, 540–545.

[Yam 2] ———, *On strong spherical index of a Sasaki metric on a tangent bundle*, Ukrain. Geometr. Sbornic, **35** (1992), 150–159.

[Ya 1] Yang C. T., *Division algebras and fibrations of spheres by great spheres*, J. Diff. Geom., **16** (1981), 577–593.

[Ya 2] ———, *On the Blaschke conjecture*, Ann. Math. Stud., Sem. in Diff. Geometry **102** (1982), 159–171.

[Ya 3] ———, *Smooth great circle fibrations and an application to the topological Blaschke conjecture*, Trans. A.M.S., **320** (1990), 507–524.

[Ya 4] ———, *Any Blaschke manifold of the homotopy type of CP^n has the right volume*, Pacific J. Math., **151** (1991), 379–394.

[YI] Yano K. and Ishihara S., *Harmonic and relatively affine mappings*, J. Diff. Geom., **10(4)** (1975), 501–509.

[Yoo] Yoo H., *Existence of complete metrics on Riemannian foliations*, Math. J. Toyama Univ., **15** (1992), 35–38.

[Yor 1] Yorozu S., *Behavior of geodesics in foliated manifolds with bundle-like metrics*, J. Math. Soc. Japan, **35** (1983), 251–272.

[Yor 2] ———, *The nonexistence of Killing fields*, Tohoku Math. J., **36** (1984), 99–105.

[Yor 3] ———, *A_ν-operator on complete foliated Riemannian manifolds*, Israel J. of Math., **56** (1986), 349–354.

[YT] Yorozu S. and Tanemura T., *Green's theorem on foliated Riemannian manifolds and its applications*, Acta Math. Hungar., **56** (1990), 239–245.

[Ze 1] Zeghib A., *Sur les feuilletages geodesigues continus des varietes hyperboliques*, Invent. Math., **114 (1)** (1993), 193–206.

[Ze 2] ———, *Feuilletages geodesigue appliquess*, Math. Ann., **298 (4)** (1994), 729–759.

[Ze 3] ———, *Ensembles invariants des flots geodesiques des variétés localement symétriques*, Ergodic Theory and Dynamical Systems, **15** (1995), 479–512.

[Ze 4] ———, *Geodesic foliations in Lorentz 3−manifolds*, preprint.

[Ze 5] ———, *Isometry groups and geodesic foliations of Lorentz manifolds*, a paraitre.

[Zh] Zheng F., *Isometric embedding of Kähler manifolds with nonpositive sectional curvature*, Math. Ann., **304** (1996), 769–784.

[Zi 1] Zimmer R.J., *Ergodic theory, semisimple Lie groups, and foliations by manifolds of negative curvature*, Inst. Hautes Etudes Sci. Publ. Math., **55** (1982), 37–62.

[Zi 2] ———, *Curvature of leaves in amenable foliations*, Amer. J. Math., **105** (1983), 1011–1022.

[Zei] Zeitler H., *Clifford-Parallelen – Was ist das?*, Mitt. Math. Ges. Hamburg., **11** (1987), 415–440.

[Zel] Zelikin M., *On Theory of matrix Riccati equation*, Math. Sbornik, **182(7)** (1991), 970–984. (Russian)

[Žu 1] Žukova N., *On the stability of leaves of Riemannian foliations*, Ann. Global Anal. Geom., **5** (1987), 261–271.

[Žu 2] ———, *Submersions with Ehresmann connection*, Soviet Math. (Izv. VUZ), **32** (1988), 27–38.

[Žu 3] ———, *Stability criterion of leaves of Riemannian foliations with singularities*, Russian Mathematics (Izv. VUZ), **36(4)** (1992), 86–89.

[Žu 4] ———, *Graph of foliation with Ehresmann connection and stability of leaves*, Russian Mathematics (Izv. VUZ), **38(2)** (1994), 76–79.

INDEX

MIX
Papier aus verantwortungsvollen Quellen
Paper from responsible sources
FSC® C105338

If you have any concerns about our products,
you can contact us on
ProductSafety@springernature.com

In case Publisher is established outside the EU,
the EU authorized representative is:
Springer Nature Customer Service Center GmbH
Europaplatz 3, 69115 Heidelberg, Germany

Printed by Libri Plureos GmbH
in Hamburg, Germany